TO

John Shickluna

with personal

compliments of

Roy L. Donahue

Forsyth,

missouri

October 23, 1973

PRENTICE-HALL VOCATIONAL AGRICULTURE SERIES

BEEF PRODUCTION. Diggins and Bundy

CROP PRODUCTION. Delorit and Ahlgren

DAIRY PRODUCTION. Diggins and Bundy

EXPLORING AGRICULTURE. Evans and Donahue

FRUIT GROWING. Schneider and Scarborough

JUDGING LIVESTOCK, DAIRY CATTLE, POULTRY, AND CROPS. Youtz and Carlson

LEADERSHIP TRAINING AND PARLIAMENTARY PROCEDURE FOR FFA. Gray and Jackson

LIVESTOCK AND POULTRY PRODUCTION. Bundy and Diggins

MODERN FARM BUILDINGS. Ashby, Dodge, and Shedd

MODERN FARM POWER. Promersberger, Bishop, and Priebe

PROFITABLE FARM MARKETING. Snowden and Donahoo

PROFITABLE SOIL MANAGEMENT. Knuti, Korpi, and Hide

SWINE PRODUCTION. Bundy and Diggins

USING ELECTRICITY Hamilton

FOURTH EDITION

EXPLORING AGRICULTURE

An Introduction to Food and Agriculture

EVERETT F. EVANS

Consultant on Environmental Conservation

ROY L. DONAHUE

Professor Emeritus, Michigan State University

Staff Associate—Ecology and Environment
Midwest Research Institute

PRENTICE-HALL, INC., ENGLEWOOD CLIFFS, N. J.

EXPLORING AGRICULTURE: An Introduction to Food and Agriculture
Fourth Edition, Everett F. Evans and Roy L. Donahue

ISBN 0-13-296004-4

Cover photograph by Grant Heilman

10 9 8 7 6 5 4 3 2 1

PRENTICE-HALL INTERNATIONAL, INC. London
PRENTICE-HALL OF AUSTRALIA, PTY. LTD., Sydney
PRENTICE-HALL OF CANADA, LTD., Toronto
PRENTICE-HALL OF INDIA PRIVATE LTD., New Delhi
PRENTICE-HALL OF JAPAN, INC., Tokyo

ABOUT THIS BOOK

The Fourth Edition of Exploring Agriculture, like the previous editions, is intended primarily for the study of general and vocational agriculture. The book is also a reliable source of information for anyone who wants to know more about farm life.

In updating Chapter 1, Careers in Agriculture and Agribusiness, the authors tried to give proper emphasis to the trend toward greater employment opportunities for young people who complete one or two years of technical education programs below the degree level.

Many new illustrations are included in the Fourth Edition. Throughout the book, the authors point out the international character of agriculture and the role of United States farmers and ranchers in world food production. Recognition is also given to the relation of land and water management to the improvement of our environment and the reduction of pollution.

The authors appreciate the recommendations which several State Directors of Agricultural Education made for the Fourth Edition. We are also grateful to the many individuals, agencies, and companies that generously provided photographs and other illustrations.

To our wives and other members of our families, and to our many friends in the field of agriculture, the authors again extend their thanks for counsel and encouragement.

EVERETT F. EVANS

ROY L. DONAHUE

CONTENTS

1 CAREERS IN AGRICULTURE AND AGRIBUSINESS

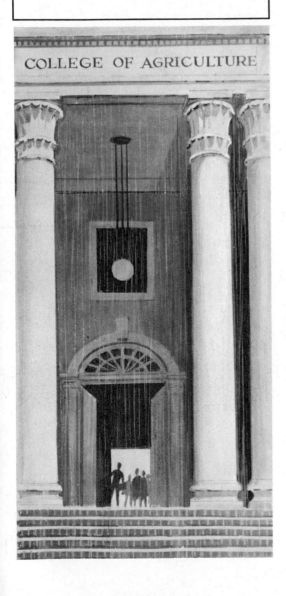

COLLEGE OF AGRICULTURE

How have agricultural occupations changed in the twentieth century? In this chapter we shall answer this question. We shall also consider some of the job opportunities and careers in the growing number of enterprises related to agriculture.

There are now less than 4.7 million family workers and 1.8 million hired workers on 2.9 million farms and ranches. Only about five million workers receive most of their income from farm work. The man-hours used for farm work are now about half the total man-hours used just after World War II. These trends reflect the rapid growth of agricultural technology. One reason that the United States is the world's leading agricultural producer is that our farmers are among the most productive in the world. In 1910, each farm worker supplied farm products for seven people. Now one farm worker can provide farm products for about 50 consumers.

In the past three decades, farm employment declined by four million workers. The average size of farms increased from 174 acres in 1940 to 389 acres in 1971 and is estimated to be 545 acres in 1980. The family farm is still the major production unit in agriculture. Family farms account for approximately 70 percent of all production.

Machines have greatly increased the area of land one worker can cultivate or harvest. A man with a team of horses could plow about two acres per day. With power equipment, one man can plow 20 to 200 acres per day, depending on the kind of equipment used. Family-size farms permit savings in cost per unit of output.

During most of the first half of this century, the traditional steps in acquir-

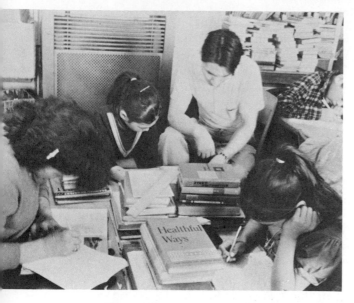

These Eskimo students in Alaska are preparing for job opportunities and careers. (*Courtesy* International Telephone and Telegraph Corporation)

Job opportunities in agriculture may mean preparing an educational exhibit to show clients the services a company offers. (*Courtesy* Allied Chemical Corporation)

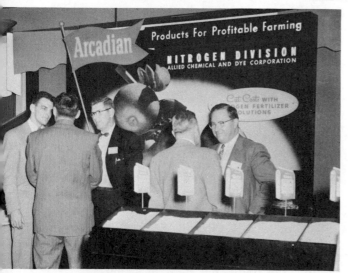

ing a farm were (1) working on the parents' farm, (2) working on other farms to earn money to become a tenant, (3) operating a rented farm, and (4) buying and eventually paying for a mortgaged farm. For most farm youth today, this "ladder" method of acquiring a farm is not successful.

Today about 80 percent of farm-reared youth must find jobs outside farming. Fortunately, there are many opportunities for youth in agriculture that are not being pursued as much as is possible. Occupations closely associated with agriculture now employ almost as many workers as there are on farms and ranches. The agriculture-related fields are often referred to as agri-business. The anticipated growth of employment in these enterprises is good news for young men and women interested in working to serve farmers who work on the land.

A college education is helpful in any agricultural occupation, although not required for all such occupations. Education for related agricultural occupations does not require as large an investment as is necessary to start a farm operation.

FINDING YOURSELF IN A CAREER

Success in a career in agriculture means more than knowing technical subject matter. Personality traits must be compatible with the demands of the work. These demands may bring out different personality traits. The trick is to find a job or career that is agreeable with the best traits in a person, and then the person must do his best to adapt his personality to the work.

Discuss personality traits necessary to a successful career. Appendix A lists 268 careers in agriculture. Do any of them appeal to you? If so, note that nearly all of them require some education beyond high school. Appendix A also contains a section on "Post-Secondary Education Curriculum in Agriculture and Natural Resources in the United States." If you prefer a 4-year course that leads to the degree of bachelor of science, see the list of 70 colleges that offer this degree.

Sixty-six Federal Job Information Centers are listed in Appendix A. The 39 field offices of the United States Department of Commerce also are given in this appendix.

AGRI-BUSINESS TRAINING

A good education for agri-business includes specialized courses in agriculture. A background in agricultural mathematics, biology, chemistry, and English is a distinct advantage. All state agricultural colleges offer a wide choice of courses for anyone interested in agriculture-related occupations.

Some special programs in agri-business training have been started, and others will be established as needed.

For information on the broad and general aspects of agribusiness selling, teachers may write to Sales and Marketing Executives International, 630 Third Avenue, New York, New York 10017.

Four examples of less-than-degree level education programs in agriculture in California, Michigan, North Carolina, and New York are described in the following paragraphs. (See Appendix A for a list of other colleges that offer similar programs.)

According to the 1971-72 catalog of Shasta College, Redding, California, the Associate in Arts degree is granted in several subjects, including natural resources. Completion of work for this degree usually requires two years. The program in natural resources includes the following required subjects: discussion skills, report writing, environmental conservation, soils, fish and wildlife conservation, forest practices, natural resources recreation, natural resources surveying, range management, natural resources mechanics, conservation administration, work experience seminar, work experience, and several suggested electives.

The environmental technology program at Michigan State University is offered as an 18-month program by the Institute of Agricultural Technology. Required courses include biochemistry, basic plant science, soil analysis and environmental quality, study and reading, writing and speaking, environmental pest management, measurement and calculations, soil chemical analysis, applied entomology, herbicides and production systems, placement training, personnel practices, crop production, application equipment, agricultural economics, financial credit practices, soils and man's environment, irrigation and erosion, urban and rural pest control systems, microbiology and public health, retail merchandising, lawn and turf management, soil classification and mapping, salesmanship, plant diseases, insecticides and their use, basic soil science, and soil management and fertilizer use.

Fayetteville Technical Institute, Fayetteville, North Carolina, offers an 18-month program of agricultural business. The following courses are listed in the 1970-72 catalog (Volume III, pages 44 and 45): grammar, chemistry, business mathematics, animal science, composition, business accounting, soil science and fertilizers, agricultural economics, oral communication, business finance, farm business management, agricultural mechanization, livestock diseases and parasites, sales development, agricultural marketing, agricultural chemicals, American institutes, business law, office machines, crop production, animal production, applied psychology, supervision, and taxes.

The Agricultural and Technical College at Alfred, State University of New York, offers a program in Agricultural Science. This is one of several options within the Division of Agricultural Technologies. The 1970-71 college bulletin contains the following list of courses in agricultural science leading to the degree of Associate in Applied Science: technical agriculture, professional practices, job-finding techniques, soils, botany, dairy cattle industry, livestock industry, agricultural zoology, agricultural microbiology, genetics, microbiology of foods, humanities, social sciences, mathematics, chemistry (three courses), and histology. Many electives are permitted in several sequences of plant or animal sciences.

SPECIAL JOBS IN AGRICULTURE

In this chapter we shall consider the opportunities and requirements for employment in some of the major occupations in agriculture and related fields.

Environment

Many positions in environmental technology are found throughout the field of agriculture and natural resources. Some positions also exist in natural sciences, engineering, and social sciences. In the broadest sense, programs in the environment include all subjects, because all subjects relate to man's physical, economic, social, and psychological environment.

Pollution is sometimes defined as a resource out of place, as a weed is a plant out of place. For example, mercury is a useful resource when used in thermometers, but it is a pollutant when it occurs in fish. Lead is an important resource when used in solder, but in gasoline it is a pollutant which gets into the air. Farmyard manure is a valuable resource when applied on fields to make plants grow, but it is a pollutant when it produces objectionable odors or when seepage gets into a water supply.

Positions in environmental technology are available through many professions. The college of your choice may offer programs in this field; if not, write to one of the colleges listed in Appendix A.

Soil Conservationist

Employment opportunities for well qualified soil conservationists have been increasing rapidly for more than 35 years. This profession probably will continue its expansion because government agencies, industries, business enterprises, and organizations are doing more work which requires conservationists.

The principal service of soil conservationists is technical assistance to owners and operators of farms and ranches. This includes planning effective land use, maintaining high fertility of good soils and improving poor soils, controlling erosion, stabilizing runoff, and developing vegetative cover. Soil conservationists also deal with problems in irrigation, drainage, flood control, forestry, and wildlife management. Soil conservationists are very useful in many of the technical aspects of agricultural pollution and its control.

The starting point for better land use may be a soil survey and the preparation of soil maps to show soil series, vegetation, water supply, topography, erosion, and other details for a conservation plan. Management practices and treatments are recommended for each part of the farm or ranch. A conservation plan may include recommendations for all farms and ranches in a watershed. To assure that conservation practices are a good investment, the relative cost of land use and treatment is compared with estimated financial benefit.

The largest employers of soil conservationists are the Soil Conservation Service in the United States Department of Agriculture and the Bureau of Indian Affairs in the Department of the Interior. There are additional opportunities for employment by other federal agencies, state and local governments, banks, utility companies, insurance companies, many agri-business enterprises and urban development agencies.

A professional soil conservationist must have at least a bachelor's degree. A degree in one of the following specialties is usually required: agronomy, soil science, general agriculture, range management, agricultural engineering, forestry, or biology. Some positions also require field training in land-use planning and farm and ranch conservation practices.

More information about soil conservationists may be obtained from: The United States Civil Service Commission, Washington, D. C. 20415; Employment Division, Office of Personnel, United

The loss of soil by erosion has increased from 3 billion to 4 billion tons annually since 1939. This means job and career opportunities in abundance for many years. (*Courtesy* Soil Conservation Service)

A career in agriculture may mean helping people to make better use of land for more food production and less pollution (Makaha Valley, West Oahu County, Hawaii). (*Courtesy* University of Hawaii)

States Department of Agriculture, Washington, D. C. 20250; or any office of the Soil Conservation Service.

Soil Conservation Aid

A college degree is not required for certain kinds of conservation personnel who assist soil conservationists in applying conservation practices after a plan of action has been prepared. An example is the position of soil conservation aid. High school graduates with a background in farming or ranching and certain kinds of experience are eligible to apply for this position in the Soil Conservation Service.

There is no written examination for this position. The United States Civil Service Commission rates applicants on the basis of education, training, and experience.

Detailed requirements and application forms for the position of soil conservation aid may be obtained from the Board of United States Civil Service Examiners, or the Department of Agriculture, Soil Conservation Service, Washington, D. C. 20250.

Soil Scientists

Working in field and laboratory, soil scientists study the physical, biological, and chemical nature of soils. A national system of soil classification is used to identify soil mapping units, and the expected result of management practices is described. Also, soils are examined to determine their behavior as engineering materials. Aerial photographs provide much of the information for the preparation of soil maps on which soil series, land use, roads, field boundaries, soil erosion, slope, and other features are indicated. A physical and chemical analysis of soil reveals much of its history, including the parent rock materials and other factors that affect soil formation.

A soil scientist may specialize in any of the following: soil classification and mapping, soil chemistry, soil physics, soil fertility, soil microbiology, soil management, or soil geography. Education and experience in soil science are excellent professional preparation for land appraisers, land managers, positions in the fertilizer industry, and other positions in enterprises related to agriculture such as urban land development.

Well-qualified soil scientists should find abundant employment opportunities during the next decade. There is a growing need for soil scientists to help

A career in agriculture may mean being a professional in animal husbandry. (*Courtesy* State of Maine Department of Agriculture)

in the scientific classification and evaluation of our nation's soil resources. A major objective of the Soil Conservation Service is to complete the soil survey of all rural and urban land. This will require research, soil classification, and interpretation of research and classification so that available information can be used by engineers, land planners, and specialists in agriculture.

Federal agencies, colleges of agriculture, and state experiment stations employ most of the soil scientists. Other sources of employment are private research laboratories, insurance companies, lending agencies, land appraisal boards, fertilizer companies, real estate firms, state conservation departments, state highway departments, farm management agencies, and park departments.

A few soil scientists are self-employed as independent consultants. Also,

there is a gradual increase in employment of these specialists as agricultural managers, consultants, and research advisers in foreign countries.

A bachelor's degree is the minimum educational requirement for a soil scientist. A doctor's degree greatly improves one's chances for advancement to the better positions in research, teaching, and extension. An aptitude for chemistry, physics, biology, and mathematics is important in the study of soil science. The land-grant colleges and state universities offer a good education for this profession.

Many colleges and universities offer fellowships and assistantships for graduate study in soil science. Some graduate students are employed for research or part-time teaching.

More information is available from the United States Civil Service Commission, Washington, D. C. 20415; Office of Personnel, United States Department of Agriculture, Washington, D. C. 20250; and any office of the Soil Conservation Service. Information also is available from your state agricultural university.

AGRICULTURAL EXTENSION SERVICE POSITIONS

Educational work in agriculture and home economics is the main activity of the Cooperative Extension Service, a federal-state partnership of the United States Department of Agriculture and the land-grant colleges. Knowledge of effective teaching methods and of many kinds of subject matter is required in this field.

Increasing the efficiency of agricultural production and marketing is a broad objective of the county agricultural agents. The development of new market outlets is a special goal. Nutrition and home management are two major fields in which county home demonstration agents work with women. County agricultural agents and home demonstration agents also work with boys and girls in 4-H Clubs. A 4-H Club agent is employed in some counties.

With the help of agricultural extension workers, families in local communities solve problems by combining practical experience and research. Some problems which cannot be approached by group methods require individual assistance to farmers and homemakers. Almost every agricultural county has an agricultural agent and a home demonstration agent. Counties with a large rural population may have several extension agents. In such counties the extension work may be assigned to specialists

in livestock, poultry, dairying, and crops. In all states the Agricultural Extension Service works with other agencies and organizations in annual community improvement programs.

State extension specialists in many subject matter fields support the county extension staff. Examples of these special fields are home economics, agricultural economics, agronomy, livestock, marketing, entomology, and horticulture.

A county agricultural agent must have a bachelor's degree in agriculture, and a county home demonstration agent must have a bachelor's degree in home economics. A master's degree is required for most positions on the staff of a state Extension Service. A Ph.D. is required in many states.

The following are good sources of information about the Extension Service: Federal Extension Service, United States Department of Agriculture, Washington, D. C. 20250; State Director of Extension at each state college of agriculture; and County Extension Service offices.

AGRICULTURAL RESEARCH

The past three decades have brought a rapid growth in the number of research programs in agriculture. The leading agencies in this work are the United States Department of Agriculture and the state experiment stations at the land-grant colleges. Independent research is conducted by many non-profit organizations, as well as by companies that make farm equipment and supplies, market farm products, and finance farm operations.

The major fields of agricultural research include bacteriology, entomology, animal breeding, plant breeding and pathology, agricultural engineering, animal and poultry husbandry, soils research and conservation, parasitology, nutrition, financing and marketing, statistical analysis, and production economics.

The United States Department of Agriculture employs a large number of agricultural research workers. Many are assigned to work in Washington, D. C., or at the Agricultural Research Center, Beltsville, Maryland. Four regional research laboratories are at Albany, California; Peoria, Illinois; Wyndmoor, Pennsylvania; and New Orleans, Louisiana. Many Department of Agriculture researchers cooperate in research programs at the land-grant colleges and overseas.

Some agricultural research is performed by other federal departments and state agencies. Much cooperation and coordination is involved in the study of interdependent resources such as soil,

Foresters with many years of experience may have a unique opportunity to work overseas. Here a forester is teaching students in Ecuador, South America, how to cut this mature tree most efficiently. (*Courtesy* United Nations)

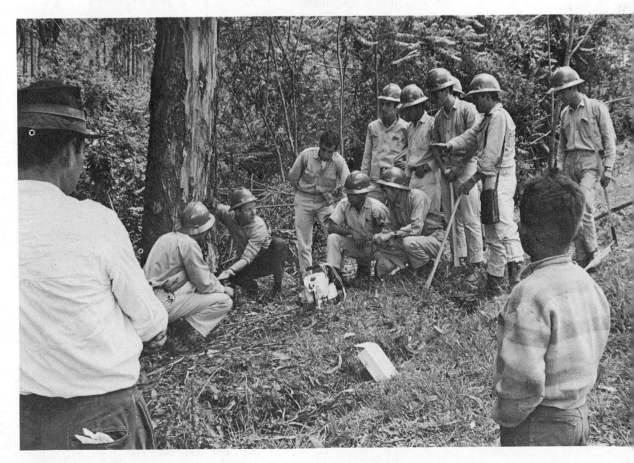

forests, and wildlife.

A bachelor's degree is the minimum educational requirement for most research positions in agriculture. The trend is toward more specialization, which requires at least five years of college study. A good background in science and mathematics is desirable for anyone who plans to pursue a career in agricultural research. A Ph.D. degree is a requirement for most research positions.

Two good sources of information on federal employment in this field are the Office of Personnel, United States Department of Agriculture, Washington, D. C. 20250; and the United States Department of Agriculture recruitment representatives at the land-grant colleges.

FORESTRY CAREERS

As the name of his profession indicates, a forester works with forests. He is concerned with the management of forest land to produce tree crops for lumber, pulpwood, poles, piling, railroad crossties, boxes, cooperage, and other wood products. As research develops new uses for wood, the role of the forester will become even more important than it is today. Foresters are also concerned with the management of forests for wildlife, watersheds, erosion control, windbreaks and shelterbelts, recreation, the management of the range, and pollution control.

A forester needs a good background in zoology, botany, plant taxonomy, plant pathology, soils, and entomology. Courses in mathematics, chemistry, and physics should also be part of a forester's education.

Employment opportunities are most numerous in the United States Forest Service, state forestry departments, and forest industries. A bachelor's degree in forestry is the minimum educational requirement. A master's degree provides additional opportunities for advancement to more responsible positions at higher pay. A doctor's degree is preferred and usually required for teaching forestry in a college or university.

More information about employment for professional foresters can be obtained from the United States Forest Service, Department of Agriculture.

The vocational agriculture graduate may qualify for nonprofessional work as a technician, aid, clerk, skilled worker, or laborer for the United States Forest Service. This agency tries to place each employee in the kind of work for which he is best qualified and which

A district ranger uses an increment borer to determine the age and growth rate of a tree. (*Courtesy* U.S. Forest Service)

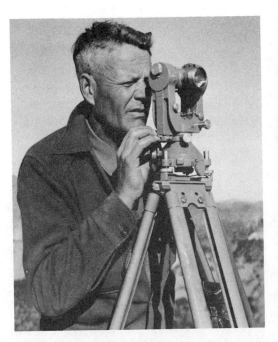 

Left: A cartographic aid uses a theodolite to measure angles on a mountain peak in the Flathead National Forest in Montana. (*Courtesy* U.S. Forest Service) *Right:* A forestry aid uses the radio in a Forest Service jeep. (*Courtesy* U.S. Forest Service)

will provide the best chance for growth in occupational skills.

Technician. In forestry, as in many other fields, there are certain kinds of work formerly done by professional personnel that can be assigned to technicians. The Forest Service has done this very successfully. Foresters have turned over to technicians such exacting jobs as supervising timber sales operations, research that requires practical skills and experience, and the management of forest areas designated for wildlife and recreation.

The forestry technician is required to have two years' general experience and one year of specialized experience. For the forest and range fire control technician, the required experience includes 1½ years general experience and one year specialized experience. An engineering technician must have general experience for 2¼ years and specialized experience for three-fourths of a year.

Acceptable general experience for forest and range fire technicians could be gained in any field of forestry, or in a related field, in which the applicant can show that he has acquired and can use basic knowledge and skills required in fire control. Specialized experience for this position would include the mastery of knowledge and skills required for fire prevention, dispatching of fire-fighting crews, or a combination of both, depending on the available position.

High school or other education may be substituted for parts of the general or specialized experience. The nature and length of subjects studied are factors in determining the amount of substitution allowed.

Aid. Three kinds of aids are (1) forestry aid, (2) forest and range fire control aid, and (3) engineering aid. The work of the aids is similar to that of the technicians but is usually somewhat less difficult and requires less responsibility. These positions require two seasons or one year of general experience. No specialized experience is required.

A diploma from a 4-year high school in which an applicant completed 6 one-semester courses in mathematics or science may be substituted for two seasons of general experience. The general experience required for the position of forest and range fire control technician would be satisfactory for the positions of aids.

Skilled worker. The Forest Service employs such skilled workers as carpenters, welders, cooks, bulldozer operators, parachute repairers and packers, and others experienced in specific trades or crafts. Skilled workers are used for construction, maintenance, fire control, and other projects.

A helper or apprentice position usually requires 6 months experience assisting a journeyman worker. (The Level 5 rating for helper or apprentice should not be confused with the GS-5 Civil Service rating.) Each month of attendance in day classes of a trade school may usually be substituted for one month of experience.

Laborers. The Forest Service employs laborers for many projects in the national forests. Unskilled workers perform such tasks as killing and removing undesirable trees, pruning trees to improve their quality, eradicating bushes that transmit blister rust to white pines, building firelines, digging ditches, planting seedlings, loading and unloading tools and equipment, and other supervised work.

Further information may be obtained by writing to the Regional Forester of the region in which you want to work. Addresses are given in Appendix A.

VOCATIONAL AGRICULTURE TEACHING

Vocational agriculture teachers supervise the farm programs of students, teach farm mechanics in school shops, and serve as advisers to local chapters of the Future Farmers of America. Agriculture teachers also provide organized instruction to help young farmers become successful farm operators and community leaders. In addition, organized instruction and individual consultations on the farm help adult farmers keep abreast of modern farm technology.

The vocational agriculture teacher must meet the same general requirements as those for agricultural extension workers. In addition, the teacher must have special courses and practice in teaching. A bachelor's degree from an approved agricultural college is the minimum educational requirement for teaching vocational agriculture.

State requirements for vocational agriculture teaching may be obtained from the head teacher trainer in agricultural education at the land-grant college in your state, or from the State Supervisor of Agricultural Education in the State Department of Education.

VETERINARIANS

The profession of veterinary medicine is concerned with the care and breeding of animals, prevention of diseases in livestock and pets, the treatment of sick or injured animals, and many special problems such as animal nutrition and meat inspection. Veterinarians may be self-employed or they may work for federal or state agencies, private nonprofit organizations, or companies. Many veterinarians in urban areas specialize in the care of pets. There are approximately 18,000 veterinarians in the United States. Approximately 5 percent are women. Almost two-thirds of the veterinarians are in private practice.

A degree in veterinary medicine is required in this profession. Two or three years of preveterinary college work followed by four years of professional study in a school of veterinary medicine are the minimum requirements for the degree of doctor of veterinary medicine (D.V.M.). All states and the District of Columbia require a license for the practice of veterinary medicine.

Further information may be obtained from the department of veterinary medicine at any university or land-grant college which offers a degree in this field, or from the United States Department of Agriculture, Washington, D. C. 20250.

AGRICULTURAL ENGINEERS

The scope of agricultural engineering is much broader than it was twenty-five years ago. No longer is this field limited mainly to the design of farm structures and equipment. In addition to the important field of design, the agricultural engineer is contributing to soil and water conservation, processing of farm products for market, better utilization of electricity, and control of pollution. Among the special kinds of work performed by agricultural engineers are production engineering, research, sales engineering, education, maintenance, testing, and combinations of these fields. Basic engineering principles and practices are applied to each specialty to improve the efficiency of the farm worker and the quality of farm products.

The United States Department of Agriculture employs more agricultural engineers than any other federal agency. Within the Department, the Soil Conservation Service and the Agricultural Research Service lead in the use of engineers. The United States Department of the Interior, the United States Department of Defense, and other government agencies employ a smaller number. Other sources of employment are state departments of agriculture, engineering departments of state colleges and universities, state highway departments, and state agencies concerned with natural resources and conservation. A few agricultural engineers are employed by cities, counties, and special districts for work on irrigation, flood control, drainage, and urban development.

There are more than forty state colleges and universities that offer professional courses leading to a bachelor of science degree in agricultural engineering. Graduate study for the master of science degree is provided by most of these colleges and universities, and a

Agricultural engineers design, test, and develop machines such as this one that moves across the field digging a trench and laying plastic drain pipe in one operation. (*Courtesy* Kraft, Inc.)

few offer the doctor of philosophy degree. At least the bachelor's degree is required for most positions.

Information on colleges which provide education in this field may be obtained from state agricultural colleges and the American Society of Agricultural Engineers, St. Joseph, Michigan 49085. Federal employment information is available through the United States Civil Service Commission, Washington, D. C. 20415; Office of Personnel, United States Department of Agriculture, Washington, D. C. 20250; and local offices of the Soil Conservation Service.

AGRICULTURAL ECONOMISTS

The state agricultural colleges are the leading employers of agricultural economists. Here the main kinds of work are research, teaching, and work in Extension Services. Production economics, farm management, marketing, and prices are examples of broad fields of study.

The United States Department of Agriculture is the second largest employer of agricultural economists. The Department requires specialists in marketing, farm finance, merchandising methods, crop estimates, poultry production, labor requirements, livestock production, farm costs, feeds and feeding, and farm tenure.

The land-grant colleges and the United States Department of Agriculture employ approximately one-half of the agricultural economists. Most of the others are employed by farm organizations, state departments of agriculture, farmer cooperatives, Federal Reserve banks, conservation and reclamation districts, food processing companies, insurance companies, farm machinery manufacturers, and feed and fertilizer companies. Recently there has been an increase in the employment of agricultural economists by companies that provide services to farmers. Also, foreign countries are using more of this nation's economists for advisory work in rural development. Overseas employment varies from temporary assignments of a few months to careers in foreign service, such as agricultural attachés.

A bachelor's degree with courses in agriculture and agricultural economics is the minimum educational requirement for an agricultural economist. Land-grant colleges and other colleges offer such essential courses as statistics, farm management, land economics, farm finance, marketing, and production economics. Agricultural economists with advanced degrees are needed in the more technical fields of this profession.

The department of agricultural economics in the land-grant college of your state can provide additional information about opportunities in agricultural economics. The Office of Personnel, United States Department of Agriculture, Washington, D. C. 20250, and the United States Department of Agriculture recruitment representatives at the land-grant colleges are the best sources of information about Federal employment in agricultural economics.

MANAGERS OF FARMER COOPERATIVES

A farmer cooperative is a business enterprise owned and operated by its members. Farmers sell their products and buy supplies through cooperatives. Groups of farmers can operate more economically than individuals in obtaining special farm business services such as electricity, irrigation, telephones, insurance, and credit. Capable managers and other personnel are needed for new positions as more cooperatives are organized and for vacancies resulting from normal employee turnover. A management position in an established cooperative is usually filled by a promotion from one of the departments.

Farmer cooperatives often show a definite preference for college graduates to fill new positions. However, many capable employees who do not have a college education are promoted. A few cooperatives give scholarships to

select students as a method of providing education for potential leaders in this field.

Local or regional farmer cooperatives in or near your community may provide information about employment opportunities. The county agricultural agent and the agricultural economics department of the state agricultural college are other sources of information. Three national organizations which provide general information on cooperatives are the National Council of Farmer Cooperatives, the American Institute of Cooperatives (both located at 744 Jackson Place, N.W., Washington, D. C.), and the Cooperative League of the United States of America, 343 South Dearborn Street, Chicago, Illinois.

FARM SERVICE WORKERS

Farmers and ranchers require many kinds of services that can be learned and performed by workers in related fields. A person may enter one of these fields as an independent operator or as an employee. A special farm service enterprise may require a large investment or a small one, depending on the size and nature of the operation. Some services, such as custom plowing and harvesting, are seasonal; other services may be performed throughout the year. A few services can be combined with the operation of a small farm.

An association of farmers usually employs one or more workers on a monthly basis to conduct the operations of cow testing and artificial breeding. Dairy herd improvement associations employ supervisors who do cow testing. For this work a high school education is required, and a farm background is needed. Inseminators employed by artificial breeding associations must have at least a high school education. Herd improvement and artificial breeding associations provide brief periods of approximately one month of specialized training. Agricultural college training is desirable but usually not required in these occupations.

Many farmers do custom work to make full use of their equipment. Where there is a long growing season, custom services can be carried on long enough to be profitable on a full-time basis. Operators of large grain-harvesting combines follow the annual grain harvest from Texas and Oklahoma across Kansas and Nebraska to Montana and the Dakotas.

An agricultural background is helpful to anyone engaged in farm service jobs, but the required agricultural skills can be learned more easily than such specialized skills as those required of soil conservationists, soil scientists, agricultural research workers, or Agricultural Extension Service workers.

OTHER AGRICULTURE-RELATED PROFESSIONS

Because of the large number of occupations related to agriculture, it is difficult to select the ones which should be included in a brief general discussion such as this. The following are closely associated with the occupations already described: agricultural writer, bacteriologist, entomologist, forester, pathologist, and wildlife biologist. The United States Department of Agriculture and the land-grant colleges are sources of information on these occupations.

Agricultural writer. The agricultural writer belongs to a field of specialization known as "agricultural communications." This is an appropriate name because the writer's purpose is to communicate ideas and information to readers. The United States Department of Agriculture employs a staff of market reporters in field offices throughout the nation. These writers report on the movement of agricultural products from farms to markets. A large number of reporters and editors prepare news and feature articles for farm magazines, bulletins, and radio programs. Many radio stations employ farm directors. State conservation agencies, land-grant colleges, and many Federal agencies employ agricultural writers.

The educational requirements in this field vary according to the kind of work to be done. Although not required for all positions, a bachelor's degree in agriculture or journalism is helpful. A degree in agriculture should be supplemented by courses in English and journalism, and a degree in journalism or English should be supplemented by courses in agriculture.

Bacteriologist. A specialist in bacteriology may be engaged in research, teaching, laboratory testing, disease control, or quality control work which requires a knowledge of microorganisms. A bacteriologist needs a good background in microbiology, including the study of fungi and viruses.

The fermentation research of bacteriologists is necessary to produce antibiotics, vitamin tablets, amino acids, sugars, and polymers by the action of microorganisms. A special problem of the bacteriologist is to prevent the spoilage of foods, cleaning and polishing waxes, and other products that are attacked by bacteria and fungi.

Bacteriologists are employed by many federal and state agencies, colleges and universities, and companies that manufacture supplies for farms and ranches. A bachelor's degree is the minimum educational requirement. The master's and doctor's degrees are required for college teaching and the better research positions.

Entomologist. Insects are the special field of interest of the entomologist. Particularly important are methods of controlling insects that affect people, livestock, crops, forests, and marketed farm commodities. The entomologist is also interested in beneficial insects.

Sources of employment and educational qualifications are similar to those of the bacteriologist. In both professions a thorough knowledge of biological sciences is needed.

Employment opportunities are most numerous in the United States Forest Service, state forestry departments, and forest industries. A bachelor's degree in forestry is the minimum educational requirement for a forest entomologist.

Pathologist. Plant pathologists are concerned mainly with causes and control of plant diseases. Animal pathologists have the same interest in animal diseases. Both specialists must deal with diseases caused by bacteria, fungi, viruses, and physiological conditions. Nutritional disorders are a particular problem of the animal pathologist. The lack of essential elements in the soil may result in pathological conditions in plants.

The principal employers of pathologists are the United States Department of Agriculture, other federal agencies,

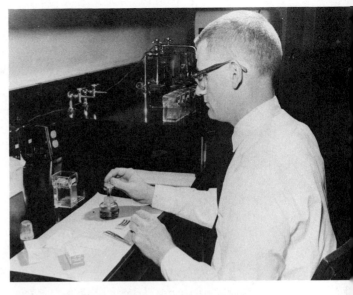

A plant pathologist makes a microscopic study to identify a plant disease. (*Courtesy* U. S. Forest Service)

and manufacturers of antibiotics, serums and vaccines, insecticides, and herbicides. A bachelor's degree is required for employment in pathology, except for some positions as laboratory technicians. Graduate degrees are required for most teaching and research positions in pathology.

Wildlife biologist. Wildlife biologists are concerned with the management of wildlife resources and the improvement of habitats for game and nongame species. The principal employers of wildlife biologists are the United States Fish and Wildlife Service and state conservation agencies. Some wildlife biologists are employed in Federal agencies other than the United States Fish and Wildlife Service.

A professional wildlife biologist must have a bachelor's or master's degree. There are some positions in wild-

Wildlife biologists help to maintain a good habitat for this white-tailed fawn and other wildlife. (Idaho) (*Courtesy* U.S. Forest Service)

This plant geneticist is doing birch tree hybridization work in the field. (*Courtesy* U. S. Forest Service)

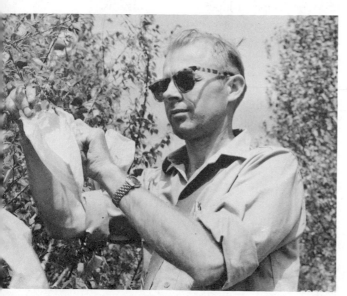

life management that do not require a degree.

Some state universities and land-grant colleges offer professional education for wildlife biologists. For information about this profession, contact your vocational agriculture teacher, county agricultural agent, or game warden; or write to the Fish and Wildlife Service, Department of the Interior, Washington, D. C., the land-grant college in your state, or your state wildlife conservation agency.

Geneticist. Geneticists study and apply principles of heredity in developing breeds, strains, varieties, and hybrids of plants and animals. The goal is to increase the production of better quality food and fiber than is presently available. High yielding hybrid corn, seedless fruit, many new roses, better broiler chickens and turkeys, and cross-bred cattle are just a few of the important accomplishments of geneticists.

Embryologist. The study of the formation and development of embryos of plants and animals is the specialty of embryologists. The embryologist is concerned with both normal embryo development and abnormal (pathological) conditions. Early diagnosis of problems sometimes permits corrective treatment. The embryologist and the geneticist have much in common in the improvement of plants and animals.

Plant and animal husbandry. Specialists in these agricultural sciences are concerned with plant and animal care and management. Their purpose is more efficient and economical production of food and fiber. Plant and animal husbandry specialists must know characteristics of breeds and varieties that best meet specific market requirements. This profession is also concerned with nutrition, prevention of diseases, insect control, sanitation, and many other related problems.

Education in plant science and animal husbandry may be obtained from a degree program in any of the 67 land-grant colleges and universities, and in three other universities, whose addresses are given in Appendix A. Animal technology programs are also listed in that appendix.

Positions in plant and animal husbandry are available in all states in the United States. Some overseas positions are available for shorter periods with the Action-Peace Corps.

Biochemist. This scientist deals with the chemical compounds and processes occurring in living animals and plants. The knowledge acquired by the biochemist is useful to the nutritionist, pathologist, plant and animal husbandry specialists, and other professional agricultural workers.

Human nutritionist. Nutrition is the science of food requirements and the processes by which the body uses food substances. The selection of balanced diets is one special aspect of nutrition. In the past three decades, human nutritionists and soil scientists have given special attention to the relation of soil fertility to human nutrition.

Rural sociologist. The customs, practices, laws, and economic conditions of rural people are the special concern of rural sociologists. These specialists work with many local, state, and federal agencies in helping farm families live better through effective use of community resources. The rural sociologist, agricultural economist, and agricultural extension agent are teamworkers in the development of better rural communities.

Public and private finance. There are many public and private lending agencies which make loans to farmers. These agencies employ men and women with a broad education in agriculture and business. In addition to practical farm experience, these finance workers are usually required to have academic education in agriculture, economics, accounting, and other subjects related to the evaluation and management of loans.

The protection of our natural environment, such as these giant cacti in Pima County, Arizona, involves many career opportunities in agriculture. (*Courtesy* U. S. Department of Agriculture Information Office)

A LOOK AHEAD

The United States Department of Health, Education, and Welfare has pointed out that there is a growing trend toward careers that do not require four years of college. By 1975 the scientific and technological changes in the United States may create more than a million new career opportunities. People with technical skills are now in great demand, and the demand may exceed the supply of new workers for many years to come.

There should also be a corresponding expansion of opportunities for young men and women with college degrees in technical fields. The income potential for skilled personnel on both the degree and nondegree levels is much higher than for the untrained.

Many kinds of technicians make up a large career field in which a year or two of technical education may be sufficient. According to a general definition by the Department of Health, Education, and Welfare, "technicians are people who work directly with scientists, engineers and other professionals in every field of science and technology." As scientists and engineers do the theoretical work, technicians put theory into practice.

Women as well as men may become technicians. Traditional barriers which once limited this field to men are breaking down. In addition to having the necessary interests and aptitudes, women must have the technical education for the specific careers of their choice.

QUESTIONS

1 How can there be an increase in agriculture and related fields when the farm population is declining?

2 What is agri-business?

3 Why is there an increase in foreign employment in agriculture?

4 Which agricultural occupations are related to conservation?

5 What agricultural occupations require skill in English?

6 Which occupations require skill in mathematics, physics, chemistry, and biology?

7 For which occupations is the United States Department of Agriculture the major employer?

8 What agricultural specialists work for state conservation agencies?

9 Compare four agricultural occupations on the basis of educational requirements.

10 What is the present trend in careers that do not require four years of college education?

ACTIVITIES

1   Ask your school counselor or librarian to suggest publications for additional reading on agricultural occupations.

2   Ask the local county agricultural agent to tell your class about the work of the Agricultural Extension Service.

3   Observe conservation work in your community to determine the occupations represented.

4   Survey the occupational interests of members of your class.

5   Examine catalogues of several agricultural colleges and report on courses, degrees, and tuition. Land-grant colleges are listed in Appendix A.

2 THE STORY OF FOOD

Assume that a pair of mature rabbits inhabits the most fertile acre of soil, planted to the most productive forage. Assume that the area is fenced against all predators, such as hawks, and that all offspring survive minor injuries, diseases, and insects. The original pair of rabbits normally would have four litters a year with six young per litter, or 24 rabbits. At seven months of age the young rabbits themselves become parents, each pair having six more rabbits, and so on.

At this rate of reproduction, how long will it be before the rabbits eat all of the forage from the acre and die from starvation?

Apply this situation to the present world population. An expanding population depends upon food grown on an essentially fixed acreage. The result is eventual starvation. In 1972, our nation supplied millions of tons of food for relief overseas.

One person in three in the world now goes to bed hungry and gets up hungrier. Death from starvation still exists, but mostly people die young because weakness from too little nourishing food makes them more susceptible to diseases.

The number of babies is increasing faster than the food supply in South America, Asia (excluding Japan), and Africa (excluding South Africa). The food deficit in countries on these continents is now approximately 16 million tons each year, and the deficit is expected to be 88 million tons by 1985.

Why don't we in the United States send our surpluses to the poorly fed and the starving people in the less fortunate countries? We do, but our reserves are now low. The anticipated food deficit of 88 million tons in 1985,

FOOD AND POPULATION GROWTH IN THE LESS-DEVELOPED COUNTRIES*

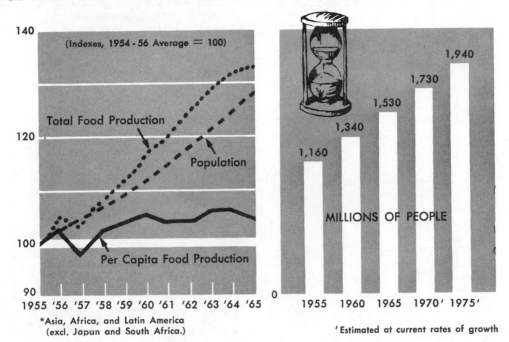

*Asia, Africa, and Latin America
(excl. Japan and South Africa.)

' Estimated at current rates of growth

Time is running out! Population growth is faster than the increase in food production in Asia, Africa, and Latin America. Can this trend be changed? (*Source:* United States Agency for International Development)

for example, compares with the 1970 *total* United States production of 148 million tons of corn and 47 million tons of wheat. We do not and will not have that much surplus. Furthermore, no country wants to be continuously dependent upon another country for food. There is the question of national pride and self-reliance and independence; and there is also the problem of finding money to pay for the food.

INCREASING FOOD PRODUCTION

Developing countries in South America, Asia, and Africa are attempting to increase food production in two contrasting ways: by an increase in acres planted and by an increase in production per acre. A few countries in South America and most countries in Africa have suitable land that can be cleared and planted for food crops. Some countries are increasing the amount of water available and irrigating more acres. Nearly all countries in Asia, however, have no new lands to be cleared and must concentrate on increasing crop yields per acre. But each succeeding year, the number of new acres that can be cleared, irrigated, and planted is reduced, until the only remaining hope of growing more food will be to increase the yield per acre.

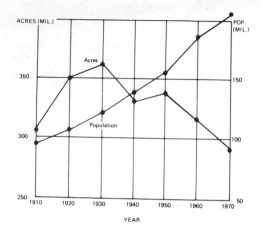

Total population in the United States from 1910 to 1970, compared with the total acres of crops harvested. Will the trends since 1950 continue? (*Source: Agricultural Science Review*, Vol. 9, No. 2, 1971, U. S. D. A.)

Yields of oats, barley, and wheat in bushels per acre from 1870 to 1970. (*Source: Agricultural Science Review*, Vol. 9, No. 2, 1971, U. S. D. A.)

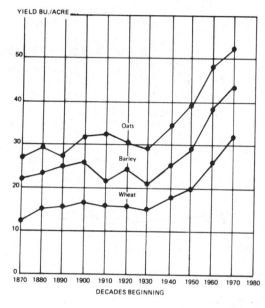

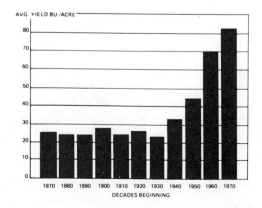

United States trends in yields per acre of corn (maize) from 1870 to 1970. What happened from 1950 to 1960? (*Source: Agricultural Science Review*, Vol. 9, No. 2, 1971, U. S. D. A.*

With modern weedicides, insecticides, and fungicides, it is not necessary to develop rice plants primarily for survival. High yields with superior management is the principal criterion now for breeding rice. Right: A common rice variety grown in the Philippines lodged and yields were low. Left: The newly bred IR8 variety did not lodge, and consequently yielded more than twice as much rice. (*Courtesy* Robert F. Chandler, Jr., Director, International Rice Research Institute)

Yields per acre in the United States remained fairly constant until plant breeders developed new crop varieties with the potential to respond to higher fertilizer applications. Then, with better seedbeds, better seed, closer spacing, more fertilizer, and effective weedicides, insecticides, and fungicides, and more timely harvesting, crop yields per acre increased rapidly.

From 1960 to 1970 the following production efficiency was achieved in the United States: 70 percent more farm output per man-hour with 46 percent less farm labor but with 13 percent more mechanized power and a 108 percent increase in tonnage of fertilizer.

1. Farm workers each produced 27 percent more food per man-hour than they did prior to 1960. Prices for food (farm value) dropped during the

1960–1970 period.

2. In 1972 one farmer was able to feed 50 other people, compared to 26 in 1960.

3. Food marketing workers turn out about 30 percent more marketing services per worker than they did 10 years ago.

4. We spend less of our take-home pay than ever for food: 16½ percent today, 20 percent 10 years ago.

5. In 1972 a farm magazine reported that in the previous year a farmer in the central United States produced more than 300 tons of milk from 68 registered dairy cows, and with only part-time family help.

As efficient as farmers in the United States are, they still do not lead the world in yields per acre of wheat and rice—only in corn. Great Britain leads in yield of wheat, and Japan in rice.

Profits on the yield per acre in the United States have increased because the cost of the production inputs has remained fairly stable while the prices that farmers receive have increased. Also, the efficient use of these inputs has increased under superior managerial skill.

Our national pattern of increasing yields per acre is being followed in the developing countries with some success. Many problems, however, prevent the full implementation of the pattern. For example, in most developing nations, fields are very small and land area per farm averages approximately three acres. Furthermore, three acres that do belong to one farmer may be scattered in ten fields at ten different locations. For the first time, in 1966 fertilizer-responsive crop varieties were available; but fertilizer prices are very high, and effec-

tive weedicides, insecticides, and fungicides may not be available, or their price may be exorbitant. But even if all production supplies happen to be available, the farmer may not be able to buy them without credit. Even with every facility available, a farmer in a developing country must still be educated to use the modern production inputs with skill, timeliness, and precision.

PROTECTING FOOD

In the developing countries it is not enough simply to increase food production; the food must be harvested at the proper time, dried or otherwise processed for storage, and stored in rat-proof, insect-proof, theft-proof, fire-proof, and water-proof shelters. In many countries as much as 20 percent of a harvested crop is destroyed by one or more of these hazards.

CONTROLLING POPULATION

Increasing food grains per acre and protecting the food from the hazards of destruction are only one side of the story of food and the sad story of starvation. Population must be controlled. Starvation will continue unless food production increases faster than the increase in babies that eat the food.

The developing countries are trying many methods of controlling populations, but obstacles to success include religious belief, illiteracy, cost, and cultural tradition.

Today there are 3.6 billion people in the world, and half of them are malnourished. Before you can complete the reading of this sentence, ten more babies will have been born. By the year 2000, there will be 7 billion people on this planet, and we will be saying, "Quit shoving! When do we eat? *What* do we eat?"

HOW THE UNITED STATES IS HELPING

The United States Government is assisting most of the developing nations to help themselves become self-sufficient in food production, and is simultaneously shipping surplus food grains to reduce starvation. Nearly one-third of the United States assistance programs to agriculture consist of sending fertilizer to developing nations. The reason that fertilizer is sent instead of sending all aid in food is that fertilizer will help a country to produce food on a self-respecting and continuing basis. For example, one million dollars spent for wheat will feed 70,000 people for a year, whereas the same million dollars spent for fertilizer and sent to a developing nation and used properly on its crops will produce additional grain to feed 200,000 people for a year.

The reserve supplies of United States grains are adequate, and the supplies of fertilizer are sufficient for our needs. We are sending both products abroad, teaching many farmers in developing nations how to use them efficiently. Fertilizer can be used as a substitute for land. The United States is helping many developing nations build fertilizer plants; technical assistance to these nations is the policy of the United States government. In the long run, each developing nation aspires to use technical inputs ef-

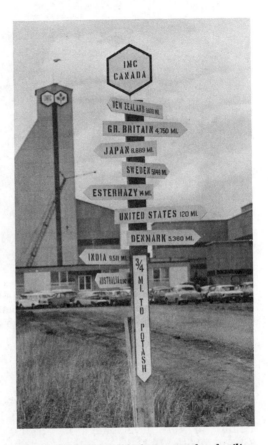

The true international character of a fertilizer company with headquarters near Chicago, Illinois, United States of America, is indicated by this sign that shows directions and distances to all of its world-wide activities. (*Courtesy* International Minerals and Chemical Corporation)

ficiently to achieve self-sufficiency as quickly as possible.

The United States Agency for International Development (USAID or just AID) was created in 1961, replacing other foreign aid agencies that had been operating since the Marshall Plan was initiated in 1948. AID has more than 3,000 persons working in 70 countries in Asia, Africa, and Latin America.

The Rockefeller Foundation has worked in Mexico continuously for more than 25 years to help develop varieties of wheat that are highly responsive to modern inputs such as fertilizer. Some of the responsive varieties were given to India, Pakistan, and other wheat-growing areas to replace low-yielding varieties. (*Courtesy* Rockefeller Foundation)

These workers provide technical assistance programs in food, agriculture, education, health, and rural reconstruction. Loans for the development of roads, schools, dams, and power and irrigation facilities are also made by AID.

Seventy-seven percent of the persons employed by AID in 1965 for work in developing nations were those working under one of the 1,308 technical service contracts. Among these AID contracts were 71 with United States universities that had overseas staffs.

Helping developing nations produce enough food for themselves is one of the most urgent needs of the world. AID personnel fully realize this problem and are helping to solve it. For example, nearly one-third of the money spent for agriculture is used for the production or purchase of fertilizer. Not only fertilizers, but better farm mechaniza-

tion, seed, insecticides, fungicides, and weedicides are required if food crop production is to increase.

An example of AID technical assistance to Ghana (Africa) can be cited. Many varieties of improved garden seeds were sent from the United States to Ghana for trials to determine the ones best adapted.

The following are examples of the best varieties of several kinds of vegetables: beans (Tendergreen), cucumber (Palmetto), lettuce (Great Lakes), Sweet pepper (World Beater), and onion (Crystal White Wax).

Some of the local Ghanaian varieties of eggplant, greens (many types), okra, hot pepper, radish, sweet potato, onion, and tomato (Improved Zuarungu) produce very satisfactorily. Each country receiving AID technical assistance has many local crops that are as good

These Indian officials are looking at a field of Sonora 63, one of the wheat varieties developed in Mexico under a program assisted by the Rockefeller Foundation. (*Courtesy* Rockefeller Foundation)

as, or better than, any that come from the United States. One function of the AID assistance then is to help the developing country to increase the supply of seeds of the best local varieties and to assist in storing, distributing, and promoting their use for increasing food production. This is an illustration of the guidelines for AID assistance to help developing nations to help themselves until they become self-sufficient. Greece, Taiwan (Formosa), the Philippines, Mexico, Korea, and Turkey are such nations that will soon be self-sufficient.

It is somewhat surprising that the United States ranks as high as it does in total crop production and gives away and sells 55 million tons of food each year from 80 million acres of cropland, yet almost no food crops are native to the United States. The genius of our people is that they have found and adapted so many "foreign" crops that we have become the "breadbasket" for the world.

FOOD AND AGRICULTURE ORGANIZATION OF THE UNITED NATIONS

The Food and Agriculture Organization (FAO) of the United Nations came into being in October, 1945. It has permanent headquarters in Rome, Italy. Work of the FAO consists of technical advisory services to member countries on agriculture, nutrition, forestry, and fisheries. Special regional projects include plant and animal disease control, and educational seminars on problems that are of vital concern to many countries. Many useful bulletins are published by FAO and are available at low cost.

A farm boy in Upper Volta, western Africa, is proud of the tomatoes he has grown with the help of experts from the Food and Agricultural Organization of the United Nations. These tomatoes had to be irrigated because of the long dry season. (*Courtesy* A. Defever, Food and Agricultural Organization of the United Nations)

THE ROCKEFELLER AND FORD FOUNDATIONS

Both the Rockefeller Foundation and the Ford Foundation assist selected countries in developing higher-yielding varieties of crops, in pursuing research and extension activities to use these crops, and in choosing the proper combination of machines, fertilizers, and tillage practices to increase yields of food. In addition, the foundations are combining their resources to establish and operate world-wide research stations for solving problems in food production. One such research station is the International Rice Research Institute in the Philippines. A second is the Tropical Research Institute at Ibadan, Nigeria (Africa). A third research station has been established in Colombia, South America. These new research facilities have developed highly-responsive wheats and rices that are readily adapted to the environments of the food-deficit nations. More food will mean less hunger.

QUESTIONS

1 On what continents is population outrunning food supply?

2 What is the estimated food deficit by 1985?

3 Which nations lead in the production per acre of rice, wheat, and corn?

4   How can fertilizer be a substitute for land?

5   What are the two principal ways in which the United States helps the developing nations?

6   Why is food considered a bargain?

7   What factors tend to prevent the increase of food production in some countries?

8   Why is population control difficult in so many nations?

9   Which of the vegetables adapted to Ghana are grown in your community?

10 How do international organizations help to solve the problem of world food shortages?

ACTIVITIES

1   Compare the number of people that can be fed a year by spending 3.5 million dollars for wheat with the number that can be fed by spending that amount for fertilizer for a developing nation.

2   Using the scale of the chart, Food and Population Growth, make a bar chart showing the approximate populations in the less developed countries in 1980 and 1985.

3   Use encyclopedias, the World Almanac, and other references to learn more about the resources and people in several developing nations.

4   Bring to class newspaper articles about the world food supply, the work of the Food and Agricultural Organization of the United Nations, and similar topics.

5   On an outline map of the world, shade or color the principal countries which do not have sufficient food. Use a different color to indicate those countries which are self-sufficient in food production.

3 AGRICULTURE ACROSS THE NATION

The agricultural regions of the continental United States are closely related to climate and topography. These two influences can be easily observed on a plane trip across the nation. Traveling from east to west, we see first the beaches of the Atlantic coast, the well-known summer vacation playgrounds. Next there is the tidewater country which supports the crab, fish, and oyster industries.

The Coastal Plain west of the inland waterways is wide in the South and narrow in the North. Much of the sandy level land is suitable for farming.

The rolling hills of the Piedmont, which the French called "foot of the mountain," have remnants of the original hardwood and pine forests. Most of the land is now in fields, pastures, and second-growth trees.

Beyond the Piedmont are the Appalachian Mountains. Here the heavy rainfall feeds the springs and streams which flow out of the mountains.

Now we cross the Mississippi Basin, through which the Mississippi River flows. The river basin is 1,200 miles wide and more than 1,600 miles long. The southern part of the basin contains the Cotton Belt. The northern part of the basin is the North Central region, also known as the Midwest, which contains much of the Corn Belt.

West of the Mississippi River the land rises gradually for a distance of 800 miles until the elevation is more than 5,000 feet. Originally covered by tall grasses, the central lowlands in the North are called prairies. The low Ozark and Ouachita mountains are in the central southern part of the basin 100 miles west of the river. Most of the Great Mississippi Basin is good farmland. The western edge of the basin is generally treeless because of the low rainfall. This is the Great Plains country, originally covered with short grasses.

The rugged Rocky Mountains begin to rise at the western edge of the basin. This western third of the nation is generally too rough and dry for agriculture. Only the high mountains have much precipitation. The rain and melting snow provide irrigation water for the valleys below.

Close to the Pacific Ocean there is a small range of mountains, the Coastal Range. The low-lying valleys between these mountains and the Sierra Nevada Range 100 miles inland from the ocean are excellent for irrigation farming.

Each kind of climate has its own type of farming and rural living. The different agricultural regions also have many common characteristics. There are numerous towns within easy reach of the farmsteads, which are scattered across the countryside. Among the farmsteads are the homes of part-time farmers who drive daily to jobs in cities and towns.

In the next chapter, and Chapter 10, on climate and weather, you will find that the agriculture of Alaska and Hawaii are also influenced by climate and topography.

FACTORS AFFECTING TYPES OF FARMING

There are many examples of how soils affect types of farming. The Corn Belt is associated with the brown silt loams, which are characteristic prairie soils. The alluvial soils of the Mississippi River Delta and the red and yellow

Adequate rainfall at the right time and fertile, fairly level, fine-textured soils that respond to fertilizers are characteristic of the Corn Belt. (*Courtesy* Allis-Chalmers)

The type of agriculture in the West is influenced mainly by the amount of available water. In the 20-30 inch rainfall belt, cattle ranching prevails. (*Courtesy* American Hereford Association)

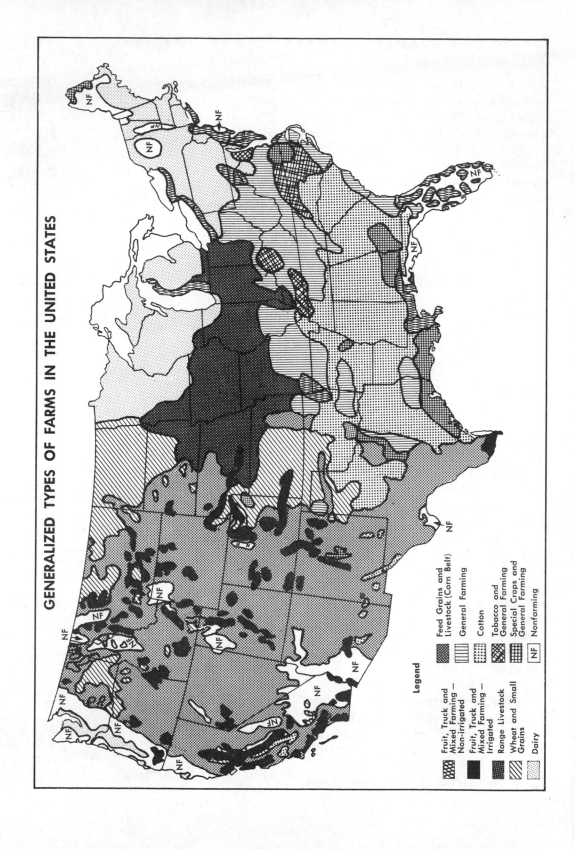

GENERALIZED TYPES OF FARMS IN THE UNITED STATES

Legend

Fruit, Truck and
Mixed Farming —
Non-irrigated

Fruit, Truck and
Mixed Farming —
Irrigated

Range Livestock

Wheat and Small
Grains

Dairy

Feed Grains and
Livestock (Corn Belt)

General Farming

Cotton

Tobacco and
General Farming

Special Crops and
General Farming

Nonfarming

NF

A modern corn picker is ready for action in the Corn Belt. (*Courtesy* Allis-Chalmers)

soils widely distributed through the South are adapted to cotton. Fruit and truck crops are best for some of the soils in the valleys of the Pacific Coast states.

The type of farming in a particular area often reflects the influence of plant and animal diseases, insects, and weeds. New varieties of crops are sometimes developed to offset these factors. The destructive influences of the cotton boll weevil in the South and of the corn borer in the Corn Belt have encouraged farmers in these regions to grow several kinds of crops in rotation instead of following the older system of one-crop farming.

Truck farming and dairying tend to be concentrated near population centers that are good consumer markets. Next in the order of nearness to cities are poultry farming, grain growing, and livestock production. Range cattle, raised in remote areas, may be fattened in the Corn Belt where the meat packing industry is concentrated.

The United States Department of Agriculture recognizes ten major agricultural regions: Feed Grains and Livestock; Cotton; Dairy; Wheat and Small Grains; Range Livestock; Fruit, Truck Crop, and Mixed Farming; General Farming; Tobacco and General Farming; Special Crops and General Farming; and Fruit, Truck, and Mixed Farming—Irrigated. There are also certain nonfarming areas.

FEED-GRAINS AND LIVESTOCK REGION (CORN BELT)

This region is best known as the Corn Belt. The topography is mainly level to gently rolling. Precipitation averages about 40 inches per year, and the rainfall is well distributed during the growing season when farm crops need abundant moisture. Corn grows rapidly during hot days and warm nights.

Originally the Corn Belt soils were very fertile, with plenty of nitrogen and organic matter. Although the region still has a high level of fertility, large quantities of commercial fertilizers are used. The beneficial results of planned conservation practices can be seen on many well-managed farms.

Corn is the principal crop in the Corn Belt farming system. Crop rotations usually include oats, wheat, grasses, and legumes. Soybeans often follow corn in the rotation. After the soybeans are harvested a winter cover crop is needed to control erosion. Winter wheat is seeded in the fall, while oats are seeded in the spring. Both these grain crops are harvested in midsummer about the time the corn has been cultivated for the last time.

Grasses and legumes, which are usually seeded with oats and wheat, are very important throughout the Corn Belt. Nutritious grass and legume pasture and high-quality hay help to reduce the cost of raising and fattening beef cattle. The soil-building grasses and legumes in the rotations also help to increase crop yields by keeping the soil soft and mellow.

Corn, oats, and soybeans. The level, fertile land of central Iowa and east-central Illinois is ideal for growing corn, oats, and soybeans to be sold for cash instead of being fed to livestock. Because of the topography, this land can be cropped heavily with little erosion

damage. Pasture and hay crops are not grown extensively enough to make beef cattle a major enterprise. Hog raising is limited by the sale of corn as a cash crop.

Landowners who do not operate farms usually prefer to sell their part of the crops as soon as possible after the crops are harvested. The operators who pay cash rent also depend upon the sale of grain rather than livestock. In this type of farming there is less investment for fences and buildings than on livestock farms.

Cattle and hogs. The feeding of cattle and hogs is the main enterprise in the subregion of the Corn Belt near the Missouri and Mississippi Rivers. Loessial or wind-blown soils are characteristic of these areas. Permanent pasture is the best use of much of the rolling land.

The cropland here is fertile enough to produce good yields of corn and other grains but must be managed carefully to prevent serious erosion. The large acreage of grasses and legumes in the cropping system favors the raising of cattle. This combination of grassland crops and cattle keeps the soil well supplied with organic matter.

The fattening of hogs is the principal use of corn in this type of farming. On many farms there is a surplus of corn, a situation which favors the feeding of cattle as well as hogs. Beef cattle are more numerous than dairy cattle because more grain and roughage can be used for fattening.

Hogs and soft winter wheat. The farms in the east-central part of the Corn Belt are smaller than in the areas used for large-scale production of corn, hogs, and cattle. Winter wheat is the main crop on the coarse-textured, well-drained soils. On most farms the entire corn crop is fed to hogs. This means that winter wheat and hogs are the principal sources of income.

Other farming. Corn is somewhat less important outside the central part of the Corn Belt. Farmers in the southern part of the region depend more upon pasture and less upon concentrates for livestock. Here the grazing of young cattle and mother cows, rather than the fattening of cattle, is emphasized. The small supply of corn also limits hog production.

On the western side of the Corn Belt there is an increase in the acreage of wheat in proportion to corn and oats. The lower rainfall reduces the carrying capacity of the pastures and the number of cattle on farms. There is not enough corn for the extensive feeding of hogs. Grain is more important as a cash crop than as feed for livestock.

In east-central Kansas and western Missouri, livestock are necessary to use the hay and pasture crops and to maintain soil fertility. The cropping system includes corn, wheat, grain sorghums, and large acreages of hay and pasture. Poor soils are one reason for this type of farming.

The broken topography of northeastern Iowa and northwestern Illinois limits the amount of land suitable for cultivation. Some of the soils have developed from glacial drift and are leached and acid. Corn and oats are the principal grain crops. Because much of the land is better adapted to forage crops, there is more roughage than concentrated feeds for livestock. This type of farming combines hog raising with dairying rather than with beef-cattle production.

The farming pattern in northeastern Indiana and northwestern Ohio includes dairying, soybeans, cash grains, and general livestock production. Soybeans are a very important crop. Corn and oats have about the same acreage. A large part of the land is used for hay and pasture. Whole-milk dairying is combined with the raising of hogs in farming areas near the large cities.

COTTON REGION

Climate is more important than soils in determining where cotton may be produced. A long growing season with high temperature is necessary for this subtropical plant. The Cotton Belt is limited mainly to states that have at least 200 frost-free days.

Except in irrigated areas, cotton does not grow well where the annual precipitation is less than 20 inches. But in most of the areas along the Atlantic Ocean and the Gulf of Mexico there is too much rain during the fruiting and harvesting seasons.

Good drainage is essential in the production of cotton. The combination of poor drainage and high rainfall makes some areas within the Cotton Belt unsuitable for this crop.

Cotton responds well to chemical fertilizers, which are used to offset the lack of minerals in some southern soils. Nitrogen, phosphorus, and potash are the critical elements most likely to be lacking.

The Cotton Belt covers most of the area of nine states in the South and parts of the other four states in the region. The black soils of Texas, Alabama, and Mississippi are famous for

A cotton picker in operation in the Mississippi Delta. Cotton was the number one cash crop in Mississippi in 1970 and the number two cash crop in Texas and Arizona. (*Courtesy* Mississippi Agricultural and Industrial Board)

In the Cotton Belt, piney woods occupy more than half of the area. In the open areas between the trees, range livestock are grazed successfully for a part of the year. (Louisiana) (*Courtesy* Southern Forest Experiment Station)

the production of cotton. Although cotton is produced on irrigated land in New Mexico, Arizona, and California, it is the Cotton Belt of the South that makes the United States the world's largest producer of cotton.

Rice and sugarcane grow better than cotton on certain areas within the Cotton Belt. Flue-cured tobacco has displaced cotton in some counties, while peanuts are better adapted to some of the coarse-textured southern soils. Markets sometimes favor peanuts, tobacco, truck crops, or wheat, rather than cotton.

Power machinery is becoming more important in the production of cotton, while hand labor is becoming less important. Such areas as the High Plains and Corpus Christi sections of Texas and the Delta lands of Arkansas, Louisiana, and Mississippi have large level fields that are ideal for the operation of heavy equipment. Machines are used for seedbed preparation, fertilizing, planting, cultivating, spraying, and harvesting.

Diversified farming, rather than one-crop farming, is the trend in the region where cotton is king. Grain sorghums are grown extensively on cotton farms in Texas. Corn, oats, and soybeans are popular crops on most of the farms in the Delta. Dairying is a growing industry in the brown-loam area of Mississippi, Tennessee, and Louisiana, and in the Black Belt of Alabama and Mississippi. There are now many fine herds of beef cattle in these states, which were once exclusively devoted to cotton.

Grassland farming is one of the most promising developments in the Cotton Belt. Grasses and legumes serve the triple purpose of providing cover to prevent soil erosion, increasing the organic matter and nitrogen in the soil, and yielding nutritious forage for livestock.

Wheat, grain sorghum, and range livestock are more important than cotton in the Panhandle of Texas and Oklahoma and in sections of central Oklahoma. Where the Cotton Belt and Wheat Belt meet in these states, market prices largely determine whether more acreage is used for wheat or cotton. As the market demand for one crop becomes stronger, production goes up. A slump in the market has the opposite effect.

In recent years there has been a gradual increase in the acreage of irrigated cotton in Texas, New Mexico, California, and Arizona. Mechanized cotton farming is a highly specialized enterprise in some places. On other areas a diversified type of farming includes alfalfa, dairying, truck crops, and fruit.

DAIRY REGIONS

The Northeast and the Lake States are two well-known dairy farming regions. This type of farming also is concentrated along the North Pacific Coast and in areas adjacent to large cities throughout the United States. The large cities in the dairying regions provide good markets for milk.

The soils and topography of the three leading dairy farming regions are generally not well adapted to the production of cereal crops. The cool climate and abundant precipitation favor the growth of grasses and legumes. Dairy cattle are efficient users of high-quality forage and can do their own harvesting of pasture crops.

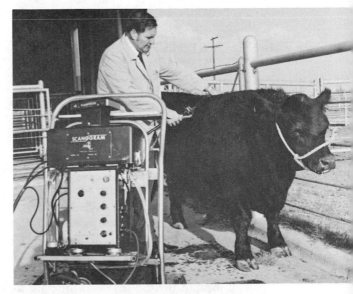

In live carcass research, an ultrasonic instrument is being used to determine the quality of meat this Angus cow will make. Cows that rate high are used for breeding. (*Courtesy*, Texas A&M Agricultural Extension Service)

A family dairy farm in one of the best dairy regions. (Wisconsin) (*Courtesy* Soil Conservation Service)

Family-size farms are typical of the highly specialized dairying areas in the central Northeast, in northeastern Illinois, in eastern Wisconsin, and along much of the Pacific Coast. On the farms the supply of hay and pasture largely determines the size of the dairy herd. Some grain is produced to supplement the forage, but the main supply of concentrates comes from other regions.

The dairy farms near the large cities specialize in the production of milk. Some of the large herds are fed forage and concentrates in dry lots and do not have access to pasture. Alfalfa hay provides about half of the nutrients in the daily ration. In rural areas that are not well located for the shipment of milk to city markets, cheese factories, powdered-milk plants, and condenseries are the best outlets for dairy products.

Southeastern Minnesota and east-central Wisconsin are two of the nation's centers for the production of creamery butter. On many farms in these areas the raising of hogs and poultry is combined with the dairy enterprise. Some of the best heifer calves are kept to replace the older and less productive cows that must be removed from the herd. The surplus calves are sold.

Creamery butter is produced in some of the northern sections of the Lake States. Here the production of feed crops is limited by the cold climate and the large amount of cutover land which is not tillable. Some farms have improved pastures and meadows that support excellent small dairy herds. Potatoes grow so well that the potato crop ranks next to dairying as a source of income.

Dairying is combined with mixed farming in central Maine and in southern New England. This type of farming, which includes poultry, fruit, and truck crops, is characteristic of the Hudson River Valley in New York. The Willamette Valley in Oregon and the Puget Sound area in Washington are centers of dairying and mixed farming near the Pacific Coast.

Dairying and general farming are carried on together in the central part of the Northeast, in Pennsylvania, and in the upper Piedmont. In these areas the farming pattern includes dairying, poultry, feed grains, truck crops, fruit, potatoes, and tobacco. A similar type of farming is found in northeastern Ohio, in northwestern Pennsylvania, on the Allegheny plateau in Pennsylvania, and in northern New England. The crops in the different localities vary with the soils and climate.

WHEAT AND SMALL-GRAINS REGION

Wheat has such a wide climatic range that it can be grown commercially throughout the United States, except in areas that have less than 15 inches of rainfall and in parts of the South where excessive rainfall encourages rust and fungus diseases.

The major wheat regions are the Great Plains in the central United States and a Pacific Northwest area in and around the Columbia Basin. The fine-textured, deep soils in these regions are brown to dark brown in color. Short grasses and mixed grasses of medium height were the original vegetation.

The level to gently rolling topography of the Great Plains is adapted to large-scale mechanized farming. Rolling land is characteristic of the Pacific Northwest wheat country. Both wheat regions have 15 to 30 inches of precipitation and growing seasons of 100 to 200 days. The warm, dry summers favor the ripening and harvesting of wheat and other small grains.

Hard winter wheat region. The greatest production of hard winter wheat is in central and western Kansas, southwestern Nebraska, eastern Colorado, and northwestern Texas and Oklahoma. Grain sorghums are important crops in these areas. Barley, oats, and corn are the other grain crops. Alfalfa and sweetclover also are grown to a limited extent.

Crop rotations, so well established in the Corn Belt, are not common in the hard winter wheat region. Wheat is usually alternated with summer fallow (plowed land which lies idle in the growing season). The large-scale farming operations are highly mechanized.

Winter wheat provides a protective cover for the soil in winter and helps to control wind erosion in the spring. When there is sufficient moisture the wheat can be grazed in winter and early spring. The winter wheat is harvested in June and July.

Spring wheat region. Montana, North and South Dakota, and western Minnesota make up the spring wheat region. Spring wheat is the main crop, although barley, rye, and flax are grown. The long, severe winters make this region unsuitable for winter wheat.

The rainfall here varies from 25 inches at the eastern border to 15 inches on the west and is well distributed through the growing season. About half of the rain occurs between March and June. Dry weather usually prevails during the ripening and harvesting season.

The major areas of wheat production are in the Great Plains where the gentle topography permits the efficient use of large combines like this to harvest wheat. (*Courtesy* John Deere)

The Northwestern part of the Wheat Region. The wheat stubble has been left on the surface to protect the soil against erosion. (*Courtesy* Soil Conservation Service)

The growing season averages 100 to 120 days over most of the region. Some areas have a growing season of 140 days.

The precipitation is higher in the Red River Valley than in the rest of the spring wheat region. The farms also are smaller and less specialized. A typical cropping system includes wheat, barley, oats, flax, and corn. Red River Valley farms produce potatoes, sugar beets, and other special crops. There is some livestock farming in this part of the wheat region.

Large-scale wheat farming is the main enterprise in northeastern Montana and in the part of North Dakota that is west of the Missouri River. On the large livestock-wheat ranches the best tillable land is used for wheat, while the rest of the land is used for pasture and hay. In some areas the raising of livestock is limited by the lack of water.

Pacific Northwest region. The Columbia River Basin of the Pacific Northwest is the third major wheat region. Both soft and hard wheats are grown in the Pacific Northwest. A growing season of 140 to 180 days permits the growing of either winter or spring wheat, although winter wheat is generally preferred. Winter wheat protects the soil against winter erosion and yields better than spring wheat.

Annual precipitation is about 20 inches in northern Idaho and in the Palouse area of eastern Washington. The Big Bend area east of the Columbia River averages about 10 inches of precipitation, and wheat is grown after the land has been summer fallowed.

The topography of this region varies from slightly undulating to rolling. Livestock raising is combined with wheat production in most of the region. Milk cows, beef cattle, hogs, and chickens are included in the small farm enterprises of the Palouse area. In the Big Bend area, livestock farming is localized in areas where there is suitable nonplowable pasture and range with water.

RANGE LIVESTOCK REGION

More than one-third of the land area of the United States is in the western rangelands, which cover approximately 700 million acres. This vast region extends from the western Dakotas and sandhills of Nebraska across southwestern Texas, New Mexico, and Arizona. The Mountain and Intermountain states and the Pacific Coast states are in the range livestock region. Also included are north-central Texas, the Arbuckle Mountain area of Oklahoma, and the Flint Hills of Kansas and Oklahoma.

The soils, climate, topography, and elevation make range livestock the most profitable farm enterprise in the range region of the West. In many areas there is not enough rainfall to produce crops without irrigation. The small amount of precipitation is not always well distributed through the growing season when crops must have moisture.

About half of the rangeland is in federal and state ownership. The stock-farm areas of Nebraska, Kansas, Oklahoma, and Texas are mainly in private holdings.

Year-long grazing is a common practice in the Southwest, except at the higher altitudes. Where rotation grazing is used, some of the land is reserved as winter range. The best grazing seasons are late summer, fall, and early winter.

Above: **Sheep on the range is a common sight in the Southwest. (New Mexico)** (*Courtesy* New Mexico State Tourist Bureau) *Left:* **Beef cattle grazing on Nevada rangeland.** (*Courtesy* Soil Conservation Service)

Forage is generally scarce in late winter and early spring.

Cattle and sheep are the main livestock in the Southwest. Goats are the major enterprise on the brushy rangeland in the Edwards Plateau of Texas, while cattle and sheep graze the open grasslands. On mixed range the three kinds of livestock make better use of the forage.

Feeder cattle and lambs are shipped to the Corn Belt for fattening. Some cattle are moved to the Flint Hills of Kansas and Oklahoma where the grasses are so nutritious that by late summer the animals can be marketed as grass-fat.

Grazing is seasonal in most of the range region outside the Southwest. Although some areas in California have a mild climate, elsewhere the extremes in temperature, precipitation, and topography do not permit year-long grazing.

Sheep may graze on some of the better open ranges throughout most of the year. Mountain ranges are grazed in summer, while the lower meadows and croplands are pastured in the fall. Winter feeding is necessary in many parts of the range livestock region. Hay from meadows of native grasses is used for wintering breeding herds.

FRUIT, TRUCK CROP, AND MIXED FARMING REGION

Fruit and truck crops are raised in localized areas throught the United States. On the map in this chapter (p. 34) the Fruit, Truck Crop, and Mixed Farming Region is divided into two regions, nonirrigated and irrigated.

Apples, pears, peaches, and other deciduous fruits are grown mainly where frost normally does not occur in late spring or early fall. Such areas are found in the Shenandoah and Cumberland valleys in Pennsylvania, Maryland, Vir-

These muskmelons in California are irrigated when the dial on the tensiometers indicate a certain dryness of the soil. (*Courtesy* the Irrometer Company)

ginia, and West Virginia; on the Ozark Plateau; on the Colorado west slope; in central Washington; in southern Oregon; and in the Sacramento and San Joaquin valleys in California. The southern shores of Lake Erie and Lake Ontario and the eastern shore of Lake Michigan are other important areas for the production of truck crops and deciduous fruits.

Truck crops and citrus fruits are grown in central Florida, in the lower Rio Grande Valley in Texas, in southern California and along the central California coast, and in southwestern Arizona. Truck farms in southern Florida specialize in growing winter vegetables.

Fruit and truck farming are mainly supplementary enterprises in dairy and general farming regions. This pattern of mixed farming is common in central

Pennsylvania, in the upper Piedmont, in the Hudson River Valley, and in southern New England. The central Virginia Piedmont and the Appalachian Mountains area are adapted to fruit growing. Truck crops are grown extensively in the upper Ohio River Valley and in some areas in the Ozark and Ouachita Mountains. Truck farming and dairying are combined in the Allegheny Plateau in Pennsylvania and in northern New England.

GENERAL FARMING REGIONS

There are several large general farming areas adjacent to the Corn and Cotton Belts and the dairy regions. General farming is centered in Kansas, Missouri, Arkansas, Oklahoma, Illinois, In-

Above: The General Farming Region in the Southeast. (North Carolina) (*Courtesy* Tennessee Valley Authority) *Below:* Hereford cattle graze lush Sudangrass in the General Farming Region. (Nebraska) (*Courtesy* Soil Conservation Service)

Tobacco is a major source of income for 625,000 farmers in 20 states in the Tobacco and General Farming Region. This crop is nearly ready to harvest. An acre of tobacco often produces more than $1,400 gross income per year. Farmers in this region receive about $1.3 billion annually. (*Courtesy* Burley Tobacco Protective Fund)

diana, and Ohio. In each of these states general farming is interspersed with other types of land use. General farming is the common pattern of agriculture in scattered areas in irrigated valleys in the West, in New Mexico and Texas, and along the Atlantic and Gulf coasts.

The climate and soils in the east-central part of the United States are suitable for many crops. Among these are corn, wheat, oats, soybeans, tobacco, fruit, and truck crops. Much of the rougher land is kept in pasture. Grasses and legumes are grown in rotation with other crops. The forage crops are fed to livestock.

Livestock and tobacco are a typical combination in parts of Tennessee, Kentucky, and Virginia. Hay, fruit, and forests are leading crops in the central Piedmont area, while truck crops are included in the general farming pattern of the upper Ohio Valley. Part-time general farming is common in the flatwoods areas along the Atlantic and Gulf coasts and in some mountainous areas, particularly in mining communities.

Many of the small general farming areas in the West are irrigated. Where the supply of irrigation water is relatively small, general farming involves less risk than the intensive production of crops that require a great amount of moisture. General farming areas are found in each of the 11 Western states. Among the crops associated with this type of agriculture are feed grains, hay, sugar beets, dry beans, potatoes and truck crops.

TOBACCO AND GENERAL FARMING REGION

There are four subregions in the tobacco and general farming region. Although a different kind of tobacco is grown in each subregion, tobacco is the dominant crop on practically all farms.

Most tobacco farms are small units that can be operated by one family. Approximately 300 to 500 man-hours are required for each acre. Harvesting and some of the other operations are not mechanized. Sharecroppers raise tobacco on some of the larger farms, giving the owners part of the crop for the use of the land.

About one fifth of the cropland in the Piedmont is used to produce tobacco. Corn is grown on approximately twice that acreage. Other crops are hay, wheat, and barley. Cotton ranks second to tobacco in the central Coastal Plains area.

The burley-tobacco area of Kentucky is also noted for bluegrass. Farming here combines livestock and tobacco. The fertile soils have developed from phosphatic limestone, and their high mineral content is beneficial to both plants and animals.

Farms in southwestern Kentucky and in northwestern Tennessee produce dark air-cured tobacco. White burley is the most popular type of tobacco in the eastern part of the area. Unfavorable markets, which occasionally result from a decline in the export demand for dark air-cured and fire-cured tobacco, have encouraged a greater production of livestock. About two thirds of the land is idle or in plowable pasture, while the other third of the acreage is used for crops.

A type of tobacco known as Maryland Leaf is grown in five Maryland counties between Chesapeake Bay and the Potomac River. In this area the farmers depend upon a single cash crop. The acreages of corn and tobacco are approximately equal. Small grains, hay, potatoes, and vegetables are minor crops.

SPECIAL CROPS AND GENERAL FARMING REGIONS

Potatoes, sugar beets, dry beans, peanuts, rice, and sugarcane are sometimes referred to as special crops. They are produced in a system of general farming or as highly specialized enterprises.

Some of the important potato-producing areas are Aroostook County,

Irrigating potatoes grown in contour strips. (Maine) (*Courtesy* Soil Conservation Service)

Maine; the Klamath Basin area of southern Oregon and northern California; the upper Snake River area; the lower Snake River section in Oregon and Idaho; the southern Yakima Valley in Washington; and the San Luis Valley and part of the South Platte River area in Colorado. There is considerable potato production in the Red River Valley in Minnesota and North Dakota. Others areas that produce potatoes are central Wisconsin and parts of Ohio, New York, Pennsylvania, and New England.

Except in Michigan, sugar beets are mainly an irrigated crop. Several centers of production are the Yellowstone River and Milk River areas in Montana; the Bighorn Basin and North Platte River areas in Wyoming and Nebraska; and eastern Colorado. In numerous areas in the West, sugar beets are grown on farms that also produce feed crops.

Left: **A red pine and white pine forest in Wisconsin.** (*Courtesy* Wisconsin Conservation Department) *Right:* **This is a part of the Nonfarming Area, but who cares about farming all the time when there is so much beauty in the yucca, the state flower of New Mexico.** (*Courtesy* El Paso Natural Gas Company)

Dry beans are a product of dry-land farming in New Mexico, Colorado, Wyoming, Montana, Idaho, and California. This crop requires less labor and fertilizer than do potatoes or sugar beets.

The lower Mississippi River Delta in Louisiana is the sugarcane country. This crop yields about three-fourths of the gross farm income in that part of the Delta. Other farm enterprises here are rice, truck crops, and beef cattle.

The United States has three major rice areas: the Gulf Coastal prairies in southwestern Louisiana and southeastern Texas; the Arkansas prairies in Arkansas; and parts of the Central Valley in California.

Rice requires large quantities of fresh water for irrigation. Impervious subsoils are necessary to prevent loss of water by seepage. Rotation of fields is necessary to maintain the physical condition of intensively irrigated land. Beef cattle make good use of the grasses and legumes during the years when rice is not grown.

Peanuts are grown commercially from Virginia to Texas. The three principal peanut areas are the North Carolina-Virginia coastal plains, the southern coastal plains in Georgia and Alabama, and the Cross Timbers section in Texas.

NONFARMING AREAS

Little farming is done in the mountainous, swampy, desert, and forested areas. Examples of nonfarming areas are the Mojave Desert, the Rocky Mountains, the Florida Everglades, and the northern woods in Maine and New Hampshire. The national parks and forests are nonfarming areas.

Parts of the range livestock region have almost no secondary types of farming. In forested regions, part-time farming is mainly the harvesting and marketing of forest products. The decentralization of industries and the development of suburban and rural communities as residential areas probably will continue to increase the amount of land classified as nonfarming.

Pupils in South Carolina learn about forestry in the woods from their teacher. (Nonfarming area) (*Courtesy* U.S. Forest Service)

Left: Forests make up a large part of our Nonfarming Area. Ponderosa pine being cut in Idaho. (*Courtesy* U.S. Forest Service) *Right:* A tractor with fire plow attached is used to clear fire lanes in this Oklahoma forest. (*Courtesy* Oklahoma Planning and Resources Board)

QUESTIONS

1 What is the principal type of farming in your community?
2 List the nine major types of farming in the U. S.
3 Locate the Corn Belt on the map.
4 Why are corn, oats, and soybeans called "cash crops"?
5 What is the principal factor that determines where cotton will grow?
6 Where is the main part of the dairy region?
7 What is the range of annual precipitation and growing season in the Wheat and Small-Grains Region?
8 Who owns the Range Livestock Region?
9 Where is the center of the General Farming Region?
10 In what farming region would you prefer to farm? Why?

ACTIVITIES

1 Use geography textbooks, encyclopedias, bulletins, and other references for additional reading on the agriculture in your state and region.
2 After reading all of this chapter, study carefully the description of the farming region in which you live. Make notes on how your community differs from the general pattern of the region.
3 On an outline map of the United States, draw the approximate boundaries of your agricultural region and two or three other regions in which you are particularly interested. Use a different color to identify each region.
4 Have a class exhibit of farm products from the local community.
5 Write a brief summary of the influences of climate, soil, insects, topography, and markets on the agriculture in your community.

4 COMPARATIVE AGRICULTURE IN THE FIFTY STATES

Agriculture in the United States continues to change in response to social, psychological, and economic pressures. Year by year, efficiency normally associated with bigness dominates the farm and ranch scenes.

The following table gives the number of farms, the average size for each decade since 1920, and the farm population as a percentage of the total population. Except for the depression year of 1930, the number of farms and farm population has decreased each 10-year period and is predicted to decrease further by 1980; farm size is predicted to continue to increase.

Year	Number of Farms	Average Size of Farms (Acres)	Farm Population as Percentage of Total Population
1920	6,518,000	147	30
1930	6,546,000	151	25
1940	6,350,000	174	23
1950	5,648,000	213	15
1960	3,962,000	297	9
1970	2,924,000	383	5
1980	1,939,000	545	3 to 4

The principal sources of information for the following state summaries are *Statistical Abstract of the United States, 1971; 1969 Census of Agriculture;* and *Agricultural Statistics,* United States Department of Agriculture.

Alabama

A little more than half of the area of Alabama is farmland. The leading farm commodities in 1970 in order of cash receipts were cotton, broilers, cattle, eggs, and corn.

In 1970, Alabama ranked fourth in

the nation in the pounds of peanuts produced. Among the states in the South, Alabama ranked fourth in the total number of chickens on farms. About one-fourth of the farms are in conservation programs.

Alabama has a long growing season of 240 frost-free days over most of the state, and an average precipitation of 50 inches per year. The commercial forests are mainly Southern yellow pine, with mixed hardwoods along the major streams.

Alaska

The farms in Alaska are relatively few in number and large in size. Three hundred and thirty farms averaging 4,823 acres account for only 0.4 percent of Alaska's land area. In 1969 farm marketings yielded approximately 3 million dollars from livestock and 1 million dollars

from crops. The leading farm products in 1970 were milk, potatoes, cattle, and eggs. The value of farm production in Alaska has more than doubled in the past fifteen years. In 1969 the annual value of farm crops in the state has exceeded 3.6 million dollars.

Most of the farms are in the Tanana Valley, Matanuska Valley, and Kenai Peninsula. Almost 90 percent of the total agriculture activity is in this area. Anchorage, Fairbanks, and military installations surrounding them are the principal markets for farm products.

In their order of production, the leading livestock products are eggs, reindeer meat, beef, pork, wool, poultry meat, lamb, and mutton. Most of the beef cattle and sheep are raised on Kodiak and adjacent islands and on the Aleutian chain. Mild weather and a long growing season characterize these areas.

Alaska is well known for large, high-

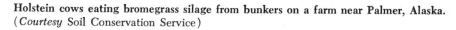

Holstein cows eating bromegrass silage from bunkers on a farm near Palmer, Alaska. (*Courtesy* Soil Conservation Service)

quality vegetables which grow well in the cool, moist summer months when there are more hours of sunshine. Intensive care and abundant use of fertilizers contribute to the high yields. Potatoes are the principal vegetable crop. Other vegetable crops are cabbage, celery, lettuce, and radishes. Most of the vegetable production is in the Matanuska-Susitna valleys near Anchorage, Tanana Valley, and Kenai Peninsula.

There is a relatively short period in which the native forage plants can be pastured or cut for hay with the maximum yield of nutritive value. Grasses and legumes bear seed and propagate readily in their native habitats. Difficulty is often encountered in germinating the seed of native grasses and legumes. Close grazing or mowing for hay or silage year after year tends to deplete the stands of these plants.

The development of farms from the cold land of Alaska requires years of hard work and a relatively large investment. The cost of developing a small vegetable farm of 20 to 25 acres may be $1,200 to $1,500 per acre. A dairy farm of 200 to 225 acres that would support 100 to 200 head of dairy cattle may require an investment of $75,000 to $100,000.

More information on farmland in Alaska may be obtained from the Alaska Department of Natural Resources, Division of Agriculture, Post Office Box 1828, Palmer, Alaska 99645.

Arizona

Arizona had 5,890 farms which averaged 6,486 acres in 1969. In comparison with Alabama's range of 44 to 66 inches of precipitation annually, Arizona farms

These bison are grazing under a special use permit, in the Kaibab National Forest in Arizona. Cattle also graze on similar Arizona rangelands. (*Courtesy* U. S. Forest Service)

and ranches get 7 to 30 inches of precipitation. The greater amounts of precipitation are in the east central (30 inches per year) and the north central (27 inches per year) parts of Arizona.

Approximately 53 percent of the state's land area is in the 38 million acres of farmland. Marketed crops yielded 236.6 million dollars in 1969; livestock brought Arizona farmers more than 372 million dollars. In 1970 Arizona ranked second in goats and fifth in cotton in the United States. There is some intensive irrigation of vegetables and farm crops.

Arkansas

Forty-seven percent of the land area of Arkansas is farmland. The average size of farms is approximately 260 acres, and more than half of the 60,000 farms are commercial (with gross sales of $2,500 or more annually). Crops brought Arkansas farmers approximately 455 million dollars in 1969, while livestock marketings amounted to approximately 515 million dollars. Arkansas ranked first in rice, fourth in soybeans and cotton, and sixth in chickens and turkeys in the United States in 1970.

California

California has ranked as the number one agricultural state for 23 consecutive years, outranking Iowa, its nearest competitor, by 600 million dollars in 1970. Farm products marketed totaled 4.4 billion dollars. Nearly all of the nation's dried prunes, raisins, dates, figs, currants,

A forest supervisor and district ranger examine the condition of cattle range in Big Whitney Meadows, Inyo National Forest. (California) (*Courtesy* U.S. Forest Service)

apricots, peaches, and pears are produced in California. The state also produces more than half of the canned fruits and one third of the canned vegetables.

Animal products in which California ranks first include cottage cheese, frozen dairy desserts, and yogurt. In meats, the state is first in numbers of sheep and lambs slaughtered and ranks third in the slaughter of beef cattle. California is also first in numbers of turkeys, market hens, egg production, beet sugar, cottonseed oil, and wines.

The high ranking of California's agriculture is accomplished on 2 percent of the nation's farms. Many of the 77,875 commercial farms in California are intensively irrigated and have high yields per acre.

Colorado

This is another state with much intensive farming and irrigation. More than three-fourths of Colorado's 27,950 farms are commercial, with an average farm size of 1,313 acres. The 37 million farmland acres account for 55 percent of the land area.

Colorado farmers marketed crops for 218 million dollars and livestock for 883 million dollars in 1969. The leading farm commodities, in the order of value, were cattle, wheat, milk, and sugar beets. In 1970, Colorado ranked second in the nation in sugar beets and seventh in number of sheep and lambs on farms and ranches. Much of the sheep production is on rangeland where there is not enough precipitation for farm crops.

There are important potato producing areas in the San Luis Valley and part of the South Platte River area in Colorado. Dry beans are a product of dryland farming in parts of the state. This crop requires less labor and fertilizer than do potatoes or sugar beets. The latter crop is well adapted to intensive irrigation farming. Irrigated fruit, truck, and mixed farming are limited to small areas in eastern, central western, and southern Colorado.

Connecticut

This is a state of small farms. The leading cash sales off the farm in 1970 were dairy products, eggs, tobacco, and greenhouse products. About 58 percent of the land area is farmland.

Dairy farming is the largest farm enterprise in the state. Mixed farming combines dairying with raising truck crops and poultry.

Information from research on movement of mineral nutrients and organic compounds through stems of slash pine seedlings is beneficial to agriculture in Florida. (*Courtesy* U.S. Forest Service)

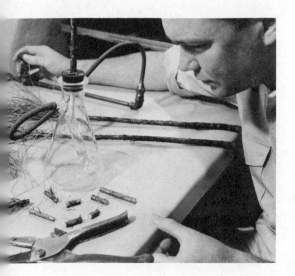

Delaware

The 740,000 acres of farmland in this state cover approximately 56 percent of the land area. More than three-fourths of the 4,700 farms are commercial, with an average of 157 acres. In comparison with Connecticut, Delaware has fewer farms, larger farms, and less total farmland. In 1969 farm marketings in Delaware amounted to 129 million dollars. This was approximately 92 million dollars from livestock and 37 million dollars from crops. Broilers, corn, soybeans, and milk were the principal farm products in 1970.

More than 1,500 Delaware farms are in agricultural conservation programs. These farms have about 41 percent of the cropland. This extent of participation in conservation is typical for the South Atlantic region.

Florida

The leading farm commodities in Florida in 1970 were oranges, cattle, milk, and grapefruit.

Florida's acreage of commercial vegetables is more than the total vegetable acreage for the other South Atlantic states.

Florida leads all other states in the production of oranges and grapefruit. This state also produces practically all the limes, tangelos, and tangerines grown in the United States.

Among the South Atlantic states, Florida ranks third in the number of chickens on farms and in the amount of whole milk sold. Conservation programs have encouraged the development of highly productive pastures which support many herds of fine beef cattle. The state had 1.8 million cattle in 1969. In

1970, Florida ranked second in the United States in vegetable production, third in sugar cane, and fifth in hens.

Georgia

Georgia's 67,431 farms averaged 234 acres in 1969, and 55 percent were commercial. Forty-two percent of the land area, or 15.8 million acres, was in farmland. The nature of farming in this state is marked by a strong three-to one division between marketings of livestock and crops in 1969, approximately 699 million dollars from livestock and 320 million dollars from crops. Broilers, eggs, cattle, and peanuts were the champion income producers.

As in the rest of the South, the trend in farming has been from the one-crop system toward diversification. Georgia ranks third in the nation in egg production. In 1965, Georgia produced more broilers than any other state.

In 1965, Georgia ranked sixth in the nation in cotton acreage. In the same year, Georgia was first in acreage of peanuts and in per-acre yield of peanuts.

Among the South Atlantic states, Georgia is second in beef cattle and second in hog production. In this region, Georgia ranks second in vegetable growing. Georgia ranked first in peanuts, second in hens, and fourth in rye grain in the United States in 1970.

Hawaii

Sugarcane, pineapple, beef cattle, and milk are the principal sources of income on Hawaiian farms. Nationwide, Hawaii ranked first in pineapple and sugarcane production in 1970.

Cane sugar and pineapple are the principal export crops. Sugarcane is grown on a little more than 5 percent of

This pineapple field is planted on the contour to reduce erosion. (Kauai Pineapple Company, Kalaheo, Kauai, Hawaii) (*Courtesy* Soil Conservation Service)

the agricultural land, and pineapple accounts for 1.5 percent of the land use.

The sugarcane and pineapple industries provide employment for a large part of the population. The pineapple industry in Hawaii normally requires about 10,000 year-round employees, with more than twice that number employed during peak periods of canning and packing.

Dry summers and wet winters are the general climatic pattern in Hawaii. The uncertainty of rainfall restricts non-irrigated agriculture, even in areas where the average annual precipitation is relatively high. Limited rainfall is partly responsible for the small acreage of well-adapted crops such as corn, wheat, oats, and legumes. In the drier zone the uncertainty of rainfall tends to discourage extensive diversified farming, with its rotation of cultivated crops and temporary pastures.

Pastures are the best use of much of the agricultural land in Hawaii. About one-third of the land is used for this purpose. The grazing industry depends almost entirely on forage plants that have been introduced. Native forage plants provide most of the grazing in the non-arable rangelands.

Irrigation and the mechanized harvesting of forage increase the production on some dairy farms. Grazing is economical only where good pasture is available. The development of more nutritious grass-legume pastures is a major need of the dairy industry in Hawaii. Nitrogen and phosphorus fertilizers are necessary on all pastures to produce high yields of nutritious forage.

Ranching depends almost entirely on grazing. Because of the high cost of imported concentrates, beef cattle are fattened on grass and locally-grown grain

sorghum rather than entirely in feedlots. Much of the rangeland is in strips which extend from seashore to mountain top to permit access to rain-fed grass for year-round grazing.

Idaho

Fourteen and one-half million farm-land acres in this state are divided into 25,475 farms which average 566 acres. Farms cover only 27 percent of the total land area, and about 80 percent of the farms are commercial. In 1969 the income from farm marketings included 296 million dollars from farm crops and 352 million dollars from livestock. In 1970 cattle, potatoes, dairy products, and wheat were the principal farm products in Idaho.

In 1970, Idaho ranked third in the United States in sugar beets and fourth in barley production. With 272,000 acres in 1969, Idaho ranked high in Irish potatoes, both in acreage and in quality.

Illinois

Farming in Illinois has much in common with that of Indiana, Iowa, and Missouri. All have family-size farms, with Indiana having the smaller average. In 1969 the 123,565 farms in Illinois averaged 242 acres. The dominant position of agriculture in Illinois is indicated by the fact that 84 percent of the total land area is in the 30 million acres of farmland.

In 1969 crops were marketed for 1.32 billion dollars; livestock for 1.29 billion dollars. Corn is the most valuable product, with hogs and cattle ranking second and third, respectively. Soybeans are next in importance.

In the United States in 1970, Illinois

In Illinois, Indiana, and Iowa, corn is the leading crop and the best farmers on the best soils may average 200 bushels per acre. (*Courtesy* Ford Motor Company)

ranked first in soybeans and second in corn and hogs.

Indiana

The farms in Indiana average 173 acres in size, which is smaller than the average in Illinois, Iowa, and Missouri. Seventy-six percent of Indiana's land area is in 17.6 million acres of farmland. The 1969 farm marketings returned 799 million dollars from livestock and 597 million dollars from crops. The leading farm commodities are hogs, corn, soybeans, and cattle. This combination of livestock and crops is typical for the Corn Belt.

In 1970 Indiana ranked third in the nation in the production of soybeans and hogs and fifth in corn. Among the five East North Central states, Indiana ranked fourth in beef cattle, with a national rank of fifteenth.

Cropping systems and conservation practices in Indiana are much like those throughout the Corn Belt. The pattern of mixed farming includes oats, barley, sorghum grains, legumes and pasture grasses, dairying, poultry, and sheep.

Iowa

Iowa is another state of family-size farms, and is best known for corn, cattle, and hogs. The average farm size is 239 acres, 89 percent of the farms are commercial, and the 33.5 million acres of farmland cover 94 percent of the total land area. Iowa had 140,354 farms in 1969.

The total value of farm marketings in 1969 was 3.65 billion dollars, including 2.77 billion dollars from livestock and 886 million dollars from crops. This comparison indicates that the grain crops, especially corn, are fed to cattle and hogs instead of being marketed directly. The leading farm products in 1970 were hogs, corn, and soybeans. In 1970 Iowa ranked first in the nation in the number of hogs and corn, second in production of soybeans, and fifth in oats and cows.

Intensive conservation practices such as crop rotations, contour farming, and terracing are controlling soil erosion and maintaining high soil fertility. As would be expected in leading grain-producing states, Iowa, Illinois, and Indiana rank high in the use of commercial fertilizers.

Although hogs, cattle, and corn dominate the Iowa farm scene, the state has a diversified agriculture. Most of the hogs and beef cattle are fattened in feedlots, as are thousands of cattle from rangelands in the Southwest. Iowa's improved pastures have a high carrying capacity, and alfalfa and red clover are high-yielding hay crops. Oats are included in many cropping systems.

Kansas

Kansas is the nation's champion wheat producer, and the average farm

Iowa ranked first in the nation in the number of hogs produced and in 1969 had 25 percent of all hogs in the United States. (*Courtesy* Hampshire Swine Registry)

size of 574 acres is roughly twice that of the leading corn states. Some large commercial wheat farms have several thousand acres. Large power machinery for plowing, planting, and harvesting can be used economically only in large fields. Sixty percent of the 86,057 farms in Kansas are commercial, covering 49.4 million acres and 94 percent of the land area.

In a state so famous for wheat, the comparison between livestock and crops is somewhat surprising. In 1969 Kansas farmers received 1.24 billion dollars from marketed livestock and 579 million dollars from crops. The annual income from cattle is greater than from wheat. In the order of value in 1969, the most important farm commodities were cattle, wheat, hogs, and sorghum grain.

In 1969 Kansas produced more grain sorghum forage than any other state. In fact, this state produced almost as much grain sorghum forage as Texas and Oklahoma combined. Kansas ranked second to Texas in sorghum grain, with Nebraska close behind. The trend toward grain sorghum replacing other grain crops has been typical in the Midwest and Southwest for about 30 years.

Kansas ranked fourth in the United States in the number of cattle marketed in 1970. Large areas of Kansas rangeland yield forage that is nutritious enough to fatten cattle without additional feeding of grain. Before they are marketed, some of the cattle from the semi-arid rangelands are sent to feedlots in Kansas, Iowa, Illinois, and Nebraska.

Kentucky

Kentucky is a state of small farms which average 128 acres in size, total 25 million acres, and cover about 63 percent of the state's land area. Approximately 49 percent of the farms are commercial. In 1969 income from crops was 343 million and from livestock 423 million dollars. Tobacco, cattle, milk, and hogs were the main sources of farm income in Kentucky in 1970.

In 1970 the production of 4.4 million pounds of tobacco in Kentucky was exceeded only by North Carolina's production of 7.2 million pounds. The soils and climate of Kentucky favor the growing of tobacco, and on the small farms much of the labor is provided by members of the farm families.

Kentucky ranks high in the nation in the amount of whole milk produced. Broiler production is important in some Kentucky communities.

Kentucky is perhaps best known for its bluegrass pastures and thoroughbred horses. Lexington is world famous as the home of the annual Kentucky Derby horse race.

Louisiana

This state's 9.8 million farmland acres represent 34 percent of the land area and are divided into about 42,269 farms. The average size of Louisiana farms is 232 acres. Marketed farm crops brought Louisiana farmers 296 million dollars in 1969; livestock, 197 million dollars. In 1970 the principal sources of farm income were, in order, cattle, soybeans, rice, and sugarcane. Louisiana ranked second in sugarcane production and fourth in rice.

In Louisiana the rotation of rice with pasture grasses is a major factor in the raising of cattle. Intensive irrigation of rice land tends to damage soil structure, and the growing of pasture grasses and

legumes for two or three years in rotation with rice helps to rejuvenate the soil. The trend toward diversified farming throughout the South Central states is reflected in the growing number of herds of beef cattle in Louisiana.

Maine

Only 9 percent of the land area of Maine is farmland. In 1969 the 1.8 million farmland acres were divided into 8,000 farms, with an average size of 221 acres. Sixty-four percent of the farms were commercial. Total farm income in 1969 was 198 million dollars. The leading farm products in Maine in 1970 were potatoes, eggs, broilers, and dairy products.

As a producer of potatoes, Maine is second only to Idaho. Maine is the leading supplier of seed potatoes. In broiler production, Maine ranks tenth in the nation.

Maryland

Almost two-thirds of Maryland's 17,181 farms are commercial. The average farm size in 1969 was 163 acres, and the 2.8 million acres of farmland covered 44 percent of the land area. In 1969 Maryland farmers received 238 million dollars from livestock marketings and 100 million dollars from crops. Broilers, milk, corn, and cattle were the leading farm products in 1970.

In the United States, Maryland is sixth in broiler production and eighth in the production of tobacco. Many of the dairy herds are Ayrshires. Aberdeen-Angus is a popular breed of beef cattle. Corn production was a little more than 38.8 million bushels in 1970.

Massachusetts

Only about 20 percent of the land area of Massachusetts is farmland, with an average size of almost 115 acres. Nearly 70 percent of the farms are commercial. Livestock marketed brought approximately 100 million dollars; crops, 75 million dollars. Milk, greenhouse products, eggs, and cranberries were the main sources of farm income in 1970.

Massachusetts produced 745,000 barrels of cranberries in 1965. Although producing half of the nation's cranberries, Massachusetts has a diversified farming pattern that includes dairy and poultry products, apples, peaches, tobacco, and maple syrup. Beef cattle, broilers, and hogs are the leading livestock products of the state.

What would Thanksgiving be without cranberries? The cranberry bogs of Massachusetts produce nearly half of the 80 million dollars worth of cranberries produced annually in the United States. This is a mechanical cranberry picker in operation. (*Courtesy* University of Massachusetts)

Michigan

Approximately 33 percent of the land area of Michigan is in a little less than 11.9 million acres of farmland. In 1969 the 77,946 farms averaged 153 acres in size. Farm marketings in 1969 totaled 829 million dollars. This included 475 million dollars from livestock and 350 million dollars from crops. In order of value, the state's leading farm products in 1970 were milk, cattle, corn, and hogs.

Michigan produces more than two-thirds of the nation's tart cherry crop and is second only to California in the production of sweet cherries and plums. The state produces nearly half of the United States production of dry beans.

In 1970 Michigan ranked third in the United States in apples and grapes, and fourth in pears and peaches. It ranked sixth in the amount of whole milk produced and in sugar beet acreage.

Minnesota

The farms in Minnesota average about 260 acres in size. The intensive character of Minnesota farming is indicated by the fact that more than 85 percent of the farms are commercial. In 1969 the state had 110,747 farms, with a total of 28.8 million acres. This is about 57 percent of the land area of the state.

In 1969 the receipts from livestock and dairying were more than 1.18 billion dollars, and the marketing of crops yielded more than 558 million dollars. That year Minnesota ranked first in the nation in the production of oats, creamery butter, turkeys, and honey; second in hay; third in flaxseed and milk; fourth in corn, sugar beets, and hogs; and seventh in the number of cattle marketed. In 1970 Minnesota was third after Wisconsin and

This Charolais calf would be at home in any state. (*Courtesy* American International Charolais Association)

New York in the number of dairy cows and heifers on farms.

Minnesota is a leading producer of soybeans, ranking sixth in the production of beans in 1970. The 1970 rank of cash receipts was cattle, dairy products, hogs, and corn. Alfalfa and other legumes are grown for the dairy herds. Good conservation practices are evident on a large percentage of the farms in this state.

Mississippi

Cotton has always been the leading crop in Mississippi. During the past three decades the trend toward diversified farming has greatly increased the production of beef cattle and broilers. Much land formerly used for the continuous production of cotton is now in improved pastures. With a total of more than 1.3 million bales, Mississippi ranked second in cotton production in 1970.

The important crops, besides cotton, in Mississippi are peanuts, sugarcane, pecans, sweet potatoes, rice, wheat, oats, soybeans, corn, and fruit. The state ranks among the nation's six leading producers of sweet potatoes. Rice is an important crop in some Mississippi counties, although the state ranks fifth in rice growing in comparison with Texas, Louisiana, Arkansas, and California.

Cotton, cattle, soybeans, and broilers were the principal commodities, in rank of cash receipts, in 1970.

Missouri

In 1969 the state had 137,067 farms. The average farm size was 236 acres. Seventy-three percent of the land area is in 32.4 million acres of farmland.

The marketing of farm products yielded 1.46 billion dollars in 1969. Livestock accounted for more than 1.0 million of this total. The leading farm products in Missouri are cattle, hogs, soybeans, and dairy products. At the beginning of 1970, the state ranked fourth in the United States in the number of hogs, goats, and turkeys on farms, and third in beef cattle.

The principal farm crops in Missouri are corn, cotton, winter wheat, and soybeans. Oats and legumes are often included in the crop rotations. Alfalfa and clovers are popular hay crops on soils to which they are adapted. Tobacco is grown in some northwest Missouri communities near the Missouri River. Cotton is a major crop in the southeastern corner of the state. Apples, peaches, popcorn, rice, and rye may be considered minor crops in Missouri.

Montana

This is a state of large farms which averaged 2,521 acres in 1969. Only Nevada, New Mexico, and Wyoming have larger averages. Montana's 62.9 million acres of farms and ranches are 68 percent of the land area in the state.

Montana livestock marketed in 1969 brought 385 million dollars; crops, 189 million dollars. Cattle, wheat, barley, and sugar beets were the main sources of income. Montana ranked fifth in the United States in wheat production in 1970 and second in barley. As of January 1, 1970, the state ranked fourth in the number of stock sheep on farms and ranches. In 1970 Montana was seventh in the number of beef cattle.

Although best known for wheat and cattle, Montana also produces apples, sugar beets, flaxseed, and potatoes.

Nebraska

Almost all of the land area of Nebraska, 94 percent, is farmland. Farms average about 634 acres in size. In 1969 livestock brought more than 1.63 billion dollars, and crops were sold for almost 536 million dollars.

At the beginning of 1970, Nebraska was fourth in the nation in the number of cattle on farms. The state's national ranking in 1970 was third in sorghum grain, rye, and corn, fifth in sugar beets, and sixth in hogs.

Nebraska has approximately 22 million acres under cultivation. Agriculture is the most important source of income and employment. Industries closely related to agriculture include chemical fertilizer, meatpacking, and food processing.

Nevada

In comparison with most states, Nevada has a relatively small number of farms. There were 2,000 in 1969, and 70 percent of these were commercial. The average farm size was 5,070 acres. About 15 percent of the state's land area is in the 10.7 million acres of farmland. Livestock sales amounted to 85 percent of the 1969 farm receipts which had a total value of almost 80.8 million dollars.

Because of the dry climate, most of the state is better adapted to grazing than to crops. Intensive farming is limited mainly to areas which have large-scale irrigation. Cattle, milk, hay, and sheep were the principal sources of farm income in 1970. Vegetables are raised on some irrigated farms.

This barley crop provides food and cover on the Edwards Peninsula Wildlife Habitat Area, Gallatin National Forest. (Montana) (*Courtesy* U.S. Forest Service)

These tree seedlings growing in the Bessey Nursery may some day be used in Nebraska shelterbelts. (Nebraska National Forest) (*Courtesy* U.S. Forest Service)

New Hampshire

Almost one-fifth of the land area of New Hampshire is in 1 million acres of farmland.

The value of farm receipts in 1970 was approximately 57 million dollars. Seventy-seven percent of the farm income was from dairy and poultry products. In rank of cash farm receipts in 1970 were dairy products, eggs, cattle, and greenhouse products. Broiler production is a commercial farm enterprise. The state produced more than 2 million pounds of broilers in 1970. The yield of maple syrup that year was 44,000 gallons.

New Jersey

The average size of farms in New Jersey is 122 acres. In 1969 there were 8,493 farms in New Jersey. The total acreage was 1.0 million. This is 22 percent of the total land area of the state.

The total farm receipts in 1969 were more than 214 million dollars. New Jersey is first in the nation in gross income per farm acre.

In New Jersey the order of cash receipts in 1970 was dairy products, greenhouse products, eggs, and tomatoes. Among the three Middle Atlantic states New Jersey ranks fourth in the quantity of whole milk sold.

New Mexico

With an average farm size of 4,019 acres, New Mexico is a state of 11,641 large farms. Seventy percent of the farms in New Mexico are commercial, and 46.8 million acres in farms are about 60 percent of the total land area.

Marketing receipts included 280 million dollars from livestock and 77 million dollars from crops. High prices for commercial vegetables were a major factor in profitable crop production that year.

New Mexico was third in the United States in the number of goats in 1970. In the number of sheep and lambs marketed, New Mexico had a rank of sixth place.

Eighty percent of the income from field crops comes from cotton, hay, and sorghum grain. New Mexico was sixth in the nation in the production of sorghum grain in 1970. Cattle, dairy products, cotton, and hay were the principal products sold from the farms of New Mexico in 1970. Intensive farming and higher crop yields are possible where there is good water for irrigation.

New York

Approximately 33 percent of the land area of New York is in 10.1 million

acres of farmland. The 51,909 farms averaged 196 acres in 1969. In 1969 receipts
from livestock products were 731 million
dollars, and the receipts from crops were
245 million dollars.

Dairying and poultry raising are the
leading agricultural industries in New
York. In 1969 New York had more than
11.5 million chickens and ranked second
in the nation in the number of milk cows,
approximately 1.7 million. In 1970 New
York was second in the quantity of whole
milk sold. That year New York was sixteenth in the number of eggs sold. Broiler
production is an important part of the
poultry industry. Principal products sold
from the farms in 1970 were, in order,
dairy products, cattle, eggs, and greenhouse products.

In other comparisons for 1970, New
York was first in the nation in buckwheat,
maple syrup, and clover; second in apples
and grapes; fourth in potatoes; and fifth
in vegetables. The grape crop supports
a large industry in New York. Other important crops are cherries, pears, peaches,
peas, beans, wheat, rye, oats, and hay.

North Carolina

The 1969 receipts from farm products in North Carolina included 594 million dollars from livestock and 812
million dollars from crops. North Carolina ranked first in the nation in tobacco
production in 1970, was second to
Louisiana in sweet potatoes, and was
third in peanuts. Cotton, corn, and soybeans are other leading crops. There is
some production of peaches, apples, barley, wheat, and oats.

The rank of cash farm receipts in
1970 was tobacco, broilers, eggs, and
hogs.

Swine numbers throughout the United States
have varied from 50 to 60 million each year
since 1954, but swine have increased steadily in
North Carolina. (*Courtesy* North Carolina
University)

North Carolina ranked fifth in the
nation in broiler production in 1970. In
January 1970 the state ranked third in
the number of chickens and turkeys on
farms. In 1970 North Carolina ranked
high among the South Atlantic states in
the quantity of whole milk sold.

Like most of the states in the South
Atlantic and South Central regions, North
Carolina has important forestry industries. Lumber, pulpwood, poles and piling, and railroad crossties are leading
sources of employment in communities
which are able to turn their forest resources into farm income.

Tobacco sales market in Wendell, North Carolina. (*Courtesy* Travel and Promotion Division, North Carolina Department of Conservation and Development)

North Dakota

Agriculture is the most important industry in North Dakota. There were 46,381 farms in 1969, and their average size was 930 acres. A total of 44 million acres, or 97 percent of the land area, is in farms.

In 1969 field crops were marketed for 480 million dollars, and livestock for 268 million dollars. In the same year, North Dakota was the nation's leading producer of spring and durum wheat, was second in the production of all wheat, and ranked first in barley and flaxseed and second in rye and oats.

North Dakota ranks third in the production of potatoes, being surpassed only by Maine and Idaho. Beef cattle and sheep are raised on the North Dakota rangelands.

Wheat, cattle, barley, and dairy products were the principal products sold from the farms in 1970.

Shelterbelts and windbreaks are a familiar part of the landscape in the Dakotas. In addition to protecting soil, livestock, and farmsteads, the shelterbelts and windbreaks are good cover for pheasant and other game.

Ohio

The typical Ohio farm is a family-size unit of about 154 acres. The state had 111,332 farms in 1969. Sixty percent were commercial. Sixty-five percent of the land area, or 17 million acres, was in farms.

For the past several years, the value of products from Ohio farms has been more than 1.2 billion dollars per year. Ohio is the world champion producer of hothouse tomatoes and ranks second in the United States in the harvesting of tomatoes for processing. In other national rankings, Ohio is fifth in popcorn, sixth in grapes, corn, and clover, and seventh in oats, soybeans, and hogs. Winter wheat, apples, peaches, and tobacco are grown commercially.

Ohio ranks high in the raising of sheep and lambs. The state is well located for shipping lambs to eastern markets. Ohio ranks high in the production of wool. The state ranks high in the quantity of whole milk sold and in the number of cattle marketed. In 1970 Ohio also ranked third in its region in the production of broilers.

The principal cash earners on Ohio farms in 1970 were, in order, dairy products, cattle, hogs, and soybeans.

Oklahoma

In 1970 Oklahoma farmers received 678 million dollars from livestock marketings and 261 million dollars from crops. The state was then third to Kansas and North Dakota in winter wheat. In 1970 Oklahoma was second in the United States in the number of beef cattle on farms and ranches.

In 1969 Oklahoma was the fifth largest producer of sorghum grain. Other commercial crops include peanuts, corn, oats, barley, pecans, and soybeans. In the South Central region, Oklahoma ranks third in the number of sheep and lambs marketed. There are large areas of rangeland which are best adapted to the raising of beef cattle.

Products sold from Oklahoma farms in 1970 ranked in this order: cattle, wheat, dairy products, and peanuts.

Oregon

Oregon had 29,063 farms in 1969, and the average size was 620 acres. The 18 million acres of farmland occupied about 29 percent of the land area.

The state has more than 2 million acres of farm woodland and is the nation's leading producer of forest products. About 56 percent of the state's industrial workers are in forest industries. The value of forest products in 1969 was approximately 6.8 million dollars. Lumber and paper are the principal forest products.

Agriculture is the second largest industry. Marketing receipts for farm products were more than 531 million dollars in 1969. Crops yielded about 260 million dollars; livestock, 264 million dollars. Cattle, wheat, and milk are the leading

Modern beef breeds have replaced the legendary longhorn cattle on the prairies of Oklahoma and Texas. (*Courtesy* U. S. Fish and Wildlife Service)

farm products.

In 1970 Oregon was first in the production of snap beans for processing, was second in pears, and ranked third nationally in the growing of hops. The state is also a leading producer of green peas and sweet corn.

Oregon ranks high in the nation in turkey production. Broilers are raised on many commercial poultry farms. Oregon is among the twelve leading apple producing states. There were more than 685,000 beef cattle on Oregon farms in 1970.

Cash earners in Oregon in 1970 were cattle, dairy products, wheat, and greenhouse products.

Pennsylvania

Pennsylvania is a state of small farms, with an average of 142 acres. There were 62,824 farms in 1969. Approximately 31 percent of the land area was in 28.7 million acres of farmland.

The east central part of the state is noted for prosperous farms. The 1969 receipts from livestock and crops were more than 945 million dollars. Much of the farm income was from dairy and poultry products.

Pennsylvania ranks first nationally in cigarleaf tobacco, sausage products, and plantation-grown Christmas trees. In the production of other farm commodities, the state was second in ice cream and in the market value of eggs, third in chickens, and fourth in dairy products and apples.

Dairy products, cattle, eggs, and greenhouse products were the principal products sold from Pennsylvania farms in 1970.

Hay of a high quality is necessary for feeding livestock in winter. Timely and rapid baling helps to improve the quality. (*Courtesy* Ford Tractor Division)

Pennsylvania forest products are valued at more than 1 billion dollars annually. The forested land in the state also adds much to the scenic beauty of the landscape.

Rhode Island

Dairy and poultry products and potatoes provided most of the 21 million dollars in farm receipts for 1969. Poultry farms in the state are well known for the Rhode Island Red breed of chickens. Rhode Island produced almost 1.5 million pounds of broilers in 1969. Potatoes and apples are the principal crops.

Rhode Island reports cash sales from farms in 1970 as having been in this order: dairy products, greenhouse products, potatoes, and eggs.

Only 1 percent of Rhode Island's labor is engaged in farming. Manufacturing and other industries, including the large textile industry, add 1 billion dollars annually to the economy of the state.

Fish and shellfish valued at approximately 3.6 million dollars annually are taken from the ocean water off the coast of Rhode Island.

South Carolina

In recent years there has been a trend toward fewer but larger farms in South Carolina.

In 1969 South Carolina farmers received 235 million dollars from crops and 166 million dollars from livestock. The state ranked third in the United States in the growing of tobacco. This crop is the leading source of farm income in South Carolina. Cotton and soybeans are the next most important crops. The diversified farming pattern also includes corn, oats, sweet potatoes, peaches, and peanuts. Some land formerly cultivated has been converted to improved pastures for beef and dairy cattle.

Poultry and eggs are the main livestock products. In 1969 South Carolina was among the top twenty states in broiler production. The state ranked almost that high in the number of eggs produced.

In 1970 South Carolina reported cash farm sales in this order: tobacco, soybeans, eggs, and cattle.

South Carolina forests yield valuable crops of sawtimber and pulpwood. Loblolly pine is the principal commercial forest species.

South Dakota

The agriculture of South Dakota is much like that of North Dakota. South Dakota had 45,726 farms in 1969. The average farm size was 997 acres. The percent of land area in farmland is also similar, 94 percent in South Dakota and 97 percent in North Dakota. South Dakota has 45.6 million acres of farmland, which is 1.6 million more than North Dakota.

In order of market value in 1970, the principal sources of farm income in South Dakota were cattle, hogs, dairy products, and wheat. Total farm receipts in 1969 were approximately 958 million dollars. South Dakota was first in rye, second in flaxseed, third in the United States in spring wheat, and third in oats. Ranking fifth nationally in the number of sheep on farms, South Dakota ranked ninth in the number of hogs and fifth in the number of beef cattle.

As in the most of the other grain states, there has been an increase in the growing of sorghum grain and forage in the past three decades.

Hogs are raised and fattened in the corn producing counties in eastern South Dakota. Beef cattle are better adapted to the drier rangelands in the western part of the state. Some cattle are fattened in the corn counties.

Among hunters, South Dakota is famous for pheasant. This game bird multiplied rapidly after its initial introduction into the James River Valley in the eastern part of the state. Mitchell, South Dakota, is widely known for its Corn Palace and annual corn festival. Like North Dakota, South Dakota has many tree shelterbelts and windbreaks.

Tennessee

The 1969 cash receipts for Tennessee farm commodities included 414 million dollars from livestock and 259 million dollars from crops. The state ranked fifth nationally in tobacco production and sixth in cotton. Cattle and milk rank ahead of tobacco as sources of farm income. At the beginning of 1970, the state had 2.3 million cattle, 7.6 million chickens, and almost a million pigs.

Tennessee ranks high in broiler production. Eggs are an important source of supplementary income on many small farms, and there is considerable commercial production of eggs.

Tennessee cash receipts in 1970 for farm products, in order, were cattle, dairy products, tobacco, and soybeans.

Receipts from forest products are about equal to receipts from the sale of livestock and crops. Tennessee is nationally known for the manufacture of hardwood flooring.

The lower Rio Grande Valley in Texas is a huge irrigated checkerboard of dark green citrus groves and lighter green vegetable fields. (*Courtesy* Texas Highway Department)

Texas

In 1969 the 213,550 Texas farms averaged approximately 668 acres. For comparison, this is almost 330 acres less than the average for South Dakota and 34 acres larger than the average for Nebraska. Eighty-five percent of the land area of Texas is in 165 million acres of farmland, and about 57 percent of the farms are commercial.

Cattle, cotton, grain sorghum, and dairy products were the leading sources of farm and ranch income in 1969. In 1969 Texas ranked third in agricultural receipts, with 1.1 billion dollars from crops and 1.8 billion dollars from livestock. On January 1, 1970 Texas had 12.5 million cattle on farms and ranches.

In 1969 the Texas cotton crop was approximately 2.86 million bales. A large acreage of Texas cotton land on the High Plains is irrigated.

Texas farms yielded 309.8 million bushels of sorghum grain in 1969. This was close to the combined production of Kansas, Nebraska, and Oklahoma. On the basis of yield of sorghum grain per acre, California, Arizona, Indiana, Iowa, New Mexico, Illinois, and Missouri were ahead of Texas in 1969.

Texas ranked first in 1970 in the production of beef cattle, sorghum for grain, cotton, sheep, and goats.

The Rio Grande Valley of Texas is well known for citrus fruits and vegetables. Peaches also are grown in the Valley. Peanuts are a leading crop in some central Texas counties. Pecans grow well in central and eastern Texas. Tyler, in East Texas, is famous for roses.

Wheat is grown extensively in the Panhandle of Texas. Corn and oats are the other major grain crops. A yield of 21.6 million pounds of rice made Texas second in the growing of this crop in 1969.

Although beef cattle are generally associated with the rangelands in western Texas, the larger numbers of cattle are found in the coastal counties of southern Texas.

Utah

Utah is a state of large farms, with an average of 867 acres per farm. In 1969 Utah had 13,045 farms in 11.3 million acres, which was more than 22 percent of the land area.

The average farm income in 1969 was 16,328 dollars, an increase from 10,075 dollars in 1964. In 1969 receipts from crops were 38.6 million dollars, and from livestock, 174.4 million dollars.

In January 1970 Utah ranked fifth in the United States in the number of goats on farms.

Barley is the leading grain crop, with a yield of 60 bushels per acre on some farms. Alfalfa, potatoes, and sugar beets are profitable crops in Utah, espe-

cially where these crops are irrigated. Utah cash receipts in 1970, in order, were cattle, dairy products, turkeys, and sheep.

Utah has large areas of dry rangeland suitable only for grazing. There is a limited production of fruits and vegetables. Peaches, apricots, and cherries grow well in the vicinity of Ogden and Brigham City.

Vermont

In 1969 Vermont farmers received 137.5 million dollars from livestock and 14.3 million dollars from crops. Dairying accounted for 80 percent of the agricultural income. With a production of 290,000 gallons of maple syrup, Vermont ranked second to New York in that commodity.

Among the six New England states, Vermont ranked third in the amount of whole milk sold. Eggs are the third ranking source of farm income in Vermont but rank far lower than milk. Potatoes, apples, and corn are considered important crops in the state.

Vermont farmers received income from these commodities in 1970 in this order: dairy products, cattle, eggs, and forest products.

Virginia

In 1969 there were 64,572 farms in Virginia, and 48 percent of them were commercial. The average size of farms was 165 acres. Approximately 42 percent of the land area of Virginia, or 10.6 million acres, was in farms.

Dairy products, cattle, tobacco, and eggs were the main sources of farm income in 1969. In 1969 the market receipts from livestock were 328 million dollars and receipts from crops were estimated at 248 million dollars. Dairy and poultry products are the main source of income from livestock.

In 1969 Virginia ranked third nationally in sweet potatoes, fourth in tobacco, and fifth in apples and peanuts. Other crops grown on Virginia farms include corn, soybeans, barley, peaches, and potatoes. Virginia ranked thirteenth in broiler production in 1969.

Forestry is an important industry in Virginia. Most important of the forest products were furniture, lumber, and paper.

Washington

The 516-acre average size of farms in Washington is above the national average of 383. In 1969 there were 34,000 farms in Washington, 35 percent were commercial, and their total acreage was 17.6 million. This was about 41 percent of the land area in the state.

The 1969 value of the products from Washington farms was 775 million dollars. Wheat, milk, cattle, and apples were the main sources of farm income. Market receipts were 420.8 million dollars from crops and 345.5 million dollars from livestock. Washington ranked fourth in the United States in the production of wheat, the state's most important crop. That year Washington was first nationally in apples, hops, late summer potatoes, dry field peas, peppermint, second in pears, apricots, and alfalfa seed, and third in winter wheat, asparagus, and strawberries. Cherries, cranberries, and grapes are also grown commercially.

Sources of cash income for farms

in 1970 were, in order, wheat, dairy products, cattle, and apples.

Washington ranked sixteenth in the quantity of whole milk sold in 1970. That year the state ranked ninth among the Western states in the number of cattle marketed. Within its region, Washington was second in broiler production.

Forests cover more than half the land area of Washington and contain about one-sixth of all the standing sawtimber in the nation. The most important commercial tree species are the Douglas fir, Ponderosa pine, western hemlocks, and red cedar. Lumber and paper are among Washington's more important industrial products.

West Virginia

In 1969 there were 23,142 farms in West Virginia. The average size was 188 acres. Farmland occupies about 28 percent of the land area and totals 4.3 million acres. About 29 percent of the West Virginia farms are commercial.

In the order of their value, the most important sources of farm income are cattle, milk, broilers, and apples. Farm marketing receipts were 106 million dollars in 1969. Within the South Atlantic region, West Virginia is second to Virginia in the number of sheep and lambs marketed. In broiler production, West Virginia is sixth in its region.

West Virginia produces a good quality of apples. Orchards can be developed on sloping and stony land not suitable for crops that require intensive cultivation.

Forests occupy large areas of the state. The forest resources include several valuable hardwood species.

Wisconsin

Wisconsin is often called the Badger State, but America's Dairyland is also an appropriate name. In 1970 the state had an estimated population of 4.37 million, and Wisconsin farms had 3.9 million dairy cows.

This is a state of intensively managed family-size farms that are much smaller than the national average. In 1969 Wisconsin had 98,973 farms, with an average size of 183 acres. Let's compare this average with the average farm sizes in New York (196 acres), Minnesota (260 acres), California (459 acres), and Pennsylvania (142 acres). All of these are major dairy states with similar production and marketing problems. The larger average size of California farms is due in part to the large areas of rangeland.

Eighty percent of the farms in Wisconsin are dairy farms. About 52 percent of the total land area is in 34.9 million acres of farmland. The characteristic appearance of the Wisconsin landscape is that of rolling pastureland interspersed with fields of alfalfa, clover, grain, and peas. Good conservation practices are part of all commercial farm operations.

Wisconsin is a state of intensively managed family farms that are dominated by Holstein dairy cattle. There are almost twice the number of dairy cows in Wisconsin than in any other state. (*Courtesy* Holstein-Friesian Association of America)

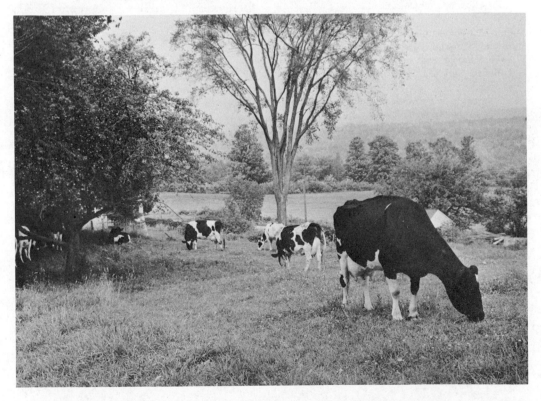

In 1969 market receipts from livestock and livestock products were over 1.25 billion dollars, and crops brought in 196 million dollars. Wisconsin was then first in the production of milk, cheese, malted milk, and peas. The state was second to Minnesota in butter, second to Massachusetts in cranberries, and fourth in oats and vegetables.

The highly diversified farming programs in Wisconsin include the production of strawberries, apples, potatoes, and honey. Clover and bees are mutually beneficial, and this combination may be seen on many farms. Wisconsin has several large vegetable processing plants. These industrial food enterprises are the main market for the state's large pea crop.

Ranking in order of cash farm receipts in 1970 were dairy products, cattle, hogs, and corn.

Wyoming

Farming and ranching in Wyoming may be described as extensive rather than intensive. In 1969 the state had 8,838 farms, with an average of 4,014 acres. This is approximately equivalent to 18 Iowa farms or 22 Wisconsin farms. Eighty-three percent of the Wyoming farms were classified as commercial. Fifty-seven percent of the land area was in farms, and the total acreage was more than 62 million.

Livestock receipts in 1969 were 211 million dollars; crop receipts, 38 million dollars. Wyoming ranks second to Texas in wool production. In January 1970 there were 1.8 million sheep and 737,000 beef cattle on Wyoming farms and ranches.

The principal crops in Wyoming are wheat, oats, sugar beets, alfalfa, corn, malt barley, and potatoes. Some of the native grasses are harvested for hay. Legumes yield well under irrigation.

In 1970 the rank of commodities to bring in cash farm income were cattle, sheep, sugar beets, and dairy products.

The Missouri, Colorado, and Columbia river systems provide water for power and irrigation in Wyoming. In recent years the construction of dams has established numerous reservoirs, and it is probable that there will be a continuing increase in the amount of water available for irrigation agriculture.

Abundant mineral resources make mining the largest industry in Wyoming. Agriculture is the second principal source of income, followed by tourism and manufacturing.

A bull moose occupies this grassy spot in the Shoshone National Forest, but similar land in Wyoming is grazed by livestock. (*Courtesy* U.S. Forest Service)

QUESTIONS

1 How many farms are there in your state? What is the average size of the farms?

2 How does the average size of farms in your state compare with the national average?

3 For what crops in your state are rainfall and length of growing season critical factors?

4 Are the farm products of your community much the same as the more important products of your state?

5 What states have the best chance for an increase in the number of farms and the area of farmland?

6 What treatment must all pastures in Hawaii have to produce high yields of nutritious forage?

7 What are the principal farming areas in Alaska?

8 Why is it difficult to develop new farms in Alaska?

9 In which states are the average sizes of farms close to the national average?

10 Which states have a highly diversified agriculture?

ACTIVITIES

1 On a blank outline map of the continental United States, write the name of each state, number of farms, number of acres in farms, and average size of farms.

2 Find the five states which rank first in each of the following: dairy cattle, beef cattle, hogs, sheep, poultry.

3 Make a list of the 50 states in descending order of the average size of farms.

4 Select five states in which the annual market value of livestock is close to the annual value of crops.

5 Using the maps in the chapter on climate, make a list of the states that have 120 to 150 frost-free days and an annual precipitation of 35 to 45 inches. Then read again the descriptions of the agriculture in these states. Does there seem to be a close relationship betwen climate and the kinds of farming in these states?

5 FARM LIFE

Many cities are becoming traffic infested, smoky, crowded, and expensive. For these and other reasons, thousands of families are moving to the country to enjoy open spaces, fresh air, lower costs of living, and freedom to live as they wish.

Science is making country living a happier experience for farm families. Rural communities are attracting other families who farm part-time but get most of their income from trades, professions, and business enterprises. The interest in country living seems to be growing rapidly. Do you know why? (See the 1971 Yearbook of Agriculture, *A Good Life for More People.*)

In your study of agriculture you will get acquainted with farm life both as a way to live and a way to make a living. You will find, too, that country living, with all its simplicity and charm, has interesting and complex problems.

If you live in a farming region or have traveled through one recently, you may have seen abandoned houses on farms or in villages. Perhaps you wondered what happened to the families who once lived in those houses. Also, you may have noticed a number of new homes on small acreages near a town or city. Some of the movement of farm families to cities and of city families to rural communities is the result of a situation much like a "squeeze play" in a baseball game. Many farmers who could not earn a good living by farming moved to the city where more job opportunities existed; while city families, looking for low-cost living, moved to the country.

Since 1900 there has been a steady decrease in the number of people needed to operate farms. As a result, several million farm families migrated to industrial centers which offered better opportunities for employment. Some went willingly because they believed they would enjoy a higher standard of living; others were squeezed off the land by circumstances they could not control, such as the high cost of machinery and labor. Although the percentage of agricultural workers in the total population is still declining, new life is being brought to country living by the development of suburban and rural communities.

Science has made the American farmer the world's most efficient producer of food and fiber. In 1900 the average production of each agricultural worker would feed and clothe about seven people. Today one farmer can provide enough food for 44 other persons, five of whom live in another country. Since 1950 the average production of each person in agriculture has increased more than during the preceding 50 years.

Science has accelerated crop yields so much that the increasing needs of America's rapidly growing population can be satisfied without farming more acres. For example, the total acreage of corn could be reduced by 25 percent if the best practices in the Corn Belt were adopted on all farms.

As the productive capacity of farm operators increases, the size of the farms they are able to tend increases. Between 1940 and 1950 the average size of farms in the United States increased from 174 to 215 acres. In 1972 the average farm covered nearly 400 acres and each year continues to increase.

THE FARM FAMILY

Many men and women credit a large part of their success in adult life to their childhood experiences as members of happy farm families. Today, even more than in the past, the farm family can enjoy a comfortable home, nutritious food, good schools and churches, and wholesome community activities. Better roads and better cars, rural electrification, and more telephones are bringing increased enjoyment of country living.

Sometimes it is said that American family life is having a hard time to survive. Particularly in the densely populated areas, too many people are looking for excitement and taking too little time for wholesome family life. As a result, many families who like to do things together are turning to farm life.

In most industrial areas, housing is a difficult problem. New families in urban communities often have trouble finding a satisfactory place to live at a cost they can afford. Sometimes houses and apartments are not available at any price. Many owners who rent property will not accept families with children, and pets are even less welcome. Here is another reason why more and more families are moving from the city to the country.

Another city problem is the lack of places for children to play. No matter how hard community leaders try to correct such situations, population growth may keep ahead of the building of homes, schools, and playgrounds. This is one more reason why families are turning to country living.

Wars and the threats of war affect every family to some extent, but the impact probably is most severe for nonfarm families. Industries must operate on a mass-production basis, and the individual worker becomes a small part of a vast and complicated organization. The demand for workers and the high

Prosperous farming is the foundation of happy country living. (West Virginia)
(*Courtesy* Soil Conservation Service)

cost of living encourage mothers as well as fathers to take their places in shops and factories. This helps to boost the family income, but it also means that the members of the family have less time together.

The following are a few of the benefits of country life that contribute to the happiness of the farm family. Perhaps you can suggest other factors that are just as important.

Space for living. Everyone likes plenty of room to work and play. This is particularly important for growing children. On a farm—even a small farm—every member of the family has space for the work or play that interests him.

A well-kept, spacious country lawn is a pleasant place for children to play. In addition to the family garden, there may be individual plots where the children can grow flowers and vegetables of their own choice. There is space for Mother's favorite roses, for Father's experiments with new varieties of plants, for Sister's pumpkin vine, and for Brother's earthworm bed. If the family outgrows the house there is space for adding new rooms or for building a new home.

Every child needs a pet, and nowhere is there a better place for pets than on a farm. A puppy, or a whole litter of puppies, can scamper across the lawn and bark without disturbing the neighbors. It is even possible for pups and kittens to be friendly playmates at a farm home. An orphan lamb or a pet pig is welcome when there is plenty of space for living. Calves, colts, goats, chickens, geese, and ducks are good farm pets. In another chapter we shall consider rabbits and other small animals that add to the pleasure of country living.

Working and playing together. Perhaps it is true that all work and no play makes a dull boy. It is also true that all play and no work makes a boy or girl lazy and unhappy. There is always work to be done on a farm.

On a livestock farm there are feed crops to be grown and harvested, animals to be cared for and marketed, and fences to be built or repaired. A grain farm requires a great deal of hard work in the planting, cultivation, and harvesting of crops. On a part-time farm which is operated mainly to provide food for the family, there are vegetables to process, fruits to pick, a garden to weed, trees to spray, eggs to gather and sell, and dozens of other things to do. All this is work, but it is twice as easy and ten times more fun when the whole family works together.

Playing together is just as important for the farm family as working together. A shady lawn is an ideal place to play croquet; the back yard may be equipped with a barbecue pit for outdoor cooking; a tennis or badminton court may be developed at little cost.

The farm family can provide wholesome recreations as by-products of the regular farming operations. For example, the pond or lake that is constructed to provide water for livestock can be made to produce fish. It may also be a good

Country living and the joy and care of farm pets go together. (*Courtesy* Western Dairy Journal)

place to swim, and if it is large enough it may be used for boating. Another recreation may be found in managing woodlands and field borders to produce small game. It is interesting, too, to help the wild birds feel at home; then they will work and sing nearby and eat harmful insects in the garden. These and countless other pleasures are among the rich rewards of country living.

Economical living. A farm family can live better on a given amount of money than can a family in any other occupational group.

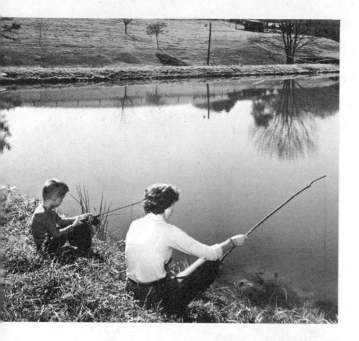

Above: Country living requires more work but also gives more fun, such as going fishing. (*Courtesy* Soil Conservation Service) *Below:* Fish taste twice as good when you catch them yourself. (*Courtesy* Soil Conservation Service)

It is true that many farm families have less cash income than do people who work in skilled trades or professions. But to get an accurate estimate of farm income, it is necessary to consider the value of the home-grown foods consumed and the cost of renting a comparable home. In general, too, taxes on farm property are somewhat lower than on city homes. And in many cases the farm family can reduce the cash outlay for home maintenance by painting, papering, and making minor repairs.

Perhaps the biggest factor in the economy of country living is the value of food grown on the farm. Three of the most expensive items in the cost of food are (1) the labor required to produce and process food crops, (2) transportation of the food to markets, and (3) the cost of marketing through wholesale and retail stores. The farm family can eliminate most of this expense, and the market value of the home-grown foods can thus be added to cash income. In addition, home-grown foods taste better because they can be harvested just when they are perfectly ripe and ready to eat.

Of course, country living has its dark days. The farm family has its ups-and-downs with drought, floods, severe winters, insects, and uncertain markets. When economic conditions are unfavorable in cities the farm family also feels the shock. The part-time farmer, like his fellow workers in the cities, is sometimes an innocent victim of unemployment. But year in and year out the farm families have more "bounce" than any other group of people in the United States.

Community teamwork. Farm families throughout the nation have demonstrated their ability to work together.

Some country living fails for various reasons. (*Courtesy* Soil Conservation Service)

Group action is bringing better schools, stronger churches, good roads, adequate medical services, and wholesome recreational programs. The same bond of common interest that holds an individual farm family together gives a whole rural community a unity of purpose that gets things done. The almost unlimited opportunity to help build happier and stronger communities is one of the most promising features of country living today.

BEGINNING FARMERS

Each year a large number of prospective farmers with a wide variety of backgrounds seek to enter farming.

Some aspire to be full-time farmers; others, part-time. Some good advice on starting a farm operation has been prepared by Kenneth H. Myers, author of *Facts for Prospective Farmers*, Farmers Bulletin 2221, U. S. Department of Agriculture, 1966. Some of the pointers given in that publication are summarized in the following paragraphs.

Farmers who start a full-time operation usually expect a farm to provide enough income to support and educate a family. Some of the prospective part-time farmers hope to become full-time farmers. Their usual plan is to continue their nonfarm work to get capital for a full-time farm business. Others who divide their time between farm and nonfarm work expect a farm to provide only supplementary income.

A large capital investment is required to buy a commercial farm and the equipment and livestock to operate it. A substantial investment is required just to rent a commercial farm. Before making such an investment the prospective farmer must know the amount of land and capital required, the type of farming to be done, and the location of the farm in respect to markets.

A commercial family farm today is a substantial business venture. It involves financial risks and has exacting requirements of capital, technical skill, and management ability. The beginning farmer also has a big problem in getting suitable land to farm.

Important farming trends. The family farm is still the dominant unit in agriculture. On a family farm the operator and his family do the managing, do most of the work, and take the financial risks. In 1972 family-size farms made up 95 percent of all farms and accounted for 70 percent of all farm production.

Farms are getting larger. The average size of farms increased from 174 acres in 1940 to about 400 acres in 1972. The increase in farm size has been accompanied by a decrease in the number of farms. In 1920 there were nearly 6.5 million farms. By 1972 the number had declined to about 2.8 million.

As the size and number of farms have changed, there has been a steady increase in the amount of capital required for farming. In 1972 the average total investment in land, buildings, livestock, machinery, and working capital was almost $90,000 per farm, more than 16 times the investment required in 1940 and three times the investment in 1960.

As farmers have become more specialized they have become more dependent on purchased feed, seed, fertilizer, equipment and other essentials. Farmers have also become highly dependent on cash markets for farm products. Because of these trends, net income can be hurt seriously in a short time by changes in prices of both farm products and purchased inputs.

Technology and economics are responsible for the increase in farm size and capital investment. Technology makes it possible for the farmer to enlarge his operation. Economics makes it necessary. Some examples of technology are labor-saving machinery, better seed, greater use of commercial fertilizer, better land use and treatment, insect and disease control, selective breeding, and more efficient feeding and management practices.

Larger farms are necessary to use modern machinery efficiently. The increase in overhead and operating costs tends to reduce profits per unit of output. This makes the improvement of net income very difficult.

These changes in farming have changed the qualifications the farmer should have. Farmers today must have a high degree of technical skill and a great deal of ability in financial and business management. It is also necessary for farmers to know about government farm programs such as allotments and market quotas.

Farm life also has changed. Modern power and equipment have taken much of the drudgery and hard work out of farming. Electricity now enables farm homes to have the same conveniences and labor-saving equipment as town and city homes.

Availability of farms. There is a steady decrease in the amount of farm land available and a continuing rise in the cost of farmland. The federal homestead laws still exist, but very little land suitable for farming becomes available. Alaska has some land which may be homesteaded and some which may be bought or leased. It is very difficult to develop a commercial farming operation in Alaska, and the risks are great because of a short growing season, limited markets, and inadequate transportation facilities. A prospective farmer considering Alaska should obtain information from reliable sources about climate, farming conditions, and markets in the area in which he is interested. One source of information is the Alaska Department of Natural Resources, Division of Agriculture, Post Office Box 1828, Palmer, Alaska 99645.

About 170,000 farms are vacated each year by operators who die, retire, or leave the farm for other reasons. It is probable that less than one-fifth of these farms are taken over by new farmers. Many of the new farmers are young men who take over the home farm from their parents. Half of the farm tracts sold each year are bought by established farmers to enlarge their farms.

Each year some land is taken out of farming for urban, industrial highway, or other nonfarm use. A smaller acreage of land is brought into cultivation by irrigation and drainage. The net result is that the total land in farms is decreasing. It is estimated that the amount of farmland is shrinking at the rate of 1 million to 1.5 million acres each year.

How much land is needed? The amount of land needed for different types of farms varies widely. A farm for producing market eggs may have

To start farming in the United States an average of almost $90,000 is required to buy land, buildings, livestock, and machinery and to provide working capital for a year. This is more than a 300 percent increase over 10 years ago. (*Courtesy* John Deere)

only 10 acres, while a sheep ranch in the Southwest may have more than 10,000 acres. In general, fewer acres of cropland are used for livestock systems of farming such as dairying than in cash-grain farming. More land is needed in areas of low production than in areas of high production. In a productive region such as the Corn Belt the cost per acre of land is high.

Beginners who do not have the capital or the experience to start a specialized farm should begin with a general type of farm. A general farm might combine a small dairy enterprise, hogs, or poultry, and perhaps a cash crop. This kind of farming requires less capital, but it also produces relatively less income than does a specialized commercial farm.

The amount of land required is related to the income level desired. The Department of Agriculture has estimated the minimum "complements of resources" needed to enable farmers to realize $2,500, $3,500, $4,500, and $5,500 a year for their labor and management on major types of farms in 30 selected farming areas in the United States. These estimates are available on request. The four income levels to which the estimates are keyed are about the median annual earnings of skilled or semiskilled workers in nonfarm employment in different regions of the country.

Getting started. Anyone seriously considering entering full-time commercial farming needs more detailed information than can be given here. Some information can be obtained from the United States Department of Agriculture in answer to particular questions. The more specific the questions, the more useful the replies can be. Information about farming in a particular state may be obtained from the State Director of the Extension Service. That office can supply up-to-date references on the state's agriculture, addresses of agencies that may be able to help beginners get started, and addresses of county agricultural agents.

Anyone interested in farming in a particular county may write to that county's agricultural agent. The county agent or a committee of farmers working with him can furnish information about farms for sale or rent in the county, soils and topography, crops suitable to the county, local farming practices, how much debt is safe, how to rent a farm, and the size of a farm necessary to support a family.

Real estate dealers in farm properties have lists of farms for sale or rent.

The United States Department of Agriculture does not maintain lists of either real estate agencies or private loan sources and is not in a position to make recommendations as to their reliability.

Always visit a farm before buying or renting it. This permits checking the soil and condition of the farm and the characteristics of the community that might affect a family's health and happiness.

Here are some points to check:

Present and planned local public services

Current and prospective tax levels

Business, commercial, and banking facilities that serve farmers' needs

General character of the community

Location of schools, churches, and social centers

Availability of medical facilities

Opportunities for supplementary off-farm work

Lending agencies usually operate through local offices. The prospective farmer should check on the kinds and amounts of credit available through federal land bank associations, Farmers Home Administration, individual lenders, production credit associations, farm equipment dealers, and other credit sources. County offices of the Agricultural Stabilization and Conservation Service (ASCS) advise on and administer commodity programs. These include allotments and marketing quotas for the basic commodities. The ASCS offices can also provide information on soil, water, timber, and wildlife conservation practices. The Agricultural Conservation Program helps farmers apply these practices on individual farms.

The ASCS offices are responsible for the local administration of price support commodity loans which are available through the Commodity Credit Corporation. These offices also administer the feed grain program, certain emergency programs in designated areas affected by drought or flood, and other farm programs.

In almost every county there is a Soil Conservation Service office that provides technical assistance and information on soil and water conservation, land-use practices, soil surveys, and resource use. These and many other free government services from many sources are available to the beginning farmer just for the asking.

PART-TIME FARMING

The United States has enough commercial farms to keep ahead of the growing demand for food and fiber. In fact, about 95 percent of the market supply of agricultural products comes from only 55 percent of the farms. Most of the remaining 45 percent of the farms are not actually needed to grow food and fiber for the market. Many of these are *part-time* farms.

Part-time farmers are those who get a major part of their income from employment other than the operation of their small farms. The United States has more than 600,000 part-time farmers. Another million Americans live on farms from which they get almost no income. Families in this second group usually raise a few chickens and have a garden for home use.

There are several factors to consider in determining whether part-time farming is good business. The quality

Above: Part-time farming and part-time work in the woods is a very good combination. (Virginia) (*Courtesy* U.S. Forest Service) *Below:* Success in farming depends upon many factors, including a farm tractor and rugged equipment such as this new sure-till harrow. (*Courtesy* Brillion Iron Works, Inc.)

of the community is extremely important. If attractive building sites are available and the homes already established are well kept, the prospective part-time farmer should then carefully determine the real market value and the probable resale value of the property he is considering.

The United States Department of Agriculture has made the following recommendations on the problem of low farm income:

1. Expand and adapt agricultural extension work to meet the needs of low-income farmers and part-time farmers.
2. Develop needed research in farm and home management.
3. Provide additional credit for low-income farmers and extend Farmers' Home Administration services to part-time farmers.
4. Increase technical assistance such as that provided by the Soil Conservation Service.
5. Expand vocational training.
6. Strengthen the Employment Service in rural areas and adapt it to the needs of this group.
7. Develop more effective programs to induce the expansion of industry in these areas.
8. Encourage farm, business, and other leadership to assume local responsibilities.

QUESTIONS

1 Give an example of how the production efficiency of the American farmer has increased.
2 How much has the size of the average farm increased recently?
3 Why are farm families usually happy families?
4 Why is it possible for farm families to live well on less cash than city families require?
5 Where would you write to get information on buying a farm?
6 What is meant by "part-time farming"?
7 What factors cause low farm income?
8 Name five steps necessary in helping low-income farmers to raise their income.
9 List all of the advantages and disadvantages of country living.
10 List five ways in which science has improved country living.

ACTIVITIES

1 Arrange a class panel discussion in which two pupils explain some of the advantages and problems of country living and two other pupils explain the advantages and problems of urban living.

2 Use geography textbooks, encyclopedias, the *World Almanac,* Census Bureau reports, or other references to obtain information about population trends in your state and region. Try to determine whether the rural population is increasing or decreasing.

3 Use American history textbooks and other references to obtain information about farming and farm life in the Colonial period. Compare that period with the nineteenth and twentieth centuries.

4 Have several voluntary class reports on the following or similar topics: part-time farming; best farm income areas in the United States; size of farms in the community; current sale prices of land in the community; or opportunities for country living.

5 If someone in the class has recently moved from the country to the city or from the city to the country, ask him to explain to the class why the move was made.

6 AGRICULTURAL SUBURBAN LIFE

The rapidly decreasing farm population and the increasing specialization required for profitable farming are depriving boys and girls of experiences that were a part of much of family life a few generations ago. A majority of boys and girls today have never seen a Hereford or Holstein cow, a Merino sheep, a Nubian goat, a Chester White Pig, a white leghorn chicken, or a riding horse of any known breed.

Many people who formerly lived in large cities are moving to the edge of cities to have more space and more freedom for family activities. Some continue to work full time in the city; others hold their city jobs but farm a little on weekends, days off, and holidays. Employed people doing "shift" work are often able to have the equivalent of three days of daylight a week for part-time farming.

In contrast, many full-time farmers on small farms discover that, with mechanization, farm work is not a full-time job. To supplement their income they obtain part time employment in the city.

One result of these shifts is that two-thirds of all farmers now earn over one-half their income off the farm. This combination of city and farm life has been called by several names, such as part-time farming. In reality it is full-time multiple job holding. The authors prefer to designate it "suburban life." The word "suburban" means at the edge of the city; but with the rapid expansion of city limits into the country, much of the farmland is near the boundaries of cities.

Whether a farmer gets a part-time job in the city or a city dweller moves to the country to do part-time farming while holding his city job, both ways of life require mobility, adaptability, and

Left: **Suburban life permits a retired man to have more space and facilities for raising tomatoes and other plants in his own greenhouse.** (*Courtesy* Boston *Herald-Traveler*) *Right:* **Getting acquainted.** (*Courtesy* Farm Vacations and Holidays)

flexibility. This means that it is usually the younger persons who make the change. A report on one study in Michigan revealed that most of those who chose agricultural suburban life were 45 years old or less and were in better health than those who did not choose this dual life. The suburban dwellers also had more children and held full-time nonfarm jobs for an average of almost 12 years. When the dual workers were asked if they would be willing to move into a city of 100,000 or larger to be closer to their jobs, nine out of ten said they would not move into a city under any circumstances.

ADVANTAGES OF AGRICULTURAL SUBURBAN LIFE

Some of the desirable features of suburban life have been described by Orlin J. Scoville in *Part-Time Farming,*

Farmers Bulletin 2178, United States Department of Agriculture.

A farm provides a wholesome and healthful environment for children. It gives them room to play and plenty of fresh air. The children can do chores adapted to their age and ability. Caring for a calf, a pig, or some chickens develops in children a sense of responsibility for work.

Part-time farming gives a measure of security if the regular job is lost, provided the farm is owned free of debt and furnishes enough income to meet fixed expenses and minimum living costs.

For some retired persons, part-time farming is a good way to supplement retirement income. It is particularly suitable for those who need to work or exercise out of doors for their health.

Generally, the same level of living costs less in the country than in the city. The savings are not as great, however, as is sometime supposed. Usually, the cost of food and shelter will be somewhat less on the farm and the cost of transportation and

Agricultural suburban life permits children to enjoy relaxing and useful club activities, such as building a bird house and riding horses. (*Courtesy* Texas Agricultural Extension Service)

utilities somewhat more. Where schools, fire and police protection, and similar municipal services are of equal quality in city and country, real estate taxes are usually about the same.

A part-time farmer and his family can use their spare time profitably.

Some persons consider the work on a farm as recreational. For some white-collar workers it is a welcome change from the regular job, and a physical conditioner.

DISADVANTAGES OF AGRICULTURAL SUBURBAN LIFE

In the same publication, Farmers Bulletin 2178, the author points out that agricultural suburban life has some disadvantages. To provide a fair comparison, these conclusions are quoted here.

Farming is confining. The farmer's life must be arranged to meet the demands of crops and livestock.

Livestock must be tended every day, routinely. A slight change in the work schedule may cut the production of cows or chickens.

Even if there are no livestock, the farmer cannot leave the farm for long periods, particularly during the growing season.

The worker who lives on a farm cannot change jobs readily. He cannot leave the farm to take work in another locality on short notice; such a move may mean a loss of capital.

Hard physical labor and undesirable hours are a part of farm life. The farmer must get up early, and, at times, work late at night. Frequently he must work long hours in the hot sun or cold rain. No matter how well work is planned, bad weather or unexpected setbacks can cause extra work that must be caught up.

It may not be profitable for a part-time farmer to own the labor-saving machinery that a full-time farmer can invest in profitably.

Production may fall far below expectations. Drought, hail, disease, and insects take their toll of crops. Sickness or loss of some of the livestock may cut into the owner's earnings, even into his capital.

Returns for money and labor invested may be small even in a good year.

The high cost of land, supplies, and labor make it difficult to farm profitably on a part-time basis. Land within commuting distance of a growing city is usually high in price, higher if it has subdivision possibilities. Part-time farmers generally must pay higher prices for supplies than full-time farmers because they buy in smaller quantities. If the farm is in an industrial area where wages are high, farm labor costs will also be high.

A part-time farmer needs unusual skill to get as high production per hen, per cow, or per acre as can be obtained by a competent full-time farmer. It will frequently be uneconomical for him to own the most up-to-date equipment. He may have to depend upon custom service for specialized operations, such as spraying or threshing, and for these he may have to wait his turn. There will be losses caused by emergencies that arise while he is away at his off-farm job.

The farm may be an additional burden if the main job is lost. This may be true whether the farm is owned or rented.

If the farm is rented, the rent must be paid. If it is owned, taxes must be paid, and if the place is not free of mortgage, there will be interest and payments on the principal to take care of.

HOW MUCH LAND FOR SUBURBAN LIVING?

A suburban family may find that three-fourths of an acre to an acre of land is enough for raising fruits and

Above: A big pumpkin like this 100-pound Big Max variety is a thrill to grow on a farm. (*Courtesy* Burpee Seeds) *Below:* A pony cart is good entertainment for city visitors taking their vacation on a farm. (*Courtesy* Farm Vacations and Holidays, Inc.)

The cost and size of a tractor must be considered in relation to the kinds and market value of the crops to be grown. (*Courtesy* U.S.D.A.)

vegetables for home use. Perhaps there can be a small flock of chickens, a cow, and two pigs; but a plot this small will not produce feed for the livestock.

The amount of land needed depends on the extent to which the family plans to supplement its income. Also there is the problem of how much farming the family can do in addition to other work off the farm. The cost of land and the prospects for an increase in value are other factors to consider. Considerable financial risk is incurred by buying more land than is needed in anticipation of suburban development.

A desired acreage may be offered for sale only as part of a larger tract. If the cost is not too great, the purchase of the whole tract is usually better than buying a plot so small that the development of adjacent land might depreciate the residential value of the farm.

HOW MUCH EQUIPMENT WILL BE NEEDED?

Only hand tools are needed to produce some of the food for home use. If plowing is to be done, it will be necessary to hire someone to do it.

Larger plantings require some kind of power equipment for plowing, harrowing, disking, and cultivating. For planting and cultivating a half acre or more, a small garden tractor may be a good investment. Local equipment dealers can provide information about the cost of these tractors. The small garden tractors are not satisfactory for plowing clay soils, so it is better to hire someone to do such plowing.

The small-scale farmer often finds that the cost of power machinery is a serious problem. This factor must be carefully considered by anyone who plans to farm for extra cash income on a part-time basis. Sometimes it is possible to buy good secondhand equipment or to share purchase and operation costs with a neighbor. If several suburban farmers can have the same kind of work done in a short span of time, it may be possible to obtain better rates from a service company that has special equipment. Commercial spraying, harvesting, and other specialized work is usually less expensive then buying the equipment.

HOW MUCH EXTRA LABOR WILL BE REQUIRED?

One man who had a full-time job had moved to a farm a year before he was asked this question: "How do you like farm life?" His answer was, "I now work twice as hard but have four times the pleasure." Some persons who choose the dual city-farm life do not adjust this well.

Anyone with a year-round, full-time job can not expect to grow more food than the family can use. Part-time farmers usually hire very little help. In deciding on the farm enterprises that can be managed by family labor, the amount of labor that the family can supply should be compared with the labor needs of various enterprises. The accompanying table will help in making such a comparison. In estimating labor requirements, the farm operator should not include all of his or his family's spare time. Only the time actually available for farm work should be considered.

People in the 70's who like agricultural suburban life insist on all modern conveniences, most of which are not possible without electricity. (*Courtesy* National Rural Electric Cooperative Association)

WHERE IS THE BEST PLACE TO LIVE?

It is often best to rent a place for a year or two before buying. This gives a family a chance to know whether a location is satisfactory and also whether the success of the farming enterprise seems probable. Some of the main things to look for in selecting a place for suburban living are suggested in the following paragraphs.

The suburban home should be within easy commuting distance of the regular job and other employment opportunities. This might prevent the necessity of selling the property if the owner were to change jobs. The availability of alternate employment would also make the suburban home easier to sell if that became necessary.

If farm produce is to be sold, it is desirable to be near good markets. This is particularly important for the sale of fresh vegetables and fruits or whole milk.

A desirable suburban neighborhood has well-kept homes. Remember that there are slums in the country as well as in the city. Few rural areas are protected by zoning. The value of residential property can be downgraded by the establishment of a tavern, junk yard, service station, rendering plant, or other business nearby.

It is always a good idea to check on the quality of schools in the community and the provision for transportation to and from the schools. Nearness to a good school helps to increase the value and salability of a home.

The availability of fire protection, sewage systems, gas, water mains, and electrical lines is essential. If these are not available, the cost of getting them may be prohibitive. It may be necessary for the property owner to pay for these facilities or get along without them.

It is not possible to get along without an adequate supply of pure water. Where water must be obtained from a well, the prospective buyer should find out whether there is a well or how much it would cost to drill one. The county health officer can give directions for having water tested. A pond may provide water for livestock.

A household of five persons needs 125 gallons of water per day. A horse needs 10 gallons, a milk cow 25, and a hog 2. Lawns and gardens need about 8 gallons per 100 square feet at each application if the water is properly applied when needed.

A HOUSE TO MAKE A HOME

Many country farm homes for sale were built when families were large and firewood for fireplaces was plentiful. In the southern United States such large houses often make good agricultural suburban homes. In the north they are not practical because now the fuel costs are too high. Today, probably very few families would be happy with fireplaces as the only source of heat. (One of the co-authors of this book once lived for a few winter months in the Northeast in an agricultural suburban home that had 14 fireplaces.)

It is well to consider whether the house on any part-time farm will make a satisfactory full-time residence. The prospective purchaser should calculate the cost of modernizing and redecorating a residence, wiring or rewiring for electricity, and installing or repairing the plumbing. The cost of central heating may be an important consideration. Air conditioning is essential for summer comfort in the warmer states.

SUITABLE SOIL

Many soils are suitable for small-scale farming, but certain hazards should be avoided. In arid parts of the

ENTERPRISE	HOURS AND SEASON OF LABOR REQUIRED
1 acre garden	500 hours a year during an 8-month period
10 acres of field corn	240 hours a year during a 9-month period
1 dairy cow *	240 hours a year (20 hours each month)
12 laying hens *	72 hours a year (6 hours each month)
2 colonies of bees	20 hours a year
rabbits (1 buck and four does) *	120 hours a year (10 hours each month)
2 milk goats *	165 hours a year (10 to 20 hours each month)

* cows, hens, rabbits, and goats require daily care.

United States some of the soils are so full of salts or sodium that crops grow poorly or not at all. Irrigation water must be satisfactory in both quality and quantity. In humid parts of the United States some soil is too wet to permit a planting date early enough for crops to mature before frost. If the soil is poorly drained, there is the problem of draining it by tile drains or ditches, depending on which method is best. It is desirable to know whether the soil is acid, neutral, or alkaline and whether the level of available plant nutrients is satisfactory.

Almost every county has a county agricultural agent and a county home demonstration agent who can help find answers to most of the questions on homemaking and soils.

BUYING A FARM

The following suggestions on buying a farm are adapted from information published by the United States Department of Agriculture.

The value of the farm depends on (1) its worth as a place to live, (2) the value of the products you can raise on it, and (3) the possibilities of selling the property later on for suburban subdivision.

Decide first what the place is worth as a home in comparison with what it would cost to live in town. Take into account the difference in city and county taxes, insurance rates, utility rates, and the cost of travel to work.

Estimate the value of possible earnings of the farm. Set up a plan for operating the farm and list the kind and quantity of things the farm can be ex-

pected to produce in an average year. Estimate the value of the produce at normal prices. The total is the probable gross income from farming.

To find estimated net farm income, subtract estimated annual farming expenditures from probable gross income from farming. Include as expenditures an allowance for depreciation of farm buildings and equipment. Also count as an expense a charge for the labor to be contributed by the family. It may be hard to decide what this labor is worth, but charge something for it. Otherwise, you may pay too much for the farm and get nothing for your labor.

To figure the value of the farm in terms of investment income, divide the estimated annual net farm income by the percentage that you could expect to get in interest if the money were invested in some other way. If, for example, you can reasonably expect to get 5 percent on the money you invest and the estimated annual net farm income would be $100, divide $100 by 5 percent (0.05). The answer, $2,000, is the value of the farm as an income source.

Whether you can buy a farm for the amount of its value to you as a home plus its value as an income producer will depend, among other things, on its nearness to a city.

A farm within commuting distance of a city, approximately 30 miles, is likely to be one-half to two-thirds more expensive than one farther out. A farm near one of the larger cities will cost more than one near a smaller city.

Land located on a hard-surfaced road is likely to cost 10 to 30 percent more than comparable land served by a dirt road. You can expect to pay more for location on a good road in the eastern part of the country than in the Midwest.

Prices asked for small farms suitable for part-time farming reflect primarily the size and condition of the house. Compare the price asked with the probable cost of a comparable house in town, making allowances for differences in heating system, water supply, and sanitary facilities.

The alternative to buying a place that already has a house is to buy a piece of land and build the kind of house you want. When this is done the cost of the land is likely to be only a small part of the total cost of setting up the part-time farm.

If you can pay cash for the land, it is usually easier to obtain financing for the cost of a new house on it than to finance the purchase of a place that already has buildings; a new house is a better security for a loan than an old one.

Many savings and loan associations and local banks provide financing for new residences on part-time farms. The Farmers Home Administration also has a special credit program for farm housing that is well adapted to the needs of part-time farmers; this agency also makes production loans to part-time farmers. Other possible sources of loans for part-time farmers are the Farm Credit Administration and some insurance companies.

Part-time farming should be considered only if the family has a genuine interest in country living. Families with a source of income that is likely to be permanent, and families with assured retirement incomes that will provide a satisfactory standard of living, will find part-time farming a challenging opportunity. It is not a dependable source of income, but it is a good way to supplement another income and to live economically in a wholesome country environment.

Some hard realities are that if full-time farming is being considered, the products to be sold and the investment required should be considered in comparison with some averages for the United States. In 1969 the average size farm was 377 acres, and increasing at the rate of about 7 acres per year; and the capital required for owning and operating the average farm was $85,400. This capital requirement is increasing rapidly every year (up from $17,200 in 1950).

One publication that will provide information about buying a farm is "Where and How to Get a Farm," *Leaflet* No. 432, United States Department of Agriculture, Washington, D. C. 20250, August

Agricultural suburban life is living at its best, because that is where the action is. (*Courtesy* Bob Taylor)

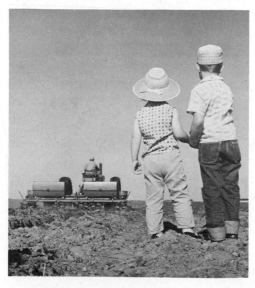

1971. Questions on buying public lands can be answered by the Bureau of Land Management, Department of the Interior, Washington, D. C. 20240. Information on loans from the Farm Credit Administration may be obtained from the Farm Credit Administration, Washington, D. C. 20578. Further information on farm loans may be obtained from the Farmers Home Administration, Department of Agriculture, Washington, D. C. 20250.

STARTING A VACATION FARM OR RANCH

As the 4-day work week becomes popular, 3-day weekends for the growing city population probably will increase the demand for vacation farms and ranches. Many young families who want to move to the country where they can see the moon without the haze of city smog may consider buying a farm with buildings suitable to start a vacation business. Most popular with tourists who want to spend their vacation on a farm or ranch are horses, ponies, and burros for riding.

Information on how to start a vacation farm or ranch may be obtained from Farm Vacations and Holidays, Inc., 36 East 57th Street, New York, New York 10022. Information may also be obtained by writing to the Director of Extension in your state, whose address is given in Appendix A.

In its Leaflet No. 301, the Extension Service at Kansas State University, Manhattan, Kansas 66502, suggests the following factors to consider before starting a vacation farm or ranch:

Factors to Consider

1. You must have attractive and comfortable quarters for your guests. Extra bedrooms in your home, cabins, trailers, or some combination of such facilities is needed. Each visiting family will prefer to have its own bathroom.
2. The entire family must be interested in meeting and sharing their home and life with new people.
3. The whole family should be energetic and in good health. Your children should help entertain the visiting children, as well as doing other work.
4. You should have recreational facilities nearby, as well as activities on the farm to keep guests busy and happy.
5. The new enterprise should not adversely affect your farm business. Consider how the extra time spent with guests will interfere with your normal farming operation.

Advantages

Many kinds of satisfaction can be gained through providing a farm vacation experience for urban families, couples, and individuals, including:

1. Additional income.
2. The opportunity to meet and learn from interesting people.
3. Helping city people to learn about country living.
4. Making new friends.

Disadvantages

There can be disadvantages in sharing your home and farm life with others:

1. Interference with the routine of farming operations and related activities.
2. The lack of any real privacy. Guests

will be around constantly, wanting to talk with you and the rest of the family when you might prefer to relax or to do something else.

3. The farm wife will have many extra hours of work. She may find herself washing dishes at 9 P.M. or otherwise making major changes in her work routine.

4. Considered on a strict business basis, the family's net income per hour may be low.

Liability Insurance

Liability insurance is very important to anyone considering developing a recreational facility. It should be investigated early in the planning stages. Liability insurance is not discussed in this publication because of the complexity of the subject.

It is essential that you check with local insurance agents on the cost and coverage of liability insurance while you are planning your recreational facilities.

SUBURBAN WATERSHEDS

Erosion and flood control is the concern of suburban residents as well as farmers. The United States Department of Agriculture has made another contribution to agricultural suburban living by publishing "Tips for City and Suburban Dwellers—Soil Conservation at Home," *Agricultural Information Bulletin* 244, 1963. A few paragraphs from this publication are quoted here to help students and teachers interpret water conservation principles and practices for small watersheds.

Many of the erosion and flood problems that plague urban and suburban residents can best be solved on a group of adjoining properties rather than on each separate lot. Some can be solved in no other way.

Farmers have found the "watershed" approach the practical way to control runoff water. The same principles apply in suburban areas, because runoff is governed by watershed boundaries and drainage patterns, not by property lines.

An interesting project for a Sunday afternoon is to try to find the boundaries of your watershed.

If you have any problem of runoff water crossing your land or flooding from a nearby stream, this search may lead you to its origin.

What is a watershed? It is all the area from which water drains to a particular point of interest. One slope of your roof is the watershed of the gutter that catches its runoff. A rill through your garden may receive water from a fraction of an acre; that area is the watershed of the rill. The depression into which the rill empties may drain water from an entire city block or several acres; this is a still larger watershed of which the rill's watershed is a part. Each larger drainageway has a successively larger watershed embracing the separate watersheds of all its tributaries. Thus, the stream that runs past your property may have a watershed of several city blocks or even several square miles.

You may be able to interest a neighbor in helping to locate the boundaries of a larger watershed that affects mutual flooding and sedimentation problems. Then, together, you can seek out the critical areas in it that contribute large amounts of flood water and eroded material. You will look for denuded construction sites, bare cuts and fills, straight-row cultivation of clean-tilled crops, eroding road ditches and streambanks, and other problem sites.

QUESTIONS

1 Why are many city families moving to the country?

2 What does the word "suburban" mean?

3 At what age are people more likely to move to the country?

4 What are four advantages of suburban life?

5 What are four disadvantages of suburban life?

6 What factors should be considered in buying a suburban home?

7 How does the kind of soil affect the kind of equipment needed?

8 What are the advantages of sharing purchase and operation costs of machinery?

9 What factors should a family consider before starting a vacation farm or ranch?

10 What factors determine the cost of suburban property?

ACTIVITIES

1 Find the "Suburban Acreage for Sale" or "Farms for Sale" columns in the advertisement section of a local newspaper for information on local land prices.

2 Write a description of the kind of suburban property you would like to own.

3 After a hard rain observe the runoff from lawns and playgrounds. Try to locate the beginning of the watershed.

4 Describe a rodeo or horse show that you have seen.

5 Visit a suburban or rural community near a city to observe how families there seem to live.

7 FARM MANAGEMENT

Agriculture is the nation's biggest industry, but unlike most nonagricultural giants, it is decentralized. No single corporation or group of companies dominates farm production. Also, the individual farmer's products have little or no identity. Unlike nationally sold, brand name, manufactured goods, one farmer's corn or wheat is about the same as another farmer's. (Source: *Statistical Abstract of the United States*, U. S. Department of Agriculture, 1971.)

Because of this decentralization, and the producer's anonymity, the production decision of just one farmer cannot markedly affect national agricultural production. National farm production does not adjust to meet consumer demand. In contrast, the steel industry reduces production to 50 or 60 percent of capacity when the demand for steel is low; automobile manufacturers try to produce only enough cars to match estimated sales; oil production and shipment is controlled and limited by state and federal laws.

For an adjusted industrial production to have the desired effect on the stability of the market and to protect each producer, all producers must limit production. But in order to maintain profits, farmers generally do not limit production. Commercial family farmers often increase production regardless of prices. The unit costs of production tend to remain about the same or may be lower as production increases. This usually means an increase in total income. When too many commercial family farmers try to maintain income by increasing production, they may weaken prices. This may cause a vicious cycle of increased production, lower prices, and depressed income.

Lower farm prices do not greatly stimulate consumption of many foods. Food distribution systems require an increasing proportion of the consumer's food dollar, and reduced farm prices have little effect on retail prices. Even if lower farm prices would reduce retail prices, people can eat just so much food. Therefore, demand tends to remain fairly constant.

THE MARKET

The trend from a rural to a city-town-industrial nation has caused many changes in farm production. A vast, complex operation is involved in channeling food and fiber from farms to more than 200 million American consumers. Farmers in this nation now produce one acre in five for the international market.

National chain stores and other food industries need large amounts of farm products for millions of customers. The food market needs quality and uniformity to back up sales promotion and advertising. Food processors, merchandizers, and farmers are making some progress in coordinating their activities. Many farmers, either alone or through their cooperatives, contract for production, processing, and marketing.

Through improved marketing methods, farmers may reduce market risks, obtain better prices, and secure more working capital. Cooperative effort is the only access to some markets, or it may reduce the cost of supplies and services. In the search for better markets, farmers may encounter pressure to expand output or to reorganize the farm operation to compete with specialized low-cost producers. Farmers may even

Farm labor has become scarce and costly, so this farmer has mechanized the filling of the fertilizer hopper. (*Courtesy* New Idea Machine Company and Allis-Chalmers)

experience some loss in the freedom of managing their individual farm enterprises.

The farmer's ability to maintain his bargaining power in national markets depends to a large extent on (1) competition among contracting firms, (2) cooperatives that can compete and perform price-setting functions, (3) developing farmer bargaining associations, and (4) having information on prices, supply, and demand conditions.

THE FARM

The wise management of a farm requires careful planning, hard work, and good business judgment. A farmer needs a rare combination of ability and experience, for he is operating in a world that changes rapidly.

Right decisions this year may be the wrong ones in next year's markets. For example, in some years potatoes may make the farmer a profit, but in other years the potato crop may cause a financial loss. And in some years the prices a farmer gets for eggs, meat, and milk may not even pay feed costs.

Whatever the cost-price relationship, the farmer can be happier, more prosperous, and less tired if he will stop before any new action and plan his next move very carefully. Here are some questions to which farm management must supply answers:

1. Is my farm work organized to make the best use of the labor each month in the year?
2. Do I have something to sell each month?
3. Are my fields more productive now than last year?
4. Do I use recommended crop varieties, fertilizers, insecticides, and herbicides?
5. Do I have an adequate forage program, including hay, silage, and pasture?
6. Is my egg production per hen equal to that of the best in the neighborhood?
7. Do I get the highest prices when I sell my farm products?
8. What can I do to make my farmwork easier, faster, or more efficient?
9. Is my home as convenient and comfortable as it should be?
10. Do I raise as much of the family food supply as possible?
11. Should I join a national farm organization?
12. Would it be better for me to do business through the local cooperative organization?
13. Am I as familiar with the work of the county agricultural agent as I should be?
14. Should I join the local soil conservation district?

All of these questions and many others are vital to farm management. The study of farm operations, markets, and related problems is a continuous process.

BALANCE OF FARM ENTERPRISES

A farmer must choose crops and livestock to fit his farm. For example, if he has only a few acres from which to make a living he must grow crops that have a high value per acre. Bees, strawberries, poultry, and specialized truck crops can be raised on small acreages. Larger acreages are needed for dairy cattle and sheep, while beef cattle require still more land for pasture and hay crops.

If it takes 5 acres of land to support one dairy cow, and if 60 cows will yield a good family income, a family-size dairy farm must have 300 acres of pasture. Beef cattle may need two to ten times as much land per animal. In arid regions, 1,000 acres is perhaps the smallest beef-cattle farm that will support a family. (In 1971 the average size of a farm was 389 acres.)

In addition to balancing his crops and animals with sufficient land, the farmer must balance his various enter-

Is the use of the land in harmony with the ability of the land to produce? *Left:* The answer is No. *Right:* The answer is Yes. (*Courtesy* U.S.D.A.)

prises to make the fullest use of his land, labor, and capital. A wheat farmer who raises only wheat is employed about three months in the year. Many wheat farmers are going into the beef-cattle or dairy business to make more of their time count. Or a beef-cattle farmer who has only a small farm may decide to sell his beef cattle and become a dairyman so that he may use his acres more intensively. Dairymen who do not have sufficient land to support the desired number of cattle may start raising poultry to supplement their farm income.

Whatever enterprises the farmer has in mind, he must study his interests and resources carefully before starting out. Perhaps he thinks he might make money in dairying, poultry-raising, or truck-gardening—but without an interest in these types of farming, he would not be successful with them. Perhaps his land is too steep or stony to raise corn, and he thinks poultry-raising might be a better enterprise. But it requires considerable capital to build and equip a poultry house and to buy feed. If the farmer does not have the capital and does not want to go in debt to that extent, he should continue investigating to find a type of farming that is adapted both to his interests and to his financial circumstances.

In planning his farm program, a farmer will ask himself many questions. Will my land grow peanuts? potatoes? nutritious grasses and legumes? Is water available to irrigate my crops? Can I

Careful farm and home planning pays the farm family by providing a better living and giving the satisfaction of work well done. (*Courtesy* Soil Conservation Service)

3. A more productive soil.
4. Adequate pasture and feed supply.
5. Better livestock and poultry.
6. Improved crop and forest management.
7. Better marketing practices.
8. Adequate farm buildings and equipment.
9. Adequate food for the family.
10. A convenient house with adequate household equipment.
11. A well-arranged and landscaped farmstead.
12. Adequate and appropriate clothing for the family.
13. Good health for the family.
14. Educational opportunities for all the family.
15. Savings for emergency and security.
16. Recreation for the family and community.
17. A better community in which to live.

sell my farm products at a profit? These questions must be answered before a working balance of farm enterprises can be achieved.

FARM AND HOME PLANNING

Careful farm and home planning provides a better living for the farm family and gives the satisfaction of work well done. A farm and home plan should be made by the entire family, taking into account the size and condition of the land, the income necessary, the age and health of each member of the family, and each person's special interests.

The Texas Agricultural Extension Service suggests that the farm and home plan should help provide:

1. A satisfactory family income.
2. A balanced crop and livestock program.

A good farm and home plan has a list of special projects and a list of things that must be done on schedule. The things to be done during the coming year are listed on a separate page. A column for "probable cost" and a column for checking each item when completed should be provided. Sources of income should be listed and total income estimated.

In making a farm and home plan, the farm family uses information from many sources. The specific farm records of previous years are always useful in planning future operations. However, there are years in which economic conditions are so uncertain that it is difficult to estimate what the cost of labor and supplies will be and what prices can be expected for farm products. High returns for one year do not necessarily mean prosperity during the next year. The farm family must know something about the factors that influence prices and costs over a longer period of time.

The state agricultural colleges have specialists who analyze the agricultural situation and issue reliable, unbiased reports on current and future trends. Much of the research work at the state agricultural colleges is reported in free bulletins. Farmers may request that their names be placed on the mailing list to receive these materials. The State Agricultural Extension Service also distributes publications on many topics of special interest to farmers.

The Agricultural Research Administration, the Agricultural Extension Service, the Soil Conservation Service, the Forest Service, and many other branches of the United States Department of Agriculture issue technical information that is valuable in farm management. By mailing a post card to the United States Department of Agriculture and another to the State Agricultural College, the farmer can obtain lists of published aids. (See Appendix A for addresses.) He should mention the type of farming in which he is interested. After receiving the lists of publications he can order specific items.

Most of the more than 3,000 counties in the United States have county agricultural agents and county home demonstration agents. Their offices are generally located at the county seat. These local representatives of the State Agricultural Extension Service can give technical assistance in many phases of farming and homemaking.

Technicians of the Soil Conservation Service help farmers develop farm conservation plans. These technicians are assigned to Soil Conservation Service work-units to assist farmers in soil conservation districts. Some soil conservation districts own special equipment that is used in constructing terraces, farm ponds, and drainage ditches. The cooperating farmers pay the costs of operating this equipment on their land.

Farming is a business enterprise, and every farm should be operated on the basis of sound business principles. Most farmers should examine their farm business more closely and make the adjustments that are needed. Today a farm family must spend a substantial amount of money for machinery, fertilizers, feeds, seeds, buildings, fencing, home equipment, and many other items. With so much money invested, a knowledge of the basic principles of business management is just as important as a knowledge of livestock and crops.

FARM RECORDS

Every farmer should keep written records of his business transactions. The purchase price of livestock should always be recorded so that profit or loss can be determined when the animals are sold. A record of the date of purchase and the purchase price of every item of equipment is necessary to compute the amount of depreciation the farmer should claim when filing income tax returns. Permanent improvements such as buildings should be accounted for over a period of several years, as these improvements are considered capital outlay rather than operating expenses.

All farmers must file income tax reports every year, and these reports must be prepared from complete and

A farm management decision must be made on whether to invest in a precision narrow-row corn planter to operate in a system of minimum tillage. (*Courtesy* Allis-Chalmers)

accurate records. All records used in preparing an income tax report for a given year should be kept at least three years—and in a safe place where they will be readily available if needed. A farmer can get the latest available information on income tax regulations by contacting the nearest office of the Internal Revenue Service. Assistance in preparing the annual income tax report is available through this office.

Farm records can be accurate, complete, and convenient without being complex. Inexpensive printed forms and ledger books may be adapted to the individual farm. A school composition book may be used to record daily transactions which later will be summarized in permanent record books.

Cancelled checks are good records of expenditures. Receipts and sales tickets should be kept until they are added up and recorded. The sales ticket and cancelled check for an item of equipment help in maintaining an inventory of equipment and in determining how much depreciation should be deducted from income.

Besides the records of the regular farm operations, there are certain personal and documentary records that must be kept permanently. Local governments provide for recording deeds of trust, deeds of sale of property, wills, birth certificates, leases, rights of way, and easements. Such records usually are filed at county courthouses or in county office buildings where important documents are protected from fire, theft, and water-damage. Insurance policies, military service papers, and records of medical expenses should be kept in a convenient, safe place.

Three kinds of farm records that are particularly important in year-to-year accounting are: (1) a property list, or inventory, (2) records of receipts and expenses, and (3) production and incidental records. Each kind of record can be kept in a separate book and summarized for several consecutive years, or else the different kinds of records can be summarized in a single book. As a farmer acquires experience and skill in keeping records, he can devise a system to suit his needs.

Property list, or inventory. An inventory is a list of the feed, seed, and supplies on hand at a given time. This list includes the items that are produced on the farm as well as the items that are bought. Young livestock, feeders, and dairy cows which are considered short-lived livestock are usually included.

Equipment and other durable items bought for use rather than resale are considered "capital." Breeding stock usually is included in this class of property.

The annual depreciation of capital equipment is computed by dividing the original cost by the number of years that the equipment will be in usable condition. For example, if a machine cost $2,000 and is expected to last five years, the annual depreciation is $400. Equipment sometimes loses its use value and becomes obsolete before it is worn out. Obsolescence may result from the development of new equipment that makes the continued operation of older equipment unprofitable. Obsolescence and depreciation are figured in much the same way.

Permanent property records help the farmer estimate how much he will need to spend for equipment during the next year or several years. These records also help to determine which kinds of equipment can be used profitably and which should be discontinued. A labor-saving machine may not be a good investment if it does not enable the farmer to make more net profit or if he does not take advantage of it to use his time in ways that increase his income.

Records of receipts and expenses. Many farmers have found that a ledger-type book is a convenient way to summarize receipts and expenses. On large farms where the business enterprises are extensive and complex, a journal with many columns may be needed. All money received from sales of products or from services performed for pay should be listed as receipts. All money paid out for supplies and services should be listed under expenses.

Records of receipts and expenses are necessary to determine net profit. Sometimes a farmer loses money on his farm operations for a given year. Accurate records of both profit and loss are required in the preparation of income tax reports. The records of receipts and expenses also are a good guide in determining the enterprises that are best adapted to a particular farm.

Production and incidental records. Production records include the milk production of individual cows or of a herd; the number of pigs, calves, or lambs raised in a given year; the egg records of individual hens or of flocks; planting dates and harvesting dates; yields of crops; and costs of fertilizers, seed, insecticides, and similar items that are a necessary part of production. Records of feeding costs are often kept to determine the feeding methods that are best.

Incidental records may include taxes, interest, rent, and insurance.

MARKETING FARM PRODUCTS

A successful farmer must be a good salesman as well as an efficient producer. He must master the problem of getting his products to the consumer.

Marketing usually means transporting products from the farm to a buyer who delivers the products to a processor. Then the products are washed, dried, heated, cooled, or cooked to make them acceptable to the consumer. The products may be packaged, stored, transported, sold, and resold before they get to the consumer.

Marketing is an expensive business. The annual cost of marketing farm products in the United States sometimes amounts to 50 billion dollars or more. Farmers receive about one-third of the money that consumers pay for farm products; the rest goes to marketers. In recent years the farmer's share of the consumer's food dollar has been approximately 70 percent for poultry and eggs, 22 percent for cereal products, and 20 percent for processed fruits and vegetables.

It has always been true that the more processing a farmer does on his farm and the more direct selling he does, the more income he earns. For example, a farmer may greatly increase his net income by operating a roadside stand.

The stand should be located on a well-traveled road where the products can be seen at some distance. There should be plenty of parking space for customers. The stand should be neat and clean, with adequate racks and tables for displaying the products. Attractive signs telling what is available should be placed to catch the attention of motorists soon enough for a safe stop. All products displayed at the roadside stand should be fresh and of high quality.

Customers who like the service and products at a roadside stand may stop again. In this way the farmer builds up a better market for his products.

FARM PRICES

The United States Department of Agriculture offers some good information on farm market prices and why these prices vary so widely. Prices of farm products are less stable than the prices for other products and services. One reason is the greater price competition in agricultural markets. Millions of farmers compete with one another in selling their products to middlemen. The latter compete in selling farm products to consumers. In contrast, in many manufacturing and service industries competition may have little effect on changes in prices from day to day or even from year to year.

Agricultural markets are not entirely competitive. Farmers have changed the structure of markets by setting up cooperative associations. Big processors and distributors are also changing the farm market situation. More important, the government is taking an active part in agricultural markets through such activities as crop loans and marketing orders. In spite of these trends, competition is still the dominant force in agricultural markets.

Food and fiber are among the most basic needs of all people. When food is very scarce, consumers may have to pay high prices to get as much as they need. Plentiful food supplies may cause prices of farm products to drop to extremely

USUAL PLANTING AND HARVESTING PERIODS FOR SELECTED CROPS

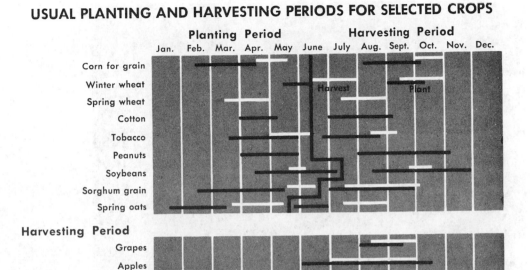

Planting dates are important because they help to determine yield. Harvesting dates are related to dates of storage or marketing. (*Courtesy* U.S.D.A.)

low levels. This is especially true in the case of foods that cannot be stored or diverted from the regular market channels. Economists say that the demand for food is "inelastic with respect to price."

Prices tend to fluctuate more at the farm level than at the retail level. One reason for this difference is the increasing amount of processing and marketing services to be included in the cost of processed foods before the foods reach the customer. The cost of such services does not fluctuate as widely as do the prices of the raw farm products going into the processed foods. Wide variations in the prices of farm products have a relatively small effect on the prices of processed retail products.

Alabama State Department of Agriculture *Circular* R-23, "Grain and Soybean Marketing in Alabama," 1970, gives indexes of seasonal variation in prices paid to farmers for selected feed grains and soybeans. The best market for oats and wheat was January through April; the best market for soybeans, February through April; grain sorghum, February through May; and corn, February through July. For all crops, statistics show that the lowest prices paid to farmers was at the time of harvest.

Marketing farm products at the right time is an essential of good farm management. However, the advantage of seasonal marketing must be weighed against the cost of handling and storage.

Dense planting, small ears, and maximum tonnage per acre when cut for silage are the result of aerial seeding of corn. (*Courtesy* The Upjohn Company)

QUESTIONS

1   Why is it necessary for a successful farmer to be a good manager?

2   Name five essentials of good farm management.

3   Give an example of the "balance of farm enterprises."

4   Why must a farmer like what he is doing if he is to be successful?

5   Name one farm enterprise that may be well adapted to steep, rocky soils, or droughty, sandy soils.

6   Name ten provisions necessary in a good farm and home plan.

7   Where can you write to get information on farm management?

8   Why is it more important than ever for farmers to keep farm records?

9   What is a farm inventory?

10   On the average, farmers receive about a third of the money paid by consumers for food. How can this fraction be increased?

ACTIVITIES

1 Select a farm enterprise in which you are interested, such as beef cattle or grain farming. List some of the requirements for success in this enterprise and some of the management problems you would encounter.

2 If you live on a farm prepare forms for keeping simple records of receipts and expenses. If you do not live on a farm prepare forms for keeping records on the kind of farm you would like to have. Use local products and prices for all sample entries.

3 Ask several farmers about the records they keep. Have a class discussion on the importance of records in farm management.

4 Make an inventory and property record for the farm on which you live, or prepare such a record for the kind of farm you would like to operate. Use local prices of feeds, seed, fertilizer, and equipment.

5 Compute the annual depreciation of: a machine purchased for $1,650 and used six years; a building constructed at a cost of $5,700 and used 15 years; a car purchased for $2,175 and used for five years.

6 Develop farm management plans for several kinds of farm enterprises such as livestock, grain, dairying, or poultry.

8 RURAL ORGANIZATIONS

Boys and girls have learned that it pays to cooperate. Whether this means joining a school club, the Boy Scouts, Girl Scouts, Future Farmers of America, Future Home Makers of America, or the 4-H Club, the object is to accomplish more through cooperation. It is more fun to work and play *with* someone.

Every organization has a purpose, sometimes written and at other times only implied. All organizations are formed for one or more of these purposes:

1. The satisfaction of doing things together.
2. A chance for companionship.
3. An opportunity to learn to share the work in completing worthwhile activities.

COMMUNITY ORGANIZATIONS

A *community* is a group of people who live close together and share mutual interests. The feeling of closeness among families in a community comes from working and visiting together and from associating with one another in schools, churches, lodges, and other organizations.

Through their local government the people of a community maintain law and order, support public schools, and provide special services such as road building and fire protection. In some communities the local government provides public utilities such as water, lights, heat, and power.

Most communities have one or more of the following organizations: Parent-Teacher Associations; Women's Clubs; service clubs such as Rotary International, Lions Club, and Kiwanis Club; youth clubs, including 4-H Clubs, Future Farmers of America, Future Home Makers of America, Boy Scouts, Girl Scouts; farm organizations; and organizations dedicated to improving the community.

Every organization needs leadership, plus a definite plan of action and active members who are willing to do their share of the work.

Many organizations fail because there are no specific plans. A good plan of action may grow out of a community problem, or it may stem from the civic pride and enthusiasm of a few individuals. Whatever the program, it should be large enough to be a challenge to the leaders and members, yet small enough to be completed successfully.

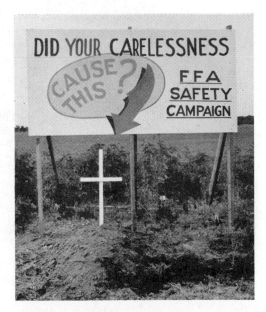

A safety program is sponsored by the Future Farmers of America. A freshly dug "grave" and a cross remind us to drive carefully. (Missouri) (*Courtesy The National Future Farmer*)

RURAL COMMUNITY IMPROVEMENT

Although every community can organize to make certain improvements, it is the rural community, varying in size from ten to 50 or more farm families, on which we now want to focus our attention. There are more pleasures in country living when there is a community organization.

Steps for improving your community. There are many ways in which your community may be improved if someone has the courage to say, "Let's do it." Here are ten suggested steps for improving your community:

1. Appeal to all families in the community through the present organizations.
2. Work out a plan for the first community-wide meeting.
3. At the meeting lead a discussion on community problems.
4. Decide as a group whether these problems can be solved faster through a community organization.
5. Form an organization if that is what the group wants to do.
6. Elect officers and permit the newly-elected president to preside.
7. As a group, study the needs of your community.
8. List these needs in the order of priority by numbering them consecutively.
9. Plan how to carry out the goals by selecting committees for each major goal.
10. Make plans for recreation as well as for work so that all work will be fun.

Left: Boys are learning to read a blueprint to follow in constructing a building. (New Mexico) (*Courtesy* F.F.A.) *Right:* A community meeting place for youth. (*Courtesy* Portland Cement Company)

These ten steps are just the beginning. The real test of community spirit comes when the people of the community tackle the problems that have been selected. Keeping alive the early enthusiasm, recruiting new leaders, and securing community-wide participation are continuous processes; but the results that can be achieved are worth the effort.

Goals for community improvement. Most communities need some help in establishing their goals. Here is a suggested list of objectives which many rural communities have used as a guide:

1. Increasing and managing family income.
   (a) Soil improvement practices.
   (b) Better crop production and management.
   (c) Better pasture and range management.
   (d) More efficient livestock management.
   (e) Improved marketing practices.
   (f) Improved forest management.
   (g) Improved family money management practices.

2. Improving health conditions and services.
   (a) Production, conservation, and use of food.
   (b) Proper selection and care of clothing.
   (c) Prevention of diseases.
   (d) Medical facilities.
   (e) Safety.

3. Improving the farm and home.
   (a) Farm buildings and equipment.
   (b) Landscaping the home grounds.
   (c) Housing and household equipment.

4. Encouraging social participation.
   (a) Recreation for the family and community.
   (b) Religious life for the family and community.
   (c) Education for all.

(d) Public facilities for community activities.

Plans for achieving some of these goals could be put in effect in a relatively short time. For example, the care and selection of clothing could be undertaken almost immediately; a room could be redecorated; a start could be made in landscaping the home grounds; or entertainment could be arranged for the next community meeting.

Most of the goals require long-range planning and years of patient work. Examples of long-range goals are soil improvement practices, better pasture and range management, improved forest management, providing adequate medical facilities for the community, improving farm buildings and equipment, and providing educational opportunities for everyone in the community.

RURAL COMMUNITIES IN ACTION

Every year thousands of rural communities compete in community improvement contests. Through such contests, a few communities receive state and national recognition. But more important is the total accomplishment of all the communities, including those that never make headlines in newspapers and magazines.

A few typical projects will show the pattern of rural communities in action. Perhaps your community has a different type of community-improvement program that is just as good.

A community center. Every community needs a place for public meetings, plays, concerts, lectures, study groups, demonstrations, exhibits, and

On school property, this chapter of the Future Farmers of America built a meeting place as one of the community goals. (Iowa) (*Courtesy The National Future Farmer*)

other activities. A school or church may be used for some of these purposes, but such use often involves difficult problems. Rural communities across the nation are establishing their own community centers. By the time this feat is accomplished the people have learned to work together and have enough confidence to undertake new projects.

Raising money for a community center usually is one of the first and most difficult problems. The cost of the desired building or buildings should be carefully estimated. Then this cost should be studied in relation to the financial resources of the community.

Since labor is such a large part of the cost of a building, the cash outlay for a community center can be greatly reduced by volunteer workers. It is not unusual for a community center to be almost completely built in this way. Individuals, organizations, and companies may contribute materials and services. Community money-raising activities include entertainments for which there is an admission charge, auctions of products and articles donated for sale, carnivals, pie and cake sales, and public meetings to solicit cash contributions or pledges of cash.

Once established, the community center becomes the hub of the community. In addition to being a source of recreation and information, it is a constant reminder that organized effort gets results.

A farmers' market. Better markets are a worthy goal for any community. A farmers' market serves the dual purpose of providing a profitable outlet for farm products and a dependable source of nutritious low-cost foods for the people in a community. The market may have an additional function in the co-

operative purchase of feeds, fertilizers, seeds, and other materials that are used on a farm.

A farmers' market should be operated on a year-round basis if the farming enterprises of the community are sufficiently diversified to make this plan successful. If highly specialized fruit or vegetable crops dominate the farming pattern, the market may be seasonal. In the mild climate of the South, inexpensive open sheds are satisfactory for the display of farm products.

Business and service organizations in cities sometimes secure the cooperation of city and county governments in establishing farmers' markets. These groups know that an increase in farm income means greater prosperity for the urban community. The availability of high-quality fresh foods at moderate prices not only contributes to the health of the community but gives the consumers more purchasing power.

Better schools through community action. The school is the center of activity in many rural communities. Although the school is financed mainly by taxation, there are many things that the people of the community can do to extend its services. Keeping the community informed about the needs of the school is a continuous process. And the planning of new school buildings should be a community-wide project. This is particularly important when bonds must be approved to finance new construction.

Most schools need certain items of equipment that have not been provided through the regular budget. For example, a projector for films or slides is a fine teaching aid which can be used not only in the instructional program of the

This rural community is definitely in action. Would you like to live here? (Texas) (*Courtesy* Ed Bryson, *The Paris News*)

school but also in community meetings. Perhaps the school has a projector but needs dark window shades so that films or slides can be used during daylight hours.

Now that electricity is available in most rural communities, more and more schools are using projectors, fans, and air conditioners. Stage curtains, folding chairs, more books for the library, and equipment for serving refreshments at community meetings are other examples of special equipment that may be provided through community effort.

Study groups. Any rural community may support a class or several classes for the study of topics of general interest. Home nursing, first aid, soil conservation, and farm management are typical courses for study groups. Veterans' classes in agriculture have been organized in communities throughout the United States since World War II. Specialists from the Extension Service, Soil Conservation Service, state agricultural colleges, and state conservation departments often serve as instructors and consultants for study groups. Teachers of vocational agriculture are especially qualified to assist.

Community recreation. The need for companionship is universal, and nowhere is this need greater than in rural areas. Typical examples of community recreation are annual festivals, Halloween and Christmas parties, square dances, plays, and activities planned especially for children and youth groups. Recreation for young people is particularly important during the summer months when schools are not in session. A community swimming pool or lake provides good recreation for all age groups.

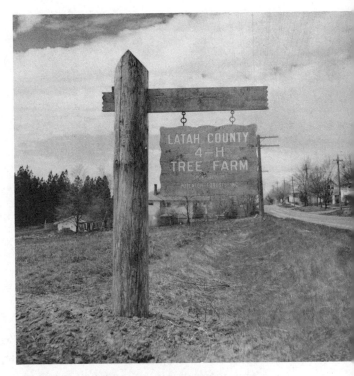

Above: The 4-H Clubs, Future Farmers of America, or other rural organizations often sponsor community-wide campaigns for the planting of trees and establishment of tree farms. (*Courtesy* National Committee on Boys' and Girls' Club Work, Inc.) *Below:* Rural Organizations may sponsor calf-roping and calf-tying contests. (*Courtesy* Benton Watson, American Quarter Horse Association)

Executive sessions of farmers help to improve the level of living for all. (Illinois) (*Courtesy* Illinois Extension Service)

NATIONAL FARM ORGANIZATIONS

Farm organizations are designed to help country people increase farm income and the satisfactions of rural living. The membership of national farm organizations is increasing as rapidly as rural people understand that united action means greater strength. Equally important is the realization that anything affecting agriculture in any part of the nation also influences farm operations in the local community.

Three major national farm organizations are the Farm Bureau Federation, the Grange, and the Farmers' Union.

The Grange. When the Grange was organized near Washington, D. C., on December 4, 1867, it was a secret society of farmers. Many of the secret phases of the organization have been maintained through the years.

The Grange Creed indicates the purpose of the organization:

We desire a proper equality, equity, and fairness; protection for the weak, re-straint upon the strong; in short, justly distributed burdens, and justly distributed power. These are American ideas, the very essence of American Independence, and to advocate the contrary is unworthy of the sons and daughters of the American republic.

The Grange has developed into a nation-wide family type of organization. Women and boys and girls over 14 years of age make up approximately 40 percent of the membership. New England has the greatest concentration of Grange members.

The Farmers' Union. This organization was started in Texas in 1902 among low-income farmers. It is the policy of the Farmers' Union to help more farmers buy the land they operate and to encourage farmers to buy and sell cooperatively.

The aim of the Union is to:

. . . attain equity and justice through maintaining a democratic . . . system and building a cooperative income system, as the practical expression of the Christian ideal of brotherhood which alone can bring lasting peace and security; to cooperate with organized groups who genuinely seek to provide economic security, preserve democratic processes, provide distribution of abundance for all the people, and maintain civil liberties.

The American Farm Bureau. Many state farm bureaus were organized soon after World War I to help farmers make post-war adjustments. In 1920 the American Farm Bureau was formally organized in Chicago. The policies of the Farm Bureau are made by farmers at their annual meetings.

The Farm Bureau works with other national leaders to help assure fair prices for farm products. This organization has always played a leading role in making the farmers of the United States conscious of the world-wide influences at work in agriculture. Farmers now realize that the price of cotton in Mississippi may vary because of prospects for a bumper crop of cotton in Brazil.

COOPERATIVES

Farmers' ownership in farmer cooperatives was 2 billion dollars in 1950 and more than 5.1 billion dollars in 1971. Farmer cooperatives are now so large that in a recent year three of them were included among the 500 largest manufacturers in the United States.

Cooperatives help farmers earn more money by marketing products efficiently, providing quality products, obtaining quality supplies, keeping down costs, and helping the local community by providing jobs and doing business in the area. There are approximately 22 million members in all types of rural cooperatives. Many rural people belong to several cooperatives. The 9,000 marketing, farm supply, and related service cooperatives have more than 7 million members, including duplication of memberships.

Other rural cooperatives include the credit cooperatives in the Farm Credit System, rural electric and telephone cooperatives, credit unions, insurance, irrigation, and similar types of service organizations. Recent trends in cooperatives show them diversifying their operations so that one association will sell, buy, and provide other services. Some cooperatives have begun to process farm products and manufacture supplies.

In a recent year, farmers marketed more than 10 billion dollars worth of farm products through cooperatives. This was about one-fourth of the total farm marketings. The farmer-members obtained about one-fifth of their farm supplies through cooperatives. These associations also manufactured most of the feed, fertilizer, and petroleum products supplied to members.

Members of marketing and supply cooperatives earn an estimated quarter of a billion dollars of extra income in refunds made by their associations. These refunds are possible because cooperatives give back to members the money left over after the costs of operation are paid and sufficient reserves are set aside.

These young men are learning to describe soils in the field. The study of agriculture in a university is often the result of earlier participation in 4-H Clubs or FFA Chapters. (*Courtesy* Michigan State University)

During the past few years there has been a decline in the number of farmer cooperatives and a slight decrease in the number of members. At the same time the dollar volume of business operations of these associations has grown. Much of the decrease in the number of co-operatives has been the result of mergers, consolidations, and acquisitions among the organizations. As long as the larger business operations prove to be more efficient, the downward trend in the number of farmer cooperatives is beneficial to farmer-members.

SOIL CONSERVATION DISTRICTS

In 1937 the first soil conservation district was organized in Arkansas. Now more than 3,000 soil conservation districts contain 99 percent of the land area of the United States. These organizations were established by farmers to improve soils and the environment. Each district is administered by a board of supervisors elected by the farmer members. More information on soil conservation districts is given in Chapter 9.

Future Farmers of America hold their annual convention to honor those who have made the best records. (*Courtesy The American Farm Youth Magazine*)

QUESTIONS

1 What are the three objectives of most organizations?

2 Name five organizations in your community.

3 Name the organizations to which you belong.

4 Give at least five suggested goals for improving your community.

5 Which goal do you think your community should work toward first?

6 What advantages are there in having a community center?

7 Is there a farmers' market in your community?

8 Does anyone lose when there is a well-operated farmers' market?

9 Does your community need a new school?

10 Name the principal farm organizations in our nation.

ACTIVITIES

1 Your class probably includes members of several organizations such as 4-H Clubs, Future Farmers of America, or Boy Scouts. Ask representatives of several organizations to describe the purposes and activities of their groups.

2 If your school has a Parent-Teacher Association, find out what it does for the school and community.

3 Interview several people in your community to obtain information about local rural organizations.

4 Conduct a class discussion on problems of the local rural community and how these problems could be solved through community organizations.

5 As a class project, plan several things that could be done to improve your community. List specific procedures to be followed, funds that would be needed and how the money might be obtained, and the approximate time that would be required to complete each activity.

9 AGRICULTURAL SERVICES

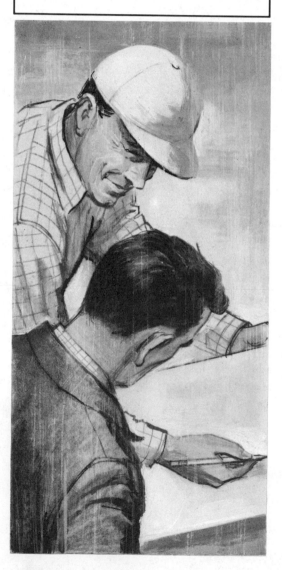

Early American farmers met and overcame great obstacles. These pioneers in agriculture were good woodsmen and expert hunters. They were resourceful in solving everyday problems that now seem almost inconceivable—hostile Indians, wild animals that killed livestock, crude farm implements, and even starvation.

During the Colonial period American farmers had no place to turn for information about better methods of farming. There was no one to tell them where good seed corn was available or how to produce enough hay to winter the livestock. There was always the question of whether there would be enough wild game for meat.

The first society for the promotion of agriculture in the United States was organized in Philadelphia on March 1, 1785. On July 4 of that year, George Washington and Benjamin Franklin were elected to membership. The same year a similar organization was incorporated in South Carolina. This society proposed to establish an experimental farm, the first in America. Other organizations with similar purposes were established in New York, Massachusetts, and Connecticut during the years 1791 and 1794. These were the humble beginnings of the first agricultural services in the United States.

Today the story of agricultural services could very well fill several large books. This chapter will introduce a few of the services that are available to farmers in your community. As you continue your study of agriculture you will find that there are many other sources of information and technical assistance. It is good business to find out where you can get help when you need it.

THE UNITED STATES
DEPARTMENT OF AGRICULTURE

The history of the United States Department of Agriculture stretches back over more than a century.

In 1836 the Commissioner of Patents began to distribute seeds and plants to farmers. Although this service was permitted, it was not officially authorized by the United States Patent Office. When the Department of the Interior was created in 1849, the Patent Office became a part of the new Department. The next step was the establishment of the Agricultural Division in the Patent Office.

The United States Department of Agriculture was created by a Congressional bill which President Lincoln signed May 15, 1862. The first Commissioner of Agriculture, Isaac Newton, did not have a seat in the President's Cabinet. In 1889 the head of the Department was given the title of Secretary of Agriculture and raised to Cabinet rank.

Today the United States Department of Agriculture is a complex organization with many diversified activities. The agencies and bureaus of the Department have offices in Washington, D. C., Beltsville, Maryland, and at approximately 4,000 locations throughout the United States.

There are several administrative offices within the Department. The Office of Information is of particular interest to farmers because it is a major source of information on many phases of agriculture. Typical activities in this office include the preparation and distribution of publications and news releases, photographic and exhibit service, production of radio and television programs, and assistance to writers and editors who request information about the work of the agencies and bureaus within the Department.

The broad objectives of the Department of Agriculture are directed toward the following problems:

1. Investigation by science of newer and better methods.
2. Economic production and distribution of essential food and fiber.
3. Wise conservation of natural resources.

Scientists in the U.S.D.A. perform hundreds of experiments to help farmers and others find answers to perplexing problems. Here some of the new herbicides (weed killers) are being tested. (*Courtesy* U.S.D.A.)

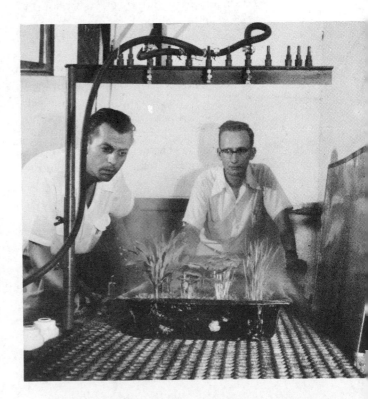

4. Sound stabilization of farm prices.
5. Regulation of markets and trade in farm products and facilities.
6. Information to farmers and the public on achievements and progress in agriculture.

The influence of agricultural research extends around the world. For example, scientists in the Department of Agriculture laboratories helped to develop methods for the mass production of penicillin, a drug which has saved many human lives. Agricultural research has been responsible for so many other scientific discoveries that it would be impossible to estimate their value to mankind. Among these are methods for the control of livestock diseases and parasites that can be transmitted to man, control of plant diseases, eradication of weeds, better methods of food processing, and new principles of human and livestock nutrition.

A brief review of a few of the agencies that serve your community will give you a better understanding of the work of the Department of Agriculture. Remember that these agencies and all others in the Department make up a highly organized team. Their organization and functions are described in the *Employees' Handbook* that is published by the Department.

Agricultural Stabilization and Conservation Service. The Agricultural Stabilization and Conservation Service operates through more than 3,000 county committees. There are 30,000 community committees, each with three members and two alternates. The County Extension Agent is an ex-officio member of the county committee.

The Agricultural Stabilization and Conservation Service administers the Rural Environmental Assistance Program (REAP), organized in 1971. Some objectives of REAP are to abate or prevent pollution of water, land, and air; reduce the loss of agricultural soil, water, woodland, and wildlife resources; encourage enduring conservation practices; and limit and direct federal cost-sharing assistance which is mutually beneficial to farmers and the public.

Agricultural Research Service. This office directs research relating to the production and use of agricultural commodities. Much of the research is in cooperation with state agricultural experiment stations which are a part of the land-grant universities. Enforcement of plant and animal quarantines, control of diseases and insect pests of plants and animals, and meat inspection are other major activities of the Agricultural Research Service.

The areas of research with which the Agricultural Research Service is concerned are: crops; farm and land management; livestock; quality of the environment; human nutrition and home economics; utilization of agricultural products; crop regulatory activities; livestock regulatory activities; administration of grants-in-aid to states; and territorial research.

Extension Service. This cooperative agency is designed to carry educational programs in agriculture and homemaking to rural and urban communities. The Extension Service coordinates all educational work under the Department of Agriculture in cooperation with the land-grant university. The Extension Service cooperates in emergency programs in connection with defense mobilization and also works with committees of business people and farmers to establish better

Left: The Extension Service furnishes specialists who work through the county agricultural agents to help farmers establish demonstrations from which other farmers may learn. (*Courtesy* New Hampshire Extension Service) *Right:* Officials of the U.S. Forest Service demonstrate to a class the proper way to plow fire lines to slow the speed of a forest fire. (Mississippi) (*Courtesy* U.S. Forest Service)

markets and improve the quality of the environment.

The staff of the state Extension Service includes specialists in agronomy, soils, livestock, landscaping, youth organizations, community improvement, wildlife conservation, home economics, and other areas. The county agricultural agent and the home demonstration agent represent the Extension Service in the counties.

Forest Service. The conservation of forest resources is the main objective of the Forest Service. Forest management, prevention and control of forest fires, forest products research, control of forest insects and tree diseases, production of trees for reforestation, soil and water conservation, and improving the quality of the environment are some of the federal

and state cooperative forestry services.

The Forest Service is the world's largest manager of rangelands. The work of the Forest Service is carried on through regional offices (see addresses in Appendix A), regional forest and range experiment stations, and the Forest Products Laboratory at Madison, Wisconsin. The Forest Service administers all national forests.

Soil Conservation Service. This agency is responsible for developing a national program of soil and water conservation—one of the first steps toward a permanent agriculture. The Soil Conservation Service administers the National Cooperative Soil Survey in cooperation with the land-grant university in each state.

Technicians of the Soil Conservation

Service work through the local soil conservation districts which are organized and controlled by farmers and ranchers. A conservation plan is developed for each farm and ranch as the operators request this service. In carrying out the prescribed conservation practices, the cooperating landowners may get additional help from the Agricultural Stabilization and Conservation Service, Extension Service, Forest Service, Farmers Home Administration, and other agencies of the United States Department of Agriculture.

Rural Electrification Administration. This lending agency was established in 1935. Loans have been made to 1,800 cooperatives, public bodies, and other qualified borrowers to help finance the construction of electric and telephone facilities in rural areas. Further information about the Rural Electrification Administration is given in Appendix A.

Farmers Home Administration. This agency in the United States Department of Agriculture makes loans to farmers and communities. Loans are made through local county offices to applicants who must show that they are unable to obtain credit elsewhere and must agree to use commercial credit resources when they are able. A borrower must operate a family-size farm and must be living on the farm. The low-to-moderate income families must live in communities under 10,000 population. Communities up to 5,000 pop-

The National Rural Electric Cooperative Association is serving this Wyoming village by providing low-cost dependable electricity. (*Courtesy* National Rural Electric Cooperative Association)

ulation may obtain loans and grants for development and improvement of water and sewer systems. Further information may be obtained by writing to the Farmers Home Administration, United States Department of Agriculture, Washington, D.C. 20250.

Other agencies. The United States Department of Agriculture also includes the Statistical Reporting Service, Office of Inspector General, National Agricultural Library, Foreign Agricultural Service, Food and Nutrition Service, Farmer Cooperative Service, Economic Research Service, Cooperative State Research Service, and Consumer and Marketing Service.

STATE DEPARTMENTS OF AGRICULTURE

Each of the 50 states maintains a department of agriculture. This division

of the state government enforces the laws relating to seed, feed, fertilizers, insecticides, fungicides, and herbicides. Usually there is a market service that checks the packaging, accuracy of weighing, and advertising claims of those who sell farm products. Some departments have a market news service so that farmers may know when to sell certain products to obtain the highest prices. A market news service also advises the public when to buy at favorable prices.

A representative of the state department of agriculture may walk into a store that sells fertilizers, show the manager his card, and take samples of fertilizers that are being offered for sale. The samples, usually about a pint each, are sent to a laboratory and analyzed to determine whether the fertilizer actually contains the plant foods that are listed on the bag. This protects the public against misrepresentation. The same in-

One function of the Soil Conservation Service is to help farmers to make farm plans. (Ohio) (*Courtesy* Soil Conservation Service)

spection service is maintained for seed, feed, insecticides and other products offered for sale to farmers and the general public. This is one of the valuable functions of a state department of agriculture.

State departments of agriculture also cooperate with Federal agencies in the administration of drought and other emergency programs.

STATE CONSERVATION SERVICES

A state conservation program usually includes forests, parks, and wildlife. All of these resources may be under a single department or commission, or there may be several separate agencies for the management of resources. For example, the Missouri Conservation Commission administers wildlife conservation, forestry, and state parks. South Dakota has a Department of Game, Fish, and Parks. In Texas, the Game and Fish Commission is responsible for wildlife conservation, and the state forestry program is operated by the Texas Forest Service.

In addition to the state conservation agency or agencies, a state may have a resources development commission that publishes reports on the resources of the state, encourages new industries, and publicizes tourist attractions. A resources development commission is concerned with promoting the economic development of the state, rather than with conservation practices.

A state forestry service administers the state laws pertaining to forest fires and to the acquisition and management of state forests. A state forestry program usually includes the prevention and control of forest fires, management assistance

to woodland owners, control of forest insects and diseases, and research. State forests are used for research and demonstration and sometimes for recreation.

Most state forestry services also maintain nurseries that produce forest tree seedlings. The seedlings are sold at cost to farmers and forest industries for planting on lands where artificial reforestation is needed.

A state wildlife conservation agency enforces hunting and fishing regulations, assists landowners in improving the wildlife habitat, and conducts research in wildlife management. Some states have fish hatcheries where fish are raised for restocking public and private waters. The state wildlife conservation agency may also manage wildlife refuges.

In some states the wildlife conservation agency determines the hunting and fishing seasons and sets limits on the amount of game that each person may take. In other states the wildlife regulations are established by the legislature.

State conservation agencies cooperate with newspapers, magazines, and radio and television stations to keep the public informed about state conservation programs. Many conservation agencies have educational specialists who assist schools, colleges, and state departments of education in the development of materials and methods for teaching conservation.

STATE SOIL CONSERVATION PROGRAMS

In 1937 Congress authorized states to provide for the organization of soil conservation districts. The purpose of this legislation was to give local communities a large share of responsibility for conserving their own soil and water.

The state soil conservation district program is directed by a committee or commission composed of farmers. The only function of this body is to help farmers and ranchers organize local soil conservation districts according to the state laws. No community is required to establish a soil conservation district.

When the farmers and ranchers in a community want to organize a district they petition the state soil conservation committee to hold public hearings. If these hearings indicate that a soil conservation district is justified, an election is held so that the farmers and ranchers can vote on the issue. When the required percentage of qualified voters approve a soil conservation district, the state committee helps to complete the organization. The supervisors of a soil conservation district are elected by the cooperating farmers and ranchers.

The Soil Conservation Service, the U. S. Forest Service, and other federal and state agencies, including the land-grant universities, give technical assistance to soil conservation districts. All district policies are established by the board of supervisors. Some districts purchase and operate terracing machines and other equipment on a cooperative basis.

VOCATIONAL AGRICULTURE

The Smith-Hughes Act of 1917, signed by President Woodrow Wilson, provided for the use of Federal funds to promote the teaching of vocational agriculture in the various states. The Federal funds, which must be matched by state and local funds, may be used

A vocational agriculture teacher shows his class how to identify cotton boll weevils and how to determine when to poison them. (Alabama) (*Courtesy Better Farming Methods*)

for salaries of supervisors, directors, and teachers of vocational agriculture in public secondary schools. These funds also are used to educate vocational agriculture teachers in colleges and universities.

During the first year after the Smith-Hughes Act was passed, more than 600 schools were given financial assistance in teaching vocational agriculture. Since that time there has been a steady increase in the number of schools offering that subject.

The program began with practical agricultural instruction in public high schools. Gradually it has been expanded to include out-of-school farm youth over 14 years of age and adult farmers in evening classes.

Each vocational agriculture student chooses and completes a farm project or program under the supervision of the teacher. A calf, a garden, poultry, an acre of corn, a forest plantation, a litter of pigs, or a similar project may be selected. Vocational agriculture students may follow a program of activities for each year of study.

The Vocational Education Act of 1963 continues and amends the Smith-Hughes and George-Barden Acts and grants funds in addition to the funds provided by the earlier Acts.

The broader new program allocates 90 percent of the basic appropriation to the states for:

(1) vocational education for persons of all ages in any nonprofessional occupation

(2) services to assure high quality vocational programs, including teacher training and supervision, program evaluation, demonstration and experimental programs, and the development of instructional materials and equipment

(3) vocational counseling and guidance services

(4) construction of area vocational school facilities.

Ten percent of the basic appropriation is for grants to colleges and universities and other public or nonprofit private agencies and institutions, State Boards for Vocational Education, and local education agencies.

Future Farmers of America. In November 1928, the Future Farmers of America was officially founded at Kansas

The sharing of the cost of constructing a farm or ranch pond is one of the optional practices permitted under the Rural Environmental Assistance Program. (*Courtesy* U. S. Department of Agriculture)

City, Missouri. It is a volunteer organization of boys who are enrolled in vocational agriculture. The main purpose of the Future Farmers of America is to provide social and educational experiences and opportunities for leadership.

Technical services and financial assistance are being supplied to farmers throughout the world by the International Bank for Reconstruction and Development (World Bank) and its affiliate, the International Development Association. *Above:* Assistance to smallholder tea growers in Uganda (western Africa). *Below:* Assistance to farmers shown here erecting trellises for growing grapes in Afghanistan (eastern Asia). (*Courtesy* United Nations)

QUESTIONS

1 On what date were George Washington and Benjamin Franklin elected to membership in the first agricultural society?

2 Under what president was the Department of Agriculture established? In what year?

3 Name three objectives of the United States Department of Agriculture.

4 What service is available from the Farmers Home Administration?

5 What does the Agricultural Research Service do?

6 Name at least one function of the Extension Service.

7 What is the main function of the United States Forest Service?

8 What is the responsibility of the Soil Conservation Service?

9 Find out the address of your own state department of agriculture. Who is head of it?

10 Approximately how many departments of vocational agriculture are there in our nation?

ACTIVITIES

1 Ask your teacher or school librarian to obtain bulletins describing the agricultural services in your state and community. Use these bulletins for additional reading for class reports or discussions.

2 Arrange a class panel discussion in which each member of the panel describes the organization and work of an agricultural service such as the Extension Service, Soil Conservation Service, or State Conservation Department. The panel should include three or four members, and reports should be five minutes or less to allow time for class discussion.

3 Ask a representative of an agricultural service to talk to your class about ways in which the different services work individually and together in the community.

4 As a class activity, plan ways to use the agricultural services that are available in the community.

5 The addresses of the 50 state Extension Service Directors are given in Appendix A. As a class project, write to the one in your state and ask for a list of available publications in agriculture.

10 CLIMATE AND WEATHER

Weather is a universal topic of conversation because we can see and feel the changes in atmospheric conditions from day to day or even from hour to hour. Rain, snow, wind, clouds, frost, dew, humidity, and temperature are the conditions that make up the weather. The average of these conditions at a given place over a period of years is known as *climate.*

The climate of a state or region can be accurately described on the basis of weather records covering a period of 40 to 50 years. Much of our knowledge of the climate of the United States is based upon records that the United States Weather Bureau has compiled since 1890. In that year an act of Congress established the Weather Bureau in the Department of Agriculture. In 1940 the Bureau was transferred to the Department of Commerce.

In 1965 the United States Weather Bureau was placed under the Environmental Science Services Administration. Another reorganization in 1970 abolished the term "Weather Bureau" and replaced it with the name "National Weather Service." Under this name it was placed under the newly organized "National Oceanic and Atmospheric Administration" in the United States Department of Commerce, Washington, D. C. 20230.

CHANGES IN CLIMATE

The earth has had several changes in climate, but these changes have occurred through millions of years. Within the short span of a human lifetime there is almost no change in the world's climate. Certain variations in climatic conditions occur, although we do not know

The Distant Early Warning System is known as the DEW line. The approach to this main station is flanked by flags of the United States and Canada. A complete characterization of the weather is one of the functions of the series of stations across northern Canada and Alaska. The ice age still exists here. (*Courtesy* International Telephone and Telegraph Corporation)

whether these changes will continue long enough to affect the long-time pattern of climate. For example, in most parts of the world the temperature has increased about 2 degrees during this century. Most of the variation is the result of slightly higher winter temperatures. Only time will tell whether this is a short-term trend or a long-term change.

Our climate today reflects the influence of the ice ages. Apparently we are approaching the end of an ice age. Perhaps that is why there have been more frequent storms, greater extremes in temperatures, and wider variations in precipitation during recent years.

The earth has had at least four major glacial periods. These glaciations have occurred at intervals of about 250 million years. In addition there have been several comparatively short glacial periods in which glaciers occurred over a small part of the earth's surface. During a glacial period the seas in polar regions remain frozen through the summer months. Today this condition exists in the arctic and antarctic regions as seen in the picture of the DEW line.

The first great glaciers may have occurred millions of years before there was any life on the earth. The glacial soils and glacial lakes in the Lake States and along the eastern edge of the Northern Great Plains are examples of glacial influence.

About one-fifth of the land area of the United States was at one time under glaciers. The ice fields extended over practically all of North America north of a line from the present site of New York City westward along the Ohio and Missouri rivers and northward almost to the border between Canada and northern Montana and Idaho.

Other glaciers occurred in Australia, India, China, Africa, Europe, and South America. The present tropical climates of parts of India, Africa, and South

America are almost unbelievable contrasts to the ice ages.

Man's knowledge of climatic changes through the ages has been obtained by studying the structure and arrangement of rocks in the layers of the earth's crust, and by reconstructing the probable life histories of fossil plants and animals. The existence of certain kinds of plants and animals during the formation of a particular layer of rocks is a clue to the climate of the region during that period in geologic history.

Scientists know that layers of limestone, sandstone, shale, and slate were formed from materials that settled to the bottoms of ancient oceans. The occurrence of oceans and glaciers on the same part of a continent is evidence that climate does change. However, these changes have occurred over periods of time that are beyond human imagination, measured in millions of years.

CLIMATE AND AGRICULTURE

Nowhere is the influence of weather and climate felt more than in the rural community. Children who ride school buses must often wait outdoors until a bus arrives. Farmers must be outside some of the time when caring for livestock. Shopping trips often require farm families to go considerable distances over poor roads.

Rural families are very weather conscious about such factors as the "chill index." A chill index chart is constructed from a combination of temperature and wind velocity in miles per hour. For example, an actual temperature of 35° F. will feel like a temperature of 21° F. when the wind velocity is 10 miles per hour, and 0° F. when the wind velocity is 50 miles per hour; an actual temperature of 10° F. will feel like a temperature of −9° F. when the wind velocity is 10 miles per hour, and −24° F. when the wind velocity is 20 miles per hour, and −38° F. when the wind velocity is 50 miles per hour. Any temperature of 25° F. or less may seem like a sub-zero temperature if the wind velocity is 20 miles per hour or more.

Climate is a major factor in determining the crops that can be grown successfully. Soil and the processes of soil formation reflect the force of climate. Both climate and weather must be considered in selecting a type of farming, planning farm buildings, developing a farm conservation plan, and managing livestock.

The daily weather forecasts indicate what the weather probably will be during the next 12 to 24 hours. Some forecasts cover 48 hours or more. The scientists who interpret the weather data for us are *meteorologists*. Sometimes the weather does not behave as predicted, but for the most part the forecasts are remarkably accurate.

There are approximately 300 primary weather stations under the supervision of the National Weather Service. These stations are located at various points throughout the nation. Observers at each station send their daily reports to a district center where forecasts are made. Teletype machines automatically record the coded weather information as it is received at the district center. Additional data is supplied by approximately 5,000 voluntary observers who cooperate without pay.

Assisting in weather forecasting is the Nimbus Weather Satellite System, which moves around the earth in a continuous orbit. It is capable of photographing cloud cover and the earth at midnight from a height of 275 miles. Storm warnings such as cyclones, the presence and movement of weather "fronts," and temperature and air pressure can be measured instantly by the Nimbus satellite. (Note the drawing of Nimbus I and the photograph which the satellite took of Italy.) The International Telephone and Telegraph Corporation Laboratories' High Resolution Infrared Radio-meter is attached to the bottom of the spacecraft. It scans at right angles to satellite travel to provide complete nighttime coverage of the earth. The radio-meter changes infrared radiation to electrical signals which are stored until a ground command causes transmission to a receiving station. There, the signals are converted to photographs. Nimbus I was launched August 28, 1964.

In the preparation of a general weather map of the United States, data from 300 to 400 observation points must be used. By using symbols to represent barometric pressure, temperature, relative humidity, precipitation, wind direction and velocity, and related information, the meteorologist can plot a great deal of information on a very small space.

The interpretation of surface atmospheric conditions at the observation points is only part of the story of a daily weather forecast. A map showing the movements of air at several elevations above the earth must be prepared. Special charts are made to show other technical information. Atmospheric conditions over the oceans also must be considered.

The daily weather forecasts serve many purposes. Farmers use weather information during planting and harvesting seasons, and in order to know when to provide protection for livestock when storms are forecast. Power companies estimate the peak load for light and heat

Left: This is an artist's rendering of Nimbus I, an earth-oriented satellite in motion. Its operating part always points toward the earth. *Right:* This infrared midnight photograph of Italy at 1. (shaped like a woman's boot) shows intensive white clouds over the Alps and Central Europe (top of photo). Scattered white clouds over the Mediterranean and North Africa are shown at 2. Yugoslavia is at 3. This infrared picture was taken by the Nimbus weather satellite at an altitude of 275 miles. (The arrow points north.) (*Courtesy* International Telephone and Telegraph Corporation)

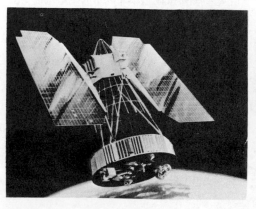
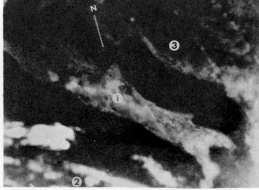

on the basis of forecasts of cloudiness and temperature. Warnings of freezing temperatures send thousands of car owners to buy antifreeze solutions for radiators. The value of increased production and the annual saving through the prevention of property damage may amount to several billion dollars. Much of this saving is in agriculture.

The National Weather Service provides special agricultural weather service for the important crop regions, including corn, wheat, cotton, and various citrus fruits.

Climate affects agriculture whenever it affects growing plants, grazing, insects, or livestock. The relation of climate to soil and to some of the major crops is discussed in other chapters.

Factors affecting growing plants. The three main climatic influences on growing plants are temperature, moisture, and light.

Different kinds of plants are adapted to extremely different temperatures. For example, certain algae plants grow in hot springs at a temperature of 200° F.; and arctic plants may survive at a temperature of −90° F. However, most plants have a much narrower range of climatic adaptation. Temperatures between 78° and 88° F. are ideal for cool-season cereal crops. A temperature range of 88° to 98° F. is best for sorghums, melons, sugar cane, and other warm-season crops.

Plants have several adaptations that reduce the chance of injury during cold weather. They may complete their life cycle before winter begins, or die down to the roots, or become dormant in cold weather. In spite of these adaptations, damage from cold is a threat almost everywhere in the United States. Moder-

ately cold winters are beneficial to deciduous fruit trees, winter wheat, rhubarb, and some other plants. Certain kinds of bulbs should be kept in cold storage before planting.

On the basis of moisture requirements, plants may be divided into three groups. Plants with great adaptation to drought are called *xerophytes*. Desert plants that do not require much moisture belong in this group. The *mesophytes* are intermediate in their moisture requirements. Most of the agricultural crop plants are mesophytes. *Hydrophytes* require a great deal of water. Some of them grow best when their roots are under water most of the time. Water lilies and cattail plants are hydrophytes.

Light is the third climatic factor that affects plant growth. The light requirements vary among the different species of plants and even among different varieties of the same plant. Some plants are affected more by the length of the growing day than by temperature. In general, plants grow in length more rapidly at night when the light intensity is very low. Daylight favors the manufacture of food in the leaves because the process of photosynthesis depends upon the energy from sunlight.

Climate and grazing. The United States has approximately a billion acres of grasslands. Most of the western rangelands receive less than 20 inches of precipitation annually. Droughts occur frequently. The high rate of evaporation and the uneven distribution of rainfall make moisture conditions even more critical. During the period of 1916–1940 the West had a lower average annual precipitation than during the period of 1886–1915.

Sugarcane is very sensitive to air temperature and grows best between 88° F and 98° F. Hawaii and Puerto Rico produce large acreages of sugarcane because of their tropical climates. (*Courtesy* Puerto Rico Economic Development Corporation)

The grazing lands in what is roughly the eastern half of the United States are generally called *pastures*. The average annual precipitation over most of this part of the United States is 30 to 50 inches, although some places in the Southeast get 60 inches of precipitation. Droughts occur less frequently here than in the western grazing regions.

Pasture and range management can be adapted to climate. By proper stocking of grazing lands it is possible to maintain a favorable balance between livestock and available high-quality forage. In the Southeast the proper rate of stocking may be one cow per acre on improved pastures, while 40 to 60 acres of arid western rangeland may be required to support one cow.

Climate and insects. Climate affects insects indirectly by influencing the kinds of plants that grow in a given locality. Weather affects insects more directly by helping to control their rate of reproduc-tion. Each kind of insect needs certain conditions of temperature and moisture. The more favorable the weather, the more generations or broods an insect can produce in one season. The coddling moth, for example, may produce three broods per year in the Ozarks but only one brood in the Northeast.

The seed-corn maggot and many species of cutworms thrive in late springs with unusually low temperatures. The boll weevil is most destructive in wet, humid summers. Hot, dry summers retard the boll weevil, and near-zero temperatures kill this insect. Grasshoppers damage farm crops most during drought years.

Climate and livestock. The nutritive value of livestock feeds depends to some extent upon climatic conditions. In general, a feed crop grown under favorable weather conditions may be expected to have a higher nutritive value than the same kind of crop grown in an unfavorable season. A high level of soil fertility may offset a deficiency of moisture to some extent, but even the best soils are more productive when temperature and moisture conditions are ideal for a particular crop.

Dairy cows usually produce more milk when temperatures are cool. Experiments have proven that milk production declines as the temperature in a dairy barn is increased from 40° to 95° F. The butterfat records of Registry of Merit Jersey cows were compared in Maine and Georgia. The cows in Maine produced more butterfat in summer, but the cows in Georgia had the best winter record. The difference in winter production may have been due in part to the greater amount of time that the Georgia cows were outdoors.

Left: **Jersey cattle are better adapted to a warm climate than most dairy cattle breeds.** (*Courtesy* Encyclopedia Britannica Films, Inc.) *Right:* **Brahman cattle are better adapted to hot, humid climates than any other beef breed.** (*Courtesy* American Brahman Breeders' Association)

Some breeds of cattle have more tolerance of hot weather than do other breeds. The Santa Gertrudis cattle, developed by crossing shorthorns and Brahmans, are particularly well-adapted to the climate in the Gulf Coast areas of the South and Southeast. Through more research in livestock breeding, answers may be found to other questions about the relation of climate to livestock production.

SOME RECORDS TO REMEMBER

The following are believed to be national or world records. It might be interesting to try to find out whether any of the records have been broken.

The highest shade temperature ever recorded in the United States was 134° F., in Death Valley, California, July 10, 1913.

The highest nongusty windspeed in the United States was 231 miles per hour in Mt. Washington, White Mountains, New Hampshire, April 24, 1934.

The nation's greatest temperature variation in a 24-hour day was from 44° to −56° F. at Browning, Montana, January 23–24, 1916. This was a 100° change in 24 hours.

The most intense rainfall in 1 minute was 1.23 inches received at Unionville, Maryland, July 4, 1956. That would dampen any celebration.

RAINMAKING

In recent years there has been much publicity about the possibility of making rain. Some communities have hired "rainmakers" who claimed that they could cause rain by seeding clouds with silver iodide crystals.

In the transition climatic zone near Cook Inlet and Anchorage, Alaska, squash can be grown without mulch (left), but with a clear plastic mulch (seen at arrow) squash yields can be increased by two or three times (right). The clear mulch acts as a greenhouse (glass house) by permitting short-wavelength rays of the sun to penetrate and warm the soil, where conversion is made to long-wavelength rays that do not penetrate outward so readily. (*Courtesy* D. H. Dinkel, Alaska Agricultural Experiment Station)

It was generally agreed that there is now so much rainmaking experimentation in the United States and other countries that an international regulatory organization is needed. The purpose of such an organization would be to prevent floods or droughts caused by "irresponsible" tampering with the weather.

A team of Israeli scientists reported that not even six years of experiments in seeding clouds with feathery silver iodide crystals projected into the sky had produced results which would be conclusive by statistical criteria. The team's report stated that cloud seeding may occasionally have very strong effects, but that on most occasions it has little or no effect.

The director-general of the United Kingdom Meteorological Service Office expressed doubt that the experimentation to date could support any claims on a realistic scientific basis. Much of the experimentation has been conducted in such a way that reliable statistical measurement cannot be applied to the results. There is much variability in natural precipitation, and at present it is not possible to predict with sufficient accuracy what would have occurred if clouds had not been seeded in the experiments.

The United States Bureau of Reclamation, Office of Atmospheric Water Resources, has experimental programs in six regions of the West. The purpose of these experiments is to determine possible practical applications of weather modification techniques to water resources problems.

CLIMATE OF THE UNITED STATES

It is difficult to describe the climate of the United States in a few paragraphs, but this summary will give you a general picture of precipitation, temperature, and growing seasons over the nation. By studying the maps in this chapter you can learn more about the climate of your state and region.

The average annual precipitation in the United States is approximately 30 inches. Rain is the principal form of precipitation over most of the country. The average annual precipitation in the states east of the Mississippi River is 44 inches. Between the Rocky Mountains and the Mississippi River the average annual precipitation is 28 inches, with a gradual decline from east to west. The area west of the Rocky Mountains averages 18 inches of precipitation, although the average in parts of the Pacific Northwest is 80 to 100 inches. Some areas in

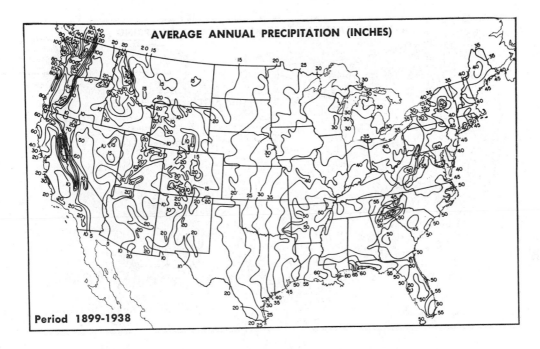

Average Annual Precipitation (inches)

Period 1899-1938

Above: What is the average annual precipitation in the area where you live? *Below:* How hot is the average day in July in your community? Is it warm enough to grow cotton? (*Courtesy* U.S. Weather Bureau)

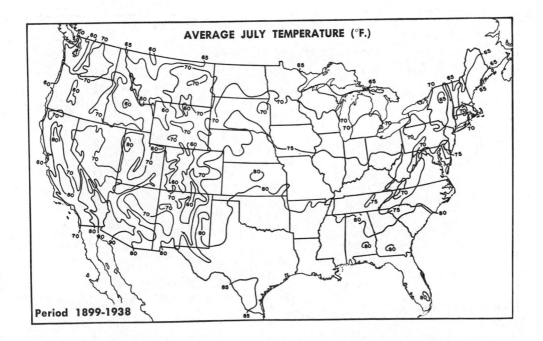

Average July Temperature (°F.)

Period 1899-1938

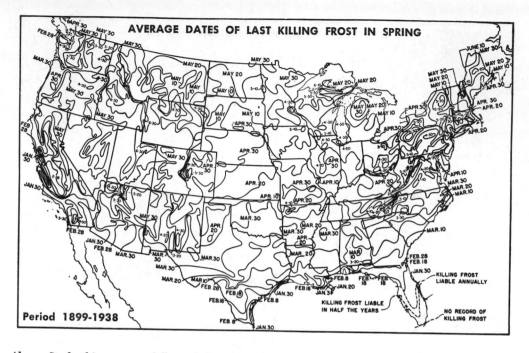

Above: Study this map carefully and then determine the best planting date for corn in your community. Below: From this map you can predict when the frost will be on the pumpkin. (Courtesy National Weather Service)

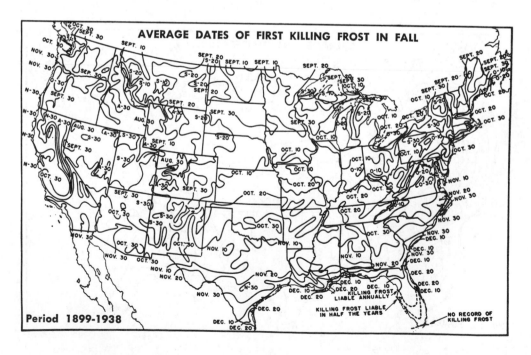

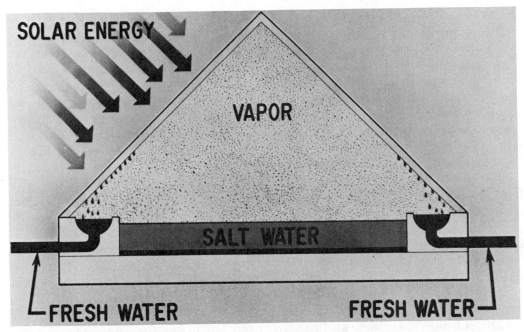

Above: Where the climate is favorable, a solar (sun) energy process of evaporation and condensation can be used to make fresh water from salty ocean water. High temperature and a high percentage of sunny days are essential to success. (*Courtesy* U. S. Department of the Interior) *Below:* No longer are the people of Alaska deprived of contact with other parts of the world because of an unfriendly climate. Television can keep them as well informed as people in more friendly climates. (*Courtesy* National Rural Electric Cooperative Association)

the West and Southwest have 5 to 10 inches of precipitation.

Most of the snowfall occurs in the West and in the Northeast. The annual snowfall in the Rocky Mountain states and the Pacific Coast states varies from 10 to 200 inches. Parts of Colorado and Oregon have 100 to 200 inches of snowfall, while in parts of Wyoming and Washington the average snowfall is 100 inches. Much larger areas in the West have averages of 30 to 40 inches of snow. In the southern part of the Northeast the average snowfall is 30 to 40 inches. Along the northern edge of the region the snowfall averages 60 to 150 inches a year.

On the basis of long-time averages, January 15 is the approximate date for the lowest temperature, while the highest temperature generally occurs about July 15. Remember that these are typical figures for the United States, not for a particular region. The precipitation averages previously mentioned and the average temperatures cited here are for a period of 40 years.

The highest average annual temperature in the continental United States is 75° F. at the tip of Florida. The average temperature in the southern parts of Texas, Arizona, and California is 70° F., although the average for the whole of each state is lower. The lowest average annual temperature in the United States is 35° F. in northeastern North Dakota near the Canadian border.

Planting dates are very closely related to temperature. The following crops are planted when the normal daily temperatures in the respective crop regions rise to the figures indicated: spring wheat, 37° F.; spring oats, 43° F.; early potatoes, 45° F.; corn 55° F.; and cotton,

62° F. The corn-planting temperature comes 10 days after the average date of the last killing frost in spring, and the cotton-planting temperature is reached approximately three weeks later.

On December 22, the shortest day of the year, the largest possible amount of sunshine varies from 10 hours and 35 minutes at the southern end of Florida to 8 hours and 10 minutes at the northern boundary of the United States from Minnesota westward. On June 21, the longest day of the year, the possible amounts of sunshine at these two locations are 13 hours and 41 minutes and 16 hours and 19 minutes, respectively. In other words, on June 21 there is a greater possible amount of sunshine in the North than in the South. In winter the South has the advantage of more sunshine because the sun is more vertical at noon.

On a 40-year average there are 340 frost-free days along the Gulf Coast of Texas and Louisiana. Florida averages 300 to 320 frost-free days, while 240 to 280 frost-free days are typical for most of California. Southern California has 320 frost-free days, and the northern part of the state has 120.

The lowest average number of frost-free days is 40 in the Rocky Mountains of Colorado. Parts of Utah, Idaho, Nevada, and California have only 80 frost-free days.

New England has a range of 100 to 180 frost-free days, with 140 being typical for the central part of the region.

The Lake States have 80 to 160 frost-free days, with 120 to 140 being typical for a large part of the region. Most of the Corn Belt has 180 frost-free days, with an average of 160 along the northern edge.

Most of the Southeast has an average of 240 to 280 frost-free days, although some mountainous areas have only 180. Much of the Southwest averages 200 to 240 frost-free days. The range of frost-free days in the Pacific Northwest is 80 to 280, but for much of the region the range is 160 to 200.

More important to you than average weather conditions for the United States are the conditions in your own county. Locate your position on several weather maps of the United States and work out your own conclusions about climate and agriculture in the area where you live.

Because the climates of Alaska and Hawaii are not well known by most of us who live in the other states, descriptions of the climates of these two states are given here.

CLIMATE OF ALASKA

This summary of the climate of Alaska is adapted from the United States Department of Commerce, Weather Bureau, publication *Climates of the States: Alaska* (September 1959), by C. E. Watson, Weather Bureau Climatologist.

There are four major zones of climate in Alaska:

1. A zone dominated almost entirely by maritime influences includes southeastern Alaska, the south coast, and the southwestern islands.
2. A zone of transition from maritime to continental climatic influences comprises a belt extending across and including the Copper River, Cook Inlet, Bristol Bay, and West Central divisions.
3. An area called the Interior Basin division is dominated by continental climatic conditions.
4. Arctic influences are dominant in the Arctic Drainage division.

Zone of dominant maritime influence. Most of the zone dominated by maritime ("ocean") influences has very rugged terrain. Islands comprise more than half of the land area of southeastern Alaska. The largest island, Prince of Wales Island, is about 120 miles long and 20 to 25 miles wide. Practically all of the narrow coastline from southeastern Alaska to the tip of Kenai Peninsula is indented by numerous tidewater bays, sounds, inlets, and fiords. There are numerous glaciers which extend through the canyons and deep valleys to the shoreline area. In southeastern Alaska the mountains are 2,000 to 5,000 feet high. Some peaks in the southeastern coastal range rise above 6,000 feet. As the coastal range approaches St. Elias Range, the mountains are higher and steeper. The 19,850-foot peak of Mt. Logan and several other peaks higher than 15,000 feet are less than 50 miles from the shoreline of Yakutat Bay. Other parts of this zone have mountains 5,000 to 7,000 feet high.

Cool summers and mild winters are typical. July and August, the warmest months, have average temperatures of 54° to 56° F. The typical temperature during January, usually the coldest month, is 25° to 32° F. Low readings of −40° F. have been recorded, but most stations have rarely had a low of −20° F. or colder.

Temperature changes between seasons are gradual. In southeastern Alaska the frost-free period averages about 145

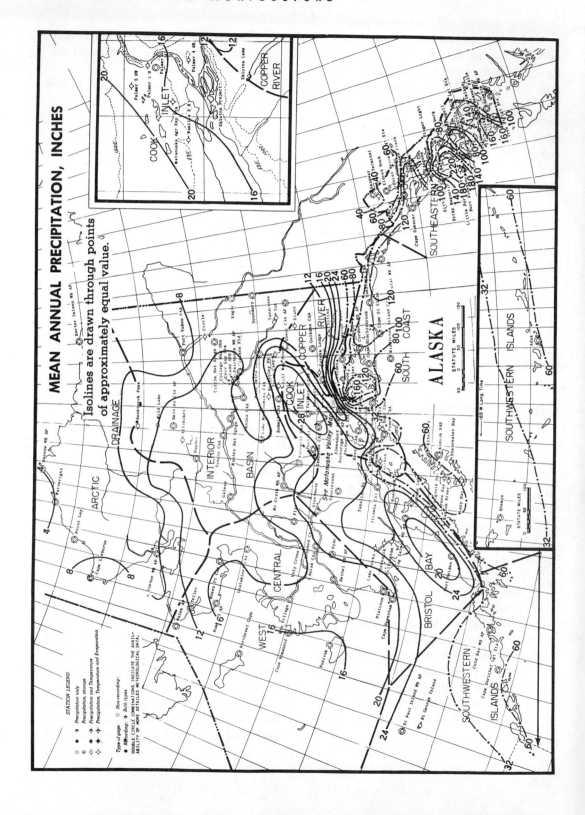

MEAN ANNUAL PRECIPITATION, INCHES

Isolines are drawn through points of approximately equal value.

days. The northern gulf perimeter has about 120 frost-free days.

Because of its exposure to the open sea, the entire maritime zone is affected by strong winds associated with intense cyclonic circulation. Areas along the gulf coast are the ones most frequently affected by strong winds. During almost every month of the year, winds of 60 to 70 miles per hour occur at exposed parts of southeastern Alaska, but these winds occur less frequently from Yakutat westward to the Kodiak area. The islands receive strong winds from practically all directions during intense storms. In winter, wind speeds of 70 to 90 miles per hour occur almost monthly at some points along the islands.

There is abundant moisture over the entire zone, particularly over southeastern Alaska. At several stations in this part of the state the average annual rainfall is more than 100 inches.

The greatest recorded average annual precipitation in Alaska is slightly more than 221 inches at Little Port Walter. About one fourth of the reporting stations in southeastern Alaska have a maximum monthly precipitation of 30 inches or more.

Along the gulf coast west of Yakutat the annual precipitation averages less than 100 inches. From Seward southwestward to Kodiak the average is 60 to 70 inches. At Chignik the annual precipitation averages a little more than 150 inches. Farther west along the Aleutian chain the heavier annual precipitation exceeds 60 inches, but the average at a few sheltered stations is only 25 to 30 inches.

Where soil conditions are favorable, the abundant moisture of the maritime zone encourages vegetation. This zone contains practically all of Alaska's forest areas of commercial importance.

Minimum temperatures in the southern portions of southeastern Alaska average near freezing. The snowfall during the winter season is variable, as small variations in temperature determine whether precipitation falls as rain or snow. There is some rain over the entire maritime zone even in winter. A permanent snow cover remains during the winter months from the Juneau area along the northern gulf to the tip of the Kenai Peninsula.

The maritime zone has some cloudy, foggy weather throughout the year. In southeastern Alaska cloudiness is greatest from September to December. At Annette heavy fog is most frequent from July through September. In the Juneau area the heaviest fog occurs in October and November. During the summer months fog occurs frequently over the Aleutians.

Transition zone. The maritime influence declines from the coastal areas toward the interior. Between the South Coast and Copper River divisions there is an abrupt change from a maritime to a semi-continental climate. This change is due to the sharp mountain ridge along the boundary. The change from maritime influence is more gradual over the Cook Inlet division. There is a gradual transition in climate in the Bristol Bay and West Central divisions.

Seasonal temperatures are more pronounced over the inland portions of the transition zone. Winter temperatures of 45° to 55° F. below zero occur in the upper Copper River Basin, upper Susitna Valley, central Kuskokwim Valley, and across the Innoko River Basin to the Yukon Valley. Some stations have record lows of 60° F. below zero or colder. Gulkana averages 103 days per year with temperatures zero or colder and about 70 days with maximum readings of 70° F. or above. The highest temperatures in the Copper River division generally vary from 90° to 95° F. Highs in the mid-80's are typical in areas more directly affected by the maritime influence.

The average annual precipitation varies widely at different places in the transition zone. At Thompson Pass, on the inland side of Chugach Ridge, there has been seasonal snowfall in excess of 600 inches during the relatively short period for which records are available. Thompson Pass had 975 inches of snow during the winter of 1952–1953.

In the Copper River division the range in maximum annual precipitation is 15 to 20 inches. Over the Matanuska Valley the annual precipitation averages 15 to 20 inches, and over the Susitna Valley the average is about 30 inches. On the western shores of Cook Inlet the average is 75 to 80 inches. Annual precipitation averages 25 to 40 inches on the southern tip of Kenai Peninsula. The lower portions of the Yukon and Kuskokwim valleys, in the West Central division, have an average annual precipitation of about 20 inches. In the Southwestern Islands the annual precipitation averages from 32 to 60 inches.

Minimum winter temperatures in the Southwestern Islands vary widely from 5° to −12° F., with recorded extreme lows of −36° F. The typical average maximum summer temperatures vary from 40° to 50° F., with a rare high of 70° F.

The entire area from the Copper River division through the Susitna Valley into the West Central division has a snow cover throughout the winter. The upper Cook Inlet has a snow cover from mid-November to about mid-April, although some thawing may occur in January.

The permafrost area (area of permanently frozen subsoil) extends southward into the northern portions of the transition zone. It is difficult to define the boundaries of the permafrost area, but it stretches from the northern slope of the Wrangell Mountains through the Glennallen area and across the upper Susitna and Kuskokwim valleys into the McGrath and Holy Cross areas. Permafrost is encountered along the inlet borders of the Cook Inlet, Bristol Bay, and West Central divisions.

Relatively light winds are characteristic of the Copper River and Cook Inlet divisions. Stronger winds occur frequently in winter. The relatively unsheltered areas of the Bristol Bay and West Central divisions frequently have strong winds, particularly during winter months. Wind speeds may average 25 to 30 miles per hour for an entire month. Winds of 60 to 70 miles per hour may persist two or three days at a time when intense low pressure areas approach the coast or enter the mainland.

Dominant continental zone. The Interior Basin division is remote from the open sea, and the surrounding land barriers hinder the inland movement of air.

The dominant continental zone is the warmest area of Alaska in summer. Almost every summer brings maximum readings of 90° F. or more. The record high for the zone is about 100° F., with several stations reporting 98° F. At Fort Yukon and Eagle, daily maximum temperatures average 70° to 75° F. during July and August. At Fort Yukon the sun remains continuously above the horizon for a month beginning about July 5. Because of the prolonged hours of daylight and abundant sunshine, plants develop rapidly during the growing season.

An extremely wide temperature range is characteristic of the Interior Basin. For example, Fort Yukon, just north of the Arctic Circle, has recorded a high of 100° F. (June 27, 1915) and a low of −75° F., a range of 175° F. Tanana has had a range of 98° to −76° F. Most of the lowland Basin area has a frost-free season from about May 20 to late August.

The Interior Basin is almost completely surrounded by high ridges. This is one of the main reasons for the light annual precipitation, which averages 10 to 13 inches. June and July are usually the months of heaviest rainfall. The normal moisture in the Interior Basin is not sufficient for profitable crop production.

Arctic region. The typical Arctic climate is confined to the Arctic Drainage division, which includes the drainage area of Kotzebue Sound. There is a prolonged period of continual daylight, but summer high temperatures rarely exceed the mid-80's. About half the weather stations have not had temperatures above 80° F. Even during the longer days in summer the sun's rays reach the earth's surface at such low angles that there is little surface warming.

Extreme low winter temperatures generally range between −45° and −60° F. About two in every five weather stations have not had a temperature colder than −50° F.

Annual precipitation averages 4 to 10 inches at most stations. Annual snowfall of 40 to 65 inches occurs at several stations in the Arctic Drainage division. Winter winds of 50 to 60 miles per hour are not unusual along the Arctic coast from Cape Lisburne to Barter Island.

CLIMATE OF HAWAII

This summary of the climate of Hawaii is adapted from the United States Department of Commerce, Weather Bureau, publication *Climate of the States: Hawaii* (February 1961), by David I. Blumenstock, Weather Bureau Pacific Area Climatologist.

Located approximately 2,100 miles west and south of California, Hawaii is the only state surrounded by an ocean and the only state within the tropics. The six major islands and their areas in square miles are Hawaii, 4,030; Maui, 728; Oahu, 604; Kauai, 555; Molokai, 260; and Lanai, 141. These islands are widely separated, with much variation in climate and topography.

Fifty percent of the State of Hawaii has an elevation of 2,000 feet, and 10 percent is above 7,000 feet. The respective maximum elevations of the island are Hawaii 13,784; Maui, 10,025; Kauai, 5,170; Molokai, 4,970; Oahu, 4,025; and Lanai, 3,370. Geologically, Kauai, to the west, is the oldest of the six major islands; Hawaii, to the east, is the youngest. The Waimea Canyon in the western half of Kauai is an example of severe erosion, and there is also much evidence of erosion in the valleys of the eastern half of the island. A long range of volcanic mountains makes up the Hawaiian Chain in the State of Hawaii.

Because the islands are relatively short and narrow, 18 to 93 miles long

MEAN ANNUAL PRECIPITATION, INCHES

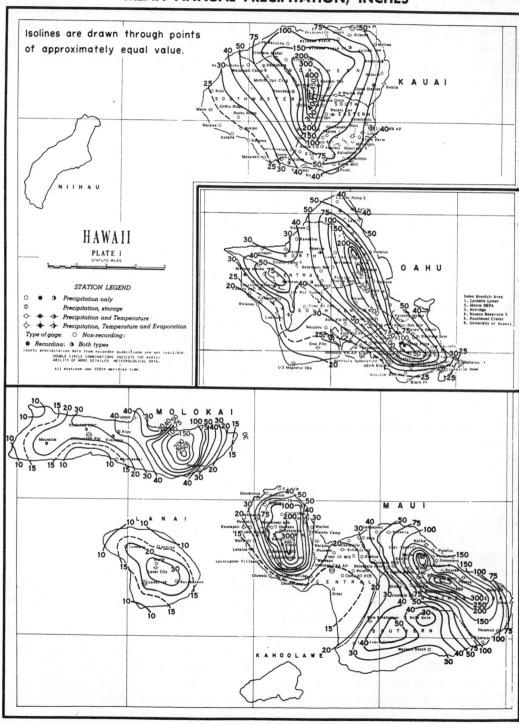

Isolines are drawn through points of approximately equal value.

HAWAII

PLATE I

STATUTE MILES

STATION LEGEND

Precipitation only

Precipitation, storage

Precipitation and Temperature

Precipitation, Temperature and Evaporation

Type of gage: Non-recording;

Recording; Both types

Hourly precipitation data from recorder substations are not available.

DOUBLE CIRCLE COMBINATIONS INDICATE THE AVAILABILITY OF MORE DETAILED METEOROLOGICAL DATA.

All stations use 150th meridian time.

and 13 to 76 miles wide, the climate reflects a strong marine influence which is modified to some extent by the mountains. Almost half of the total area is within five miles of a coast, and only a small part of the Island of Hawaii is more than 20 miles inland.

The characteristic circulation of air across the tropical Pacific Ocean is the east-to-west trade-wind flow. Trade winds are prevalent 80 to 95 percent of the time during the period of May through September and 65 to 80 percent of the time during the period of October through April.

Seldom are there cloudless skies over the islands, although the lowlands often have only scattered clouds. Areas at or near the summits of the high mountains and at elevations of 2,000 to 4,000 feet on the windward side of other mountains have cloudy skies more than 70 percent of the time. Showers occur frequently, most of them light and brief. Thunder and lightning rarely accompany any showers, even the heavy ones. Most of the major storms occur during the six months from October through March. There are usually two or three major storms per year, although in some years there may be six or seven. Heavy rains and violent winds are characteristic of these storms.

The storms may be associated with the leading edge of a mass of relatively cool air moving from west to east or from northwest to southeast. Storms may also be associated with a large atmospheric eddy, generated by moving air in much the same way that a flowing river may produce whirling water. Torrential rains and large clouds are produced when moist, warm air swirls into a large eddy of moving air.

Daylight and solar energy. There is little variation in the length of day and night throughout the Hawaiian Islands, as they have little difference in latitude. Honolulu (21° north latitude) has 13 hours and 20 minutes between sunrise and sunset on the longest day of the year, and 10 hours and 50 minutes on the shortest day, a total difference of 2 hours and 30 minutes. In comparison, St. Louis, Missouri, and Washington, D. C. (both 39° north latitude) have 15 hours between sunrise and sunset on the longest day of the year, and 9 hours and 20 minutes on the shortest day, a total difference of 5 hours and 40 minutes.

The seasonal variation in solar energy is much less in Hawaii than in the continental United States. The amount of solar energy received by the upper atmosphere over Hawaii each day is only 50 percent greater in early June than in late December. At Washington, D. C., the maximum daily receipt of solar energy is three times the minimum (a 300 percent difference), while at Anchorage, Alaska, the maximum daily receipt of solar energy is twenty times the minimum (a 2,000 percent difference).

Temperature. There is a very small seasonal temperature range in Hawaii. In downtown Honolulu the range is from 78.5° F. in August to 72° F. in January and February. No weather station in the islands has an annual variation of more than 9° F. In the continental United States, only a small zone along the coast of California has a comparable small variation in temperature between summer and winter. Here are some comparative temperature ranges: Miami, Florida, 13° F.; Seattle, Washington, 23° F.; Salt Lake City, Utah, more than

Contour planted pineapple. (Kauai Pineapple Company) (*Courtesy* Soil Conservation Service)

40° F.; Fort Worth, Texas, 30 to 39° F.; New Orleans, Louisiana, 26° F.; Chicago, Illinois, 50° F.; and Fairbanks, Alaska, 70° F.

Average temperatures decrease as elevations increase. For example, Hilo, with an elevation of 40 feet above sea level, has a mean January temperature of 71° F. and a mean August temperature of 76° F.; at Mauna Loa Observatory, where the elevation is 11,150 feet above sea level, the mean January temperature is 39° F. and the mean August temperature is 47° F.

The winter season in Hawaii is October through April, and summer is May through September. The trade winds are strongly dominant during the warmer summer season. The average monthly rainfall is also less during the summer, except on the Kona Coast (leeward coast) of the Island of Hawaii. On the Kona Coast, rainfall averages about an inch more in summer than in winter.

Rainfall. Rainfall varies widely in the Hawaiian Islands. On the leeward coastal areas and near the summits of the very high mountains of Mauna Loa, Mauna Kea, and Haleakala, the average annual rainfall is about 20 inches. Along the lower windward slopes of the high mountains and near the summits of the lower mountains of Kauai, Oahu, and western Maui, the annual rainfall averages more than 300 inches. The average annual rainfall for the entire State of Hawaii is 70 inches, while that of the continental United States is about 30 inches. It is estimated that the Hawaiian average would be about 25 inches if the area were only water. This means that the islands intercept an additional 45 inches of annual rainfall.

The lowlands receive more rain at night or in the hour immediately after sunrise than during the daytime. At Honolulu, on the average, 59 percent of the annual rainfall occurs between 8 P.M. and 8 A.M. The rainfall variations between day and night are greater in summer, when rains occur most frequently between 3 A.M. and 8 A.M.

There are occasional droughts in the Hawaiian Islands. The worst drought years are the ones in which the winter rains fail. In exceptionally dry winters there may be only one or two rainstorms, sometimes none. After a dry winter the summer rainfall is usually insufficient for crops. Although severe droughts are usually limited to areas of approximately 50 square miles, they may cause considerable economic loss even in irrigated areas.

Storms. Major storms are mainly the result of three classes of disturbance. A cold front moving across the islands may bring locally heavy showers and gusty winds. A storm eddy associated with a low pressure area may bring "kona storms," which are often accompanied by strong winds and heavy rains. The third class of disturbance includes the tropical storm and the true hurricane.

Cold front storms. Cold front storms occur in winter and may move across two or more islands. The number for a single winter may vary from one or two to eight. The Island of Kauai, located at the northwest end of Hawaii, usually has a few more cold front storms than do Oahu and some of the other islands. The encroachment of the southern edge of a mass of cool air often reaches Kauai but not Oahu.

Rains which accompany the winter storms vary from a fraction of an inch in some areas to several inches in others and may last a few hours or several days. Skies are relatively cloudless after a cold front passes. Temperatures in the lowlands may drop to about 62° F. but rarely below 60° F.

Kona storms. Kona storms are also winter storms. They get their name from the fact that they often bring winds from the "kona," or leeward, direction. If high winds accompany a kona storm, they tend to be steadier, more prolonged, and not as strong as cold-front winds. Oc-casionally there is a winter without a well-developed kona storm. Usually there are one or two a year, less frequently four or five.

Tropical storms and hurricanes. It is good to know that true hurricanes do not invade the Hawaiian Islands very often. The Weather Bureau began its observations in September 1904, and by April 1960 only four hurricanes with winds 74 miles per hour or more had come close enough to affect the islands. These were Hiki, August 1950; Della, September 1957; Nina, December 1957; and Dot, August 1959. The latter passed south of the eastern and central islands and swung north across Kauai. This island suffered most of the damage.

Tropical storms are more frequent than hurricanes. Winds of tropical storms are less than 74 miles per hour, often much lower in velocity. Because of the similarity between tropical storms and kona storms, it is difficult to distinguish one type from the other. On the average, tropical storms probably pass close enough to the Hawaiian Islands to cause heavy rains and moderate to high winds once every three or four years.

QUESTIONS

1 What is the difference between *weather* and *climate*?

2 How many primary National Weather Service stations are there in our nation? Find out where your nearest station is located.

3 Make a list of the items of weather given in your local TV or radio forecast, such as temperature and wind direction.

4 What are the three main climatic influences that affect plant growth?

5 What are xerophytes?

6 Name a hydrophyte.

7 What is the average annual precipitation in the United States?

8 What is the average annual precipitation in your community?

9 On the western range, what is the general relationship between precipitation and rate of stocking of livestock?

10 What is the length of the growing season in your community?

A C T I V I T I E S

1 Listen to radio or television forecasts of the weather every day for two weeks. Make notes about the weather predictions and the kind of weather that actually occurs.

2 Save the weather maps from a daily newspaper for a week. Use these maps in a class discussion of the recent changes in the weather, the reasons for these changes, and the way weather forecasts are made.

3 Find a tin can with straight sides, a diameter of 6 to 8 inches and a depth of 8 to 10 inches or more. Put the can in an open place where it will catch rain. Tie the can securely to two or three stakes. Measure and record the depth of water after each rain. Empty the can so that it will be ready for the next rain.

4 Release a toy balloon and watch it until it is out of sight. Observe the direction in which it moves as it rises. Note any change in direction of movement, and whether its movement is slow or rapid. Explain how air movement affects weather.

5 Use science and geography textbooks, encyclopedias, and other references for additional reading on climate and weather.

11 EARTH AND LIFE SCIENCES

There are many reasons for learning all we can about nature. We are completely dependent upon nature for life—for the air we breathe, the water we drink, and the food we eat.

In our daily living we are guided by natural laws, usually without being aware that they exist. The rotation of the earth gives us day and night, and the movement of the earth around the sun brings the seasons of the year. The sun is the unfailing source of energy for all living things. At night the moon and stars tell us that all is well in the universe. So great is the power of nature that man cannot live without the sun's energy, nor can he escape the water cycle which brings the snow and the rain.

Natural science helps us to understand and appreciate the wonderful natural environment in which we live. The stories that can be read in rocks and soil are just as exciting as any that can be found in books. Getting acquainted with plants and animals is a delightful experience for anyone. A study of earth and life sciences in an adventure, and is fun.

Conservation is the wise use of resources for better living. The best conservation practices are copied from nature's own system for keeping soil, water, plants, and animals in harmonious balance. Conservation and earth and life sciences are considered together here because they are so closely related. You will learn more about conservation in the chapters on soil, water, forests, wildlife, and grasses and legumes; it is a subject of prime importance.

KINDS OF NATURAL RESOURCES

Soil, water, plants, animals, and minerals are natural resources. Air and

sunlight are so important that they also must be considered as parts of our natural environment. Climate is an influence rather than a resource, but climate must be recognized in a study of earth and life sciences.

Indestructible resources. Air, sunlight, and water are not destroyed by use. However, the usefulness of these resources may be reduced by abuse.

One of the most difficult problems in conservation is the management of water so that it will be available when and where it is needed. Although the total amount of water in the world remains constant, floods and drought may

Once wild and part of the natural resources in southern India and Ceylon, this elephant has been trained to satisfy human wants. Here the elephant is loading heavy logs on a truck. (Ceylon) (*Courtesy* Food and Agriculture Organization of the United Nations)

Left: A dramatic illustration of the intimate relationship between earth and life sciences is shown here. A white cedar tree has grown to near-maturity by sending its roots into the cracks and solution holes of the limestone bedrock. The tree died only after a careless fire burned the 6-inch organic layer of soil that was developing. (Michigan) (*Courtesy* J. O. Veatch) *Right:* Soil can be used forever and it will not "wear out." It may blow or wash away if not used properly. The elements of fertility may also be depleted, but these can be replenished by animal manures and chemical fertilizers. (*Courtesy* Cornell University)

Left: The young mother and her son are studying this Ohio school forest which she helped to plant 15 years before when she was in the sixth grade. Forests are a renewable resource. (*Courtesy* U. S. Forest Service) *Right:* If you want to learn to grow something near a window in your home or in a classroom, try this curlicress. It will produce edible greens in two weeks. (*Courtesy* Burpee Seeds, Philadelphia, Pennsylvania)

occur at the same time at localities only a few hundred miles apart.

Plants use air, sunlight, and water in the process of photosynthesis. Animals breathe oxygen, which is abundant in the atmosphere around the earth. Oxygen makes up 20 percent of the air.

Every farmer knows that sunlight helps to keep livestock healthy. Buildings may be designed to make maximum use of the sun's energy for animals or for plants. Greenhouses are an example of this principle. The quality of natural light is important in many kinds of buildings.

Water can be changed from liquid to solid (ice) or gas (vapor) and back to liquid without affecting its chemical composition. The same water can be used several times for one purpose or several purposes. For example, the water that is used to generate hydroelectric power at one point on a river may become part of the water supply of a city downstream. An industrial plant may reuse water several times.

Renewable resources. Plants and animals are renewable because new generations can be produced to take the place of those that are harvested. Commercial forests can be managed to provide a permanent supply of timber. Ranges and pastures can be grazed without destroying the forage plants so long as there is a satisfactory balance between livestock and available forage.

The conservation of grasslands and forests depends to a great extent upon the natural reproduction of plants. Sometimes the reseeding of grasslands or the artificial reforestation of woodlands is necessary after the land has been abused.

Soil is renewable to some extent. However, if all of the topsoil on a field has been removed by erosion, the restoration of the organic matter and the fertility may be a long, difficult, and expensive process. When the best cropping systems and erosion-control practices are used, topsoil forms faster than it is destroyed.

When satisfactory habitat conditions are maintained, game can be harvested as a crop. If the surplus wildlife population is not harvested and used it will be destroyed by natural causes such as starvation, severe weather, and natural enemies.

Nonrenewable resources. The minerals, particularly the metals such as iron and the fuel minerals such as coal, are destroyed by use. Mineral fertilizers such as potash salts are also consumed in use. Nature produces these resources so slowly that they must be considered nonrenewable.

The conservation of metals and fuel minerals includes the finding of new reserves, the development of substitutes, and more efficient use of the available supply. It is not possible to *stop* using these resources, but they can be used

Oil is an example of a nonrenewable resource. This oil refinery makes gasoline, kerosene, and other products from crude oil. (*Courtesy*** Puerto Rico Economic Development Administration)**

with less waste. Some of the metals can be reused. The salvage of scrap iron is an example of this conservation practice.

THE BALANCE IN NATURE

Nature has many ways to keep plants and animals in balance with each other and with their natural environment. Certain species of plants and animals may increase to the disadvantage of other species, but sooner or later nature will reverse this trend. Man often upsets the balance in nature, although he also can help to restore it.

Before the settlement of North America began more than 300 years ago, the forests, grasslands, streams, and wildlife were in balance. The American Indians were part of the balance of nature, as they were not numerous enough to destroy the forests and wildlife over large areas. During severe droughts the bison overgrazed parts of the Great Plains, but for the region as a whole this was a temporary condition.

The aquatic plants and animals in a stream or lake are an example of the balance in nature. The green plants use carbon dioxide and release oxygen through the process of photosynthesis. The animals use the oxygen and release carbon dioxide. Most of the oxygen in water, however, comes from the air.

The balance in nature can be demonstrated with an aquarium. Algae and other water plants will live indefinitely with such aquatic animals as fish, snails, and tadpoles. If an aquarium is not available, you can demonstrate the balance in nature by putting algae and a few minnows or tadpoles in a fruit jar filled with water. Without the green

In humid regions the original balance of nature consisted of trees, shrubs, deer, birds, insects, and other living things. Man cleared the forests and plowed the soil to make new fields. The new balance consists of moles and gophers, farm crops, and insects and diseases that must be controlled by pesticides. (*Courtesy* (above) Cities Service Company and (below) Charles E. Kellogg)

Left: Bison were a part of the balance of nature before white men came. (*Courtesy* Soil Conservation Service) *Right:* A honey bee in a cotton flower. Bees increase cotton yields by spreading pollen to fertilize flowers. (*Courtesy* U. S. D. A.)

plants these little water animals would die in a few hours.

Predatory animals such as hawks, owls, weasels, and coyotes help to control rodents. The populations of the predators are controlled by their natural food supply. For example, certain kinds of hawks are most numerous where field mice and rabbits are abundant. On some of the western rangelands the population of coyotes seems to vary according to the abundance or scarcity of prairie dogs.

Predatory animals are more likely to attack poultry and livestock when the natural prey species are scarce. In wildlife management the small rodents are called "buffer species" because they provide food for birds and mammals that otherwise might destroy game birds.

Soil, water, plants, and animals are interdependent, and all depend upon air and sunlight directly or indirectly. Green plants are the primary source of food and feed. Animals that eat only flesh must depend upon other animals that eat plants. Plants get their moisture and nutrients from the soil. Plant roots, bacteria, and many other kinds of soil organisms need air. As plants and animals complete their life span, they return to the soil as organic matter.

IDENTIFYING PLANTS AND ANIMALS

Nature study affords many opportunities to get acquainted with the plants and animals in your community. Perhaps it is somewhat easier to study plants because they do not run away. With patience and practice you can develop skill in observing even the more timid animals.

Plants and animals have common names and scientific names. In nature study it is usually sufficient to know common names. However, scientific names must be used for positive iden-

Left: The fish should be enough food to last about two days for the snapping turtle.
Right: "I wish I hadn't eaten so many grasshoppers." (*Courtesy* Ralph J. Donahue)

tification, because different plants may be known by the same common name. For example, several species of oaks are known locally as "pin oak."

Latin and Greek words are used for scientific names because these languages do not change. Also, a Latin word or a Greek word has the same meaning anywhere in the world. A scientific name usually has two parts, one indicating the genus (a division in classification) and the other indicating the species (the next lower division of classification). For example, all oaks belong to the genus *Quercus.* The scientific name of the red oak is *Quercus rubra,* while the white oak is *Quercus alba.*

The four major groups of plants are:

1. Bacteria, algae, fungi, and lichens.
2. Mosses and liverworts.

This 5-spotted tomato worm will eat only a few more tomato leaves before it dies from the parasitic wasp larvae and cocoons. (*Courtesy* Ralph J. Donahue)

Left: Lichens are growing on an 800-year-old church in north central Ethiopia (eastern Africa). This church was hewn from solid volcanic rock in the form of a cross. (*Courtesy* Roy L. Donahue) *Right:* Which would farmers rather have, spiders or grasshoppers? Here a golden garden spider has killed a grasshopper. (*Courtesy* Ralph J. Donahue)

3. Ferns and their allies.
4. Seed plants.

The plants within each group have certain common characteristics by which they can be identified. It is usually easy to place a plant in one of the four major groups, but identifying the plant according to the family or species to which it belongs is a more difficult problem.

In nature study you probably will be more interested in the seed plants than any other division of the plant kingdom. The flowering plants are the largest and most important group of seed plants. The flowering plants include grasses, legumes, grains, and most of the other important agricultural crops. The broad-

leaved deciduous trees and many kinds of wildflowers also belong to this group.

COMMON PLANTS WHICH ARE POISONOUS

Few people know that parts of common garden and field plants are poisonous. Sometimes these plants are fatal if eaten. A 10-cent packet of castor bean seeds may contain enough poison to kill five children. One tulip bulb contains enough poison to kill a man. The National Safety Council estimates that about 12,000 children every year eat parts of potentially poisonous plants.

According to the National Safety Council, more than 700 kinds of plants

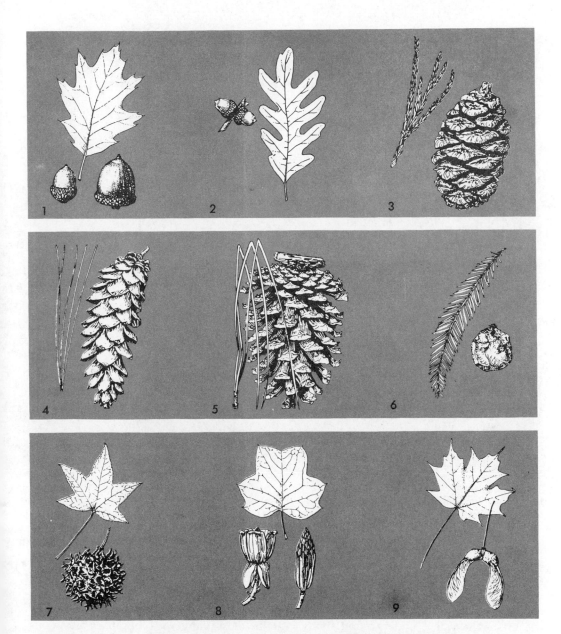

1. Northern red oak, Quercus Borealis. 2. White oak, Quercus alba. 3. Cone and twig of the giant sequoia in California. 4. Eastern white pine cone and needles. The needles are always found in bundles of five. 5. Cone and leaf cluster of loblolly pine, the most common southern pine. 6. Cone and leaf of bald cypress, one of the few cone-bearing trees that sheds its leaves each fall. 7. Sweet-gum seed ball and star-shaped leaf. 8. Tulip tree has a tulip-like flower, a cone-like seed pod, and a leaf with a notch on the end. 9. Sugar maples have seeds like a double-winged key and the leaves have many points. (*Courtesy* U.S. Forest Service)

Left: Big bluestem grass has a seed head like a turkey foot. *Right:* Sudangrass has a seed head that is shaped like a Christmas tree. (*Courtesy* U.S.D.A.)

have been known to cause accidental poisonings. Many of these grow in gardens. Among these are bulbs of daffodils, seeds of larkspur, leaves and flowers of the lily-of-the-valley, underground stems of iris, seeds and pods of wisteria, berries of jasmine, and all parts of azaleas, rhododendron, and oleander.

Some parts of the more common fruit or vegetable bearing plants are toxic. The leaves and vines of the tomato plant and the foliage of the potato plant contain alkaloid poisons that can cause severe digestive upset and nervous disorder. Rhubarb leaf blades are perhaps the most dangerous because they contain oxalic acid. This acid can crystallize in the kidneys and cause severe damage.

Plant poisonings may occur under seemingly harmless circumstances, especially among children at play. A play luncheon of daphne berries was known to cause the death of a young girl. Other children have been poisoned by using the pithy stems of plants such as the elderberry for blowguns. Most toddlers are attracted to pretty flowers and place the plants in their mouths.

The following plants are among the most common which are poisonous when eaten. All except the columbine, sweet pea, and spider lily can be fatal if taken in the quantities which a child might eat.

PLANT	POISONOUS PART
Columbine	Berry
Sweet pea	Stem
Spider lily	Bulb
Elephant ear	Any
Narcissus	Bulb
Four o'clock	Root, seed
Cyclamen	Tuber
Poison ivy	Leaves
Potato	Seed, sprouts
Pimpernel	Any
Oleander	Leaves
Lily-of-the-valley	Any
Burning bush	Leaves
Jimson weed	Any
Rhododendron	Any
Dumb cane	Any
Iris	Underground stem
Pinks	Seed
Mock orange	Fruit
Spanish bayonet	Root
Bittersweet	Berry
Castor bean	Seed
Foxglove	Leaves
Scotch broom	Seed
Bluebonnets	Seed
Tulip	Bulb
Mountain laurel	Any
Monkshood	Root

A stream has many uses. This scene is on the Blanco River, near Wimberley, Texas. (*Courtesy* Texas Highway Department)

The symptoms of plant poisoning depend on what plant is eaten. A common symptom is difficulty in breathing. Some poisonous plants burn the mouth. Parents who find their children eating a flower or other part of a plant should call the family doctor or the emergency room of a hospital. If the parents or medical personnel can not readily identify the plant, it may be necessary to send it to a university or laboratory.

Parents and other adults should avoid the common habit of chewing on a leaf or stalk. Children should be taught that it is dangerous to put leaves, stems, and especially berries, into the mouth. Many plant bulbs are poisonous and should be stored where children cannot get them.

QUESTIONS

1 Is nature study a necessity or simply a hobby?

2 How much oxygen is there in the air?

3 In what way is water unique among the natural resources?

4 "Soil is renewable to some extent." Explain.

5 Name five nonrenewable resources.

6 Give one example of the balance of nature.

7 In nature, what is responsible for the manufacture of all food and feed?

8 Name the four major groups of plants.

9 Name one kind of bird that is a permanent resident of your community.

10 How many kinds of plants have been known to cause accidental poisonings?

ACTIVITIES

1 Measure the amount of water required to fill an ice tray in a refrigerator. Place the tray in the refrigerator until ice cubes form. Remove the tray, melt the ice cubes, and measure the water. Did the freezing of the water and the melting of the ice change the volume of the water?

2 Put a potted green plant in a dark place for three or four days, then observe the change in the color of the leaves. Wrap two or three leaves of a potted green plant in tinfoil or black paper. After three or four days compare these leaves with the other leaves on the plant.

3 Make individual and class collections of insects, leaves, wood samples, flowering plants, and soil. Label each item so that it can be identified for further study.

4 With the help of your teacher, learn to identify several of the trees, shrubs, grasses, and legumes that grow near your school.

5 By observing conservation practices in your community and by interviews with farmers and conservation specialists, find out what is being done to conserve the natural resources.

6 Find out what poisonous plants grow in your community. Arrange a class activity in the identification of these plants. An exhibit of local poisonous plants would be helpful to other students. The exhibit should be labeled "Poisonous Plants," and the collection should be exhibited under glass so that the plants cannot be handled.

7 Set up a demonstration by the class or by an individual to show that most seeds, such as corn, require the right proportion of air and water to germinate. It should also demonstrate that aquatic (water) plants, such as rice, will germinate under water as well as at the surface of the water. Neither type of seed will germinate in air. (Corn seed is readily available, but obtaining rice seed may be a problem. Rice purchased in a store will not germinate because it has been polished and the germ has been injured. Unpolished (rough) rice for a demonstration may be obtained from the Rice Research Station, Crowley, Louisiana 70526.)

LEFT: CORN

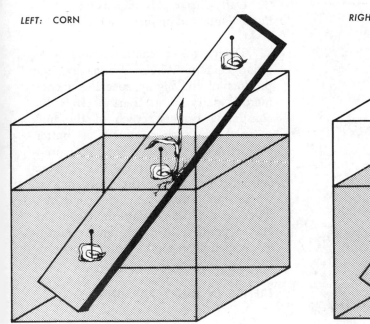

RIGHT: RICE

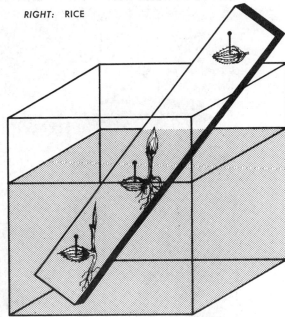

8 The following are believed to be national or world records. Use encyclopedias and other references to find other "firsts" in nature.

The world's biggest salt mine is the Morton Salt Company mine at Fairport, Ohio. The mine was opened in June 1956. It is worked to a depth of 2,025 feet, can yield 12,000 tons a day, and will last 200 years.

The largest known underground chamber is Carlsbad Caverns in New Mexico. The Big Room is 4,270 feet long, 328 feet high, and 656 feet wide.

The most extensive known cave system in the world is Mammoth Cave in Kentucky, discovered in 1799. It is 150 miles long.

12 DAIRY FARMING

Dairy farming is a highly developed enterprise in the Northeast, the Lake States, and some areas along the North Pacific Coast. In these parts of the United States the soils and topography are not suitable for large scale production of grain crops. But the cool climate and well-distributed rainfall are ideal for hay and pasture crops which can best be marketed by feeding dairy cattle. Dairy farming is increasing rapidly in the South and on farms in the irrigated West.

Dairying is concentrated near population centers. Whole milk is a major product of the highly specialized commercial dairy farms. Farther from the consuming centers, much of the milk is used for such products as butter, cheese, and ice cream. For these products the dairy enterprise can be less specialized than for the marketing of whole milk. Often dairying is part of a diversified farming pattern which includes poultry, livestock, and specialty crops such as vegetables for canning.

TRENDS IN DAIRY FARMING

The number of dairy cows has decreased sharply in recent years, and the trend is continuing. For example, from 1964 to 1971 the number of dairy cows declined 15 percent. By 1971 the number in the United States was approximately 12.4 million. The major decline was on small dairy farms. The reasons given for the decline have been that dairy farming is very confining (no golf or fishing or holidays), and alternate enterprises are attractive. Since the demand for beef has been increasing, a dairy farmer can easily switch to intensive beef cattle production and still play golf and go fishing.

As the number of dairy farms has decreased, the size of dairy farms has increased. The total value of all dairy cattle, average value of each dairy cow, and average yield of milk from each cow have been increasing also. Each year from 1950 to 1970 the average yield of milk per dairy cow has improved. In 1950 it was 5,314 pounds, and in 1970 it was 9,158 pounds. The average annual production of butterfat (milkfat) also has risen from 210 pounds to 337 pounds per cow during the same 20-year period.

The milk and milkfat production per cow in Alaska and Hawaii ranked higher than the United States average in 1970. The average production of milk per cow in Alaska was 9,889 pounds with 356 pounds of milkfat; in Hawaii it was 10,391 pounds wtih 353 pounds of milkfat. In a recent survey of the fifty states, comparison figures show that only in California, Arizona, New Jersey, and Washington did the average milk and milkfat production per cow exceed that of the production in Hawaii.

THE RETAIL COST OF MILK

Studies by the United States Department of Agriculture and the dairy industry indicate that if 50 cents were paid for a half gallon of milk the money received would be distributed as follows: the farmer, 23.5 cents; the retailer, 7.5 cents; the wholesaler, 7.7 cents; processor and packager, 7.6 cents; procurement of milk from the farm producer, 0.2 cent; dealer-processors or administrative expenses, 2.1 cents; and assembling of milk at central plants for processing and distribution, 1.4 cents.

The studies cited above indicate the following distribution of the 26.5 cents going to all middlemen combined: labor, 10.5 cents; containers and supplies, 3.2 cents; buildings, equipment, and other overhead, 3.3 cents; advertising and promotion, 1.7 cents; transportation, 2.4 cents; miscellaneous, 3.2 cents; and profits before income tax, 2.2 cents.

According to these studies, the farmer gets 17.7 cents when milk that is worth 50 cents if sold retail is sold instead for use in other dairy products such as butter, cheese, ice cream, or dried milk.

SMALL-SCALE DAIRY FARMING

Some excellent information on this topic may be found in *Raising Livestock on Small Farms,* Farmers Bulletin 2224, United States Department of Agriculture, 1966. Part of that information is summarized in the following paragraphs.

A grade cow, well fed and well cared for, produces enough milk to more than pay for her feed, even if all the feed is purchased. The cow should produce 3,000 to 6,000 quarts of milk per year. This is enough for a family of two adults and three children.

Any savings in the dairy farmer's family food bills should be charged against the cost of buying and feeding a cow. A large family may want a cow to provide an abundance of milk and milk products.

A cow will eat 20 to 25 pounds of hay a day, or 3 to 4 tons a year, if no pasture is available. In addition, the cow will need 1 to 2 tons of a concentrate grain mix. The market price of hay varies greatly over the nation, but a

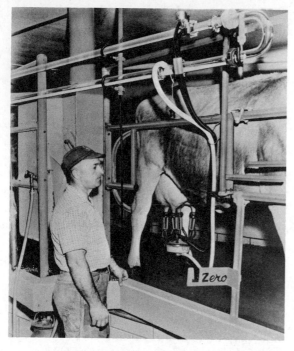

A pipeline milking system is used on many modern dairy farms. (*Courtesy* Zero Manufacturing Company)

range of $20 to $60 per ton may be typical. Depending on the protein content, the cost of concentrates may vary from $30 to $80 per ton. In one year a cow needs 800 to 1,600 pounds of straw for bedding. The average cost of feeding and bedding a cow is about $100 to $300 a year.

The cost of keeping a cow can be reduced if part or all of the feed can be grown on the farm. Two acres of good land usually will provide feed, mainly pasture, for 6 months of the year. This cuts feed costs almost in half.

Buying a cow. It is best to select a cow from one of the principal dairy breeds: Ayrshire, Brown Swiss, Guernsey, Holstein-Friesian, or Jersey. Jerseys and

Guernseys are often used for family cows because they do not require as much feed as the larger breeds. Jersey and Guernsey milk is higher in butterfat than that of some other breeds, although total milk production may be lower.

A 4 or 5 year old cow that has had two or three calves is young enough to have years of production ahead, and old enough to have shown her ability to produce milk. A good family cow is sound and healthy, easy to milk, and gentle without bad habits.

When buying a dairy cow, examine the udder to determine that it has no lumps or hardened tissue and that the teats are a good size for convenient milking. A large udder does not necessarily mean high milk production; avoid large, meaty udders that do not shrink when empty. It is desirable to see the cow milked by hand, or to milk her yourself a few times. The milk should not contain clots, flakes, strings, or blood. Draw several streams of the first milk from each teat on a close-woven black cloth stretched over a tin cup to determine whether the milk contains any undesirable material.

Do not buy a cow that kicks, or one that wears a yoke, muzzle, or nosepiece. Such devices indicate that the cow has bad habits such as breaking through fences or self-sucking.

A family cow must be free from tuberculosis, brucellosis, and leptospirosis. These diseases can be transmitted to people. Make sure that a veterinarian has tested the cow for these diseases within 30 days of the time of sale.

Summer feeding. Approximately 2 acres of pasture is required to provide grazing for each cow. Bluegrass or mixtures of grass tend to drop in production in summer and may have to be supplemented to provide a uniform feed supply. In most of the northern half of the United States, a mixture of alfalfa and ladino clover with grasses produces well during the summer. The alfalfa and clover must be reseeded every 3 to 5 years.

Sudangrass, crosses of sudangrass and sorghum, and bromegrass make excellent summer pasture in the North. A half acre of this temporary pasture planted next to permanent pasture provides grazing, and the excess may be cut and thrown into the permanent pasture for feed. CAUTION: Do not allow cows to eat sudangrass during its early growth or its regrowth after drought or frost. Sudangrass in these stages may cause prussic acid poisoning.

Sudangrass should not be grazed until it is 18 inches high or cut for hay until 2 feet tall. Tall yellowish-green sudangrass is relatively safe, but short dark-green sudangrass is likely to be dangerous.

Coastal Bermudagrass, pearl millet, carpetgrass, dallisgrass, and lespedeza make good summer pasture in the South but do not grow early in the spring. In this region part of the pasture should be planted to crimson clover or small grains, such as oats, rye, barley, or wheat in the late summer or early fall. This will provide some forage for winter and late spring grazing.

A vegetable garden can furnish a little summer feed. Cows will eat pea vines, sweetcorn stalks, cabbage leaves, and sweet potato vines.

Winter feeding. Hay and a mixture of concentrates are good winter feed for the family cow. Alfalfa, soybean, alsike clover, or early-cut grass hay are satisfactory. A Jersey or Guernsey cow needs at least 10 pounds of hay a day, and a pound of grain for each 2 to 4 pounds of milk she produces.

A good concentrate to feed with hay is a mixture of ground corn and wheat bran. Some soybean oil meal or linseed oil meal may be added to the diet of grain and hay for extra protein. Reliable ready-mixed feed can be purchased. Sixty-four pounds of concentrate furnish approximately as much nutritive value as 100 pounds of hay.

Cows should have a block of trace mineralized salt in a sheltered box. Loose salt may be added to the concentrate mix at the rate of one pound to every 100 pounds of feed.

Housing and care of cows. A cow needs a sunny, comfortable shelter or stable. She may be left untied in a box stall about 10 feet square, or confined to a smaller space and held by a stanchion, chain, rope, or strap. A cow needs about three times as much bedding in a box stall when confined in a smaller space.

If a cow is confined by a stanchion, there should be a manger in front extending beyond the stall. There should be a gutter for droppings behind the cow, with 4 or 5 feet of space to make it easy for the cow to get into the stall and to facilitate removal of manure. Feeding is easier when there is enough space in front of the manger.

The sides of the stable should be constructed to prevent drafts in cold weather. A box stall affords good protection, and it can be open on the south except in very cold weather. An arrangement that permits the sun to shine into the box stall in winter adds to the cow's comfort. A stall that is entirely closed should be ventilated by a tilting window on the side opposite the prevailing winter winds.

A cow should always be handled gently and quietly. Fences should be well constructed so the cow will not develop a habit of breaking through. A fence of four barbed wires tightly stretched and fastened to good posts is generally satisfactory.

Regular grooming is especially important for a cow confined in a stall. Daily brushing is recommended to keep manure from caking on the cow's flanks and thighs.

To keep the dairy cows comfortable and the milk sanitary, pesticides must be used on the cows and in the dairy barn. However, some pesticides may get into the milk. The following recommendations on the use of pesticides on cows and in the dairy barn have been made by Auburn University, Auburn, Alabama, 1970:

1. Use only recommended pesticides. Do not use malathion or methoxychlor sprays on milking cows.
2. Do not use pesticides in amounts greater than those recommended.
3. Keep a record of the use of all pesticides in barns, on pastures, and on forage crops.
4. Observe waiting periods after pesticides have been used. Do not graze or harvest crops until waiting periods have expired.
5. Read all labels and follow the manufacturers instructions carefully.
6. Use good santiation practices. Do not contaminate food, feed, forage, water, or milking utensils.
7. When in doubt, consult your county agent.

Cows are usually milked twice a day. Before milking, inspect the udder and flanks to be sure they are free of dirt that might fall into the milk pail. It may be necessary to wash the udder and flank, and it is always a good practice to wipe them with a clean damp cloth before milking. The milker's hands should be clean and dry.

The county agricultural agent can provide information about breeding service in the local area. Artificial insemination may be available. A cow should usually have a calf at 12-month intervals. It is desirable that a cow be dry for a month or 6 weeks before the calf is born. Cows can be made to go dry by reducing their feed and gradually discontinuing milking.

Care of milk. Immediately after the milk is drawn it should be strained through a clean cloth. Single-use cloths are best. Cloths to be reused should be washed and boiled after each use. All raw milk should be pasteurized by heating it to 142° F. and holding it at that temperature for 30 minutes. Milk can also be pasteurized by heating it to 161° F. and maintaining that temperature for 15 seconds. Small electric home pasteurizers cost about $40. A local appliance dealer may provide information about the current price of this item and the source from which it can be obtained.

After milk is pasteurized it should be cooled as rapidly as possible to 50° F. or lower. Milk should be kept in a refrigerator until needed. Surplus whole milk may be held for butter making. If the butter is to be churned, the milk should be kept in a deep container until the cream rises to the top. Skim milk may be used for cooking or for making cottage cheese. Skim milk is also a welcome addition to the menu for a pig.

Milk utensils should be rinsed in cold water immediately after use. Then they should be washed in hot water containing a dry washing powder or detergent and scrubbed with a brush. Rinse utensils with hot water, then scald them with boiling water. Sanitized utensils should be stored uncovered in a clean, airy place. Seamless milk pails and other utensils are best because they do not have crevices in which milk can lodge.

MILK AS A NEAR-PERFECT FOOD

One reason for the importance of dairying is the high nutritive value of dairy products. Milk is one of the best sources of calcium, the mineral so essential for good bones and teeth. High-quality milk also contains considerable phosphorus and iron. The protein content of milk is particularly important because several of the essential amino acid proteins are abundant. In addition, milk is a good source of vitamins A and D and a fair source of vitamin B_1. A child should consume a quart of milk a day, while the diet of an adult should include a pint of milk daily. Milk is a lifetime requirement.

During the past 35 years the food habits of American families have gradually changed. There has been a decrease in the consumption of potatoes, cereals, and other starchy foods. At the same time there has been an increase in the consumption of milk and cheese, meats, eggs, fruits, and vegetables. But about one-third of the people in this country still do not drink milk.

As more and more people under-stand the nutritive value of milk products and desire their benefits, dairying will become even more important in American agriculture. The continuing rapid growth in population also favors the expansion of the dairy industry.

DAIRYING AS A LARGE-SCALE FARM ENTERPRISE

Almost half of the dairy cows in the nation are in the North-Central states. The South-Central states and the North-Atlantic states rank next. Dairying also is an important farm enterprise in the Western states and South-Atlantic states.

The states with the greatest numbers of dairy cows are Wisconsin, Minnesota, New York, California, Pennsylvania, Michigan, Ohio, Illinois, and Kentucky. Wisconsin, Minnesota, and New York probably are better known for dairying than for any other type of farming, although each of these states has a diversified agriculture.

The principal dairy products are whole milk, cream, butter, cheese, and ice cream. Evaporated milk and condensed milk are two other products that are found in every grocery store.

Improvements in refrigeration and transportation have made fresh dairy products available to almost every community in the United States. The demand for whole milk is greatest in and near the large cities. Nearness to large population centers is somewhat less important in the processing and marketing of cheese, butter, evaporated milk, and condensed milk.

The steady demand for quality dairy products makes dairying a stable

About 70 percent of all dairy cows in the United States are of the Holstein-Friesian breed, as are the cows shown here eating silage. (*Courtesy* Holstein-Friesian Association of America)

farm enterprise. The dairy farmer is busy throughout the year, and his income is fairly regular from month to month. Dairying is a good enterprise in a diversified type of farming that includes the growing of grain crops, grasses, and legumes in rotation. Dairy cows use large quantities of forage efficiently and at the same time help to maintain soil fertility.

Dairy products help the farm family to have a balanced diet at relatively low cost. An adequate supply of high-quality fresh milk is particularly important to a large family, and many part-time farmers keep cows to provide milk for home use. The dairy industry would benefit from more home use of milk by farm families in every region.

Dairying requires a great deal of labor and also considerable money for the purchase of breeding stock, equipment, buildings, and feed. Dairy cattle are susceptible to several diseases and nutritional disorders. In addition to these problems, certain substitutes tend to reduce the market prices for dairy products. All of these factors should be considered along with the advantages of dairying as a farm enterprise.

BREEDS OF DAIRY CATTLE

All breeds of dairy cattle were introduced into the United States from other countries. About 70 percent of all dairy cattle in the United States are made up of these six breeds: Jersey, Guernsey, Holstein-Friesian, Ayrshire, Brown Swiss, and Red Danish. The Milking Shorthorn, Red Poll, and Devon are dual-purpose breeds that combine some of the characteristics of both dairy and beef animals. The dual-purpose breeds originated in England.

Jersey. The Jersey Island in the English Channel is the native land of this breed. Jerseys were brought to the United States a little more than a century ago. The smallest of the dairy breeds, the Jersey is noted for the high fat content of its milk.

In color the Jerseys vary from light fawn to black. The color may be solid or mixed with white patches. The tongue and the switch of the tail are black or white, and the black muzzle has a light-colored ring around it.

Jerseys are good grazers and efficient users of forages and concentrates. Since only a moderate amount of feed is needed for body maintenance, Jerseys produce milk economically. Although they are inclined to be nervous, Jerseys are very gentle when properly handled. The bulls are often vicious and should always be handled carefully.

Guernsey. A native of the Guernsey Island off the coast of France, Guernseys are well-known as good producers of high-quality milk. This breed is widely distributed throughout the United States.

Guernsey colors vary from light fawn to almost red. White markings occur on the face, flank, legs, and switch, and sometimes on the body. The skin is yellow; the nose is cream, buff, or smoky in color.

Guernseys are larger than Jerseys but smaller than Holsteins, and their feed requirements are intermediate.

They are active and good grazers. Because of their good dispositions, Guernsey cows are easily managed.

Holstein-Friesian. Holland is the native land of the Holsteins. The first Holsteins were brought to the American colonies more than 300 years ago, and today they are the most popular dairy breed.

Holsteins are noted for the production of large quantities of milk. Holstein milk is excellent for the manufacture of

Left: Spreading manure daily is a practice that pays in a cleaner farm environment, cleaner cows, cleaner milk, and more productive soils. Only when the soil is frozen and impervious should manure not be spread. Coarse crop residues and manure help to reduce the frequency with which the soil freezes like concrete. (*Courtesy* New Idea Machinery Company) *Right:* Modern methods for growing alfalfa for hay in arid parts of California include the use of the best variety, fertilization based upon a soil test, and irrigation. The sprinkler irrigation system is turned on when the soil reaches a certain dryness, as indicated by these tensiometers (center foreground). (*Courtesy* the Irrometer Company)

cheese. Although the fat content of the milk is generally low, the large yield of milk results in a high total production of butterfat.

Because of their large size and their high production of milk, Holsteins require large quantities of forage and other feeds. Holsteins are good grazers, particularly on good pastures.

Holsteins are black and white, with considerable variation in the comparative amounts of these colors. Only black-and-white Holsteins are eligible for registration as purebreds. Solid colors are not acceptable. The legs must be white below the knees or hocks, and the switch must be white.

Holstein cows have a quiet disposition. Some of the bulls are vicious and hard to handle.

Ayrshire. The Ayrshire breed originated in Ayr County, Scotland. These cattle are red and white. Either color may be the more conspicuous, with

A New England Ayrshire dairy herd coming to the barn to be milked. (*Courtesy* New Hampshire Extension Service)

markings of the other color. However, the red may vary from very light to very dark.

The Ayrshires are larger than Guernseys but smaller than Holsteins. A nervous disposition is characteristic of this breed. The Ayrshires are very active and make excellent use of pasture forage.

The milk production of Ayrshires is somewhat lower than that of Holsteins, and the butterfat content of the milk is slightly higher than in Holstein milk. Ayrshire calves make good veal.

Brown Swiss. This breed originated in Switzerland. Brown Swiss cattle were first imported into the United States in 1869, although much of the improvement of the breed has been accomplished in the past two or three decades.

The Brown Swiss is light fawn to almost black. The muzzle is light in color, and there is a light stripe along the backbone. The horn tips, nose, tongue, and switch are black.

Brown Swiss cattle are large. The calves are good vealers, and cows and steers are easily fattened for beef.

The Brown Swiss cattle have a good disposition. They are very active and rugged—two characteristics that make them good grazers.

The butterfat content of Brown Swiss milk is comparable to that of Holstein milk. The better Brown Swiss cows compare favorably with Jerseys and Guernseys in milk production.

Red Danish. The United States Department of Agriculture imported a few Red Danish cattle from Denmark in 1933. Much of the experimental work with this breed has been done in Michigan. The American Red Danish Cattle Association has members in many states, and the breed probably will become more popular as it becomes better known among dairymen.

DAIRY BREED ASSOCIATIONS

To obtain more information on any of the six principal dairy breeds, write to the following dairy breed associations.

The Ayrshire Breeders Association, Brandon, Vt. 05733

The Brown Swiss Cattle Breeders Association, Beloit, Wis. 53511

The American Guernsey Cattle Club, Peterborough, N.H. 03458

The Holstein-Friesian Association of America, Brattleboro, Vt. 05301

The American Jersey Cattle Club, 1521 East Broad St., Columbus, Ohio 43216

The American Red Danish Cattle Association, Rte. 3, Marlette, Mich. 48453

IMPROVING THE DAIRY HERD

The selection of good foundation stock is one of the first steps in successful dairying. Usually it is best to choose a breed that is popular in the local community so that breeding stock will be available.

The beginning dairyman must decide whether to start with purebred or grade cattle. Many people prefer to start with grade cattle. If purebreds are purchased the buyer should get the pedigree records that show the ancestry of the animals and the breed association with which they are registered. Calves are not eligible for registration unless their parents are registered.

Before buying foundation stock a dairyman should check the production records of the herd from which selections are to be made. If the owner of the herd is a member of the Dairy Herd Improvement Association, reliable records are available. Many dairymen who do not belong to an association keep accurate records on the production and ancestry of individual cows and herds. The reputation of the seller is very important to the buyer of foundation stock.

All foundation stock should have the desired characteristics of the dairy type and breed. The animals should be healthy and without physical defects that would impair their value or usefulness. The buyer should find out whether the cattle have been tested for and vaccinated against certain diseases.

If the dairyman plans to raise heifers to increase his herd or replace older and less productive cows, the selection of herd sires is particularly important. Many dairymen start with grade cows and purebred bulls. Nearly all progressive dairymen now belong to an artificial breeding association.

By selecting the best heifers as replacements and continuing to use purebred sires, the dairyman can upgrade the herd until all of the animals have characteristics that approach the standards of purebreds. The mating of purebred cows and bulls of the same breed is *pure-breeding*.

The maintenance of a productive dairy herd requires frequent replacement of cows for reasons of age, disease, injury, or poor production. The dairyman may raise heifers as replacements, or he may prefer to sell all the calves and buy heifers or cows when replacements are needed. Experience is the best guide in determining which practice to follow. Whether the replacements are

Above: **A registered Jersey herd on improved pasture in Vermont.** (*Courtesy* Soil Conservation Service) *Below:* **The parts of a dairy cow.** (*Source:* Clifford Burton, "Judging Dairy Cattle," Oklahoma State University *Circular* C-698, January 1971)

raised or purchased, they should have good ancestry and desirable dairy characteristics.

A productive herd is a healthy herd. Only disease-free foundation stock should be purchased, and diseased animals should be removed from the herd as soon as possible. Bovine tuberculosis, an infectious disease that was once a constant threat to the dairy industry, has been practically eliminated in most communities as a result of organized testing programs. All counties in the United States are now in a modified accredited area, which means that these counties are practically free of bovine tuberculosis. Every herd should be retested periodically to prevent a recurrence of the disease.

Brucellosis (Bang's disease) is an infectious disease that lowers milk production and greatly reduces the calf crop. Resistance to this disease can be increased by vaccinating calves between the ages of six and eight months. The county agricultural agent, vocational agriculture teacher, or a veterinarian can provide information on vaccination and the federal-state cooperative program for the prevention of Bang's disease. As in the control of other livestock diseases, sanitation is one of the best preventive practices.

FEEDING DAIRY CATTLE

Nutritious pasture forage, hay, and silage are the most economical sources of nutrients for dairy cattle.

Forage from well-managed pastures and meadows supplies a little of each of the essential minerals. Thirteen mineral elements are necessary for the health,

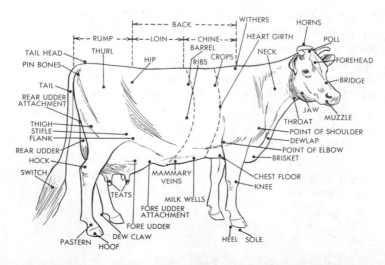

growth, and reproduction of all animals. These minerals are: calcium, chlorine, cobalt, copper, iodine, iron, magnesium, manganese, phosphorus, potassium, sodium, sulfur, and zinc. The essential minerals are a vital factor in both human and livestock nutrition.

Even the best forages do not contain enough sodium and chlorine. Salt is fed to correct this deficiency.

Some forages are deficient in calcium and phosphorus, the bonebuilding minerals. Bone meal or a commercial mineral mixture may be fed to supply additional calcium and phosphorus. Forage grown on fertile soil contains more calcium, phosphorus, and other essential minerals than does forage grown on poor soils. This is one of the many reasons why soil conservation practices are important on dairy farms.

A deficiency of iodine is a common characteristic of forages produced in the humid northern part of the United States. Iodized salt is fed to provide additional iodine. Northern forages generally are deficient in cobalt. In cobalt-deficient areas a cobalt salt, usually cobalt sulfate, is added to the mineral supplement or dairy feed.

Dairy cows must have forages high in carotene to produce milk rich in vitamin A. Green grasses and legumes from fertile pastures and meadows provide all the carotene that dairy cows need.

A Jersey cow will eat 30 pounds of dry forage daily, while a Holstein cow will eat 35 pounds if plenty of forage is available. Cows need four times as much green forage as dry. This means that a Jersey cow needs 120 pounds of green forage and a Holstein 140 pounds. These figures are based upon a 20 percent mois-

ture content of cured hay. Dairy cows on full feed will consume 30 to 35 pounds of silage and about 15 pounds of cured hay daily.

For the most economical milk production, at least 80 percent of the nutrients required by dairy cows should come from green forage, hay, and silage. On most farms a greater use of forage would reduce feeding costs and increase production. High-producing herds need grains and protein concentrates to supplement forages.

The best feeding program, however, is the one that is most profitable. Cows fed only green legumes, grasses, hay, and silage may produce only about 70 percent as much milk as they are capable of producing, but the feeding of concentrates does not always increase milk production enough to pay for the extra cost of feed. A great deal depends upon local market prices of feeds and milk products.

Alfalfa and other legume hays generally have a higher protein content than do grass hays, but cattle sometimes bloat when allowed to graze green legumes. Keeping dry hay available for cattle on legume pasture seems to help prevent bloat. Grass-legume pastures are less conducive to bloat, and the mixture is also excellent for hay.

On some dairy farms the green forage is cut and chopped with a forage harvester and hauled to the dairy barn instead of allowing the cows to graze on pastures. This is known as "zero grazing." On commercial dairy farms near large cities, especially in the Northeast, the West Coast, and Hawaii, the dairy cows are sometimes fed in dry lots and never allowed to graze. But in most dairy regions the feeding program includes pasture forage, hay, silage, and some grain or protein concentrates.

Dairy cattle feed must be high in protein. This is an expensive ingredient. A few years ago it was shown through research that a certain percentage of the protein in the ration of a dairy cow could be provided by substituting cheaper synthetic urea. This finding has helped to reduce the cost of high-quality feed for dairy cattle.

It is possible to save about $18 per cow per year in protein supplement costs by adding 10 pounds of urea to each ton of corn silage in a silo. More money can be saved by adding up to 15 pounds of urea to each ton of dairy feed concentrate. Milk production is likely to drop if more than that amount of urea

Records of milk production per cow as well as feeding and breeding records are necessary for success in operating a dairy. (*Courtesy Great Lakes Steel Corporation*)

is added to the concentrate while urea-treated silage is being fed.

THE MODERN DAIRY "FACTORY"

The traditional picture of contented dairy cows knee-deep in verdant grass-legume forage, alternately grazing and lying down, will soon be only a memory. No longer can a modern dairy farmer afford the luxury of cows ruining more standing forage than they eat or trampling the moist surface soil into hoof-pans that reduce intake of rainwater and cause more loss of water by surface run-off.

The modern dairy farm is run as a factory on an around-the-clock basis. All forage is cut and chopped in the field and immediately hauled to the cows. Some of it, along with corn or sorghum, is chopped and put into a modern silo and fed when dry weather or winter comes. Milking is done in a parlor where each cow in turn is permitted the luxury of eating concentrated feed out of spotless troughs and is milked with a gentle-action milking machine. Each cow is milked twice or sometimes three times a day. The milk from each cow goes through the milking machine directly into pipes that lead to a milk cooler. Two or three times a week a bulk tank milk truck drives up, pumps the cooled milk into the tank and delivers the milk to a bottling (cartoning) plant. Here it is usually pasteurized and homogenized and put in cartons for the retail market where we buy the fresh milk. In all this processing the milk is untouched by human hands and does not risk contamination by exposure to dirt.

Is this a good way to get rich? Hardly, because it takes approximately one million dollars to get started in a dairy business big enough to compete successfully for the next 50 years. The one million dollars would be sufficient to buy 1,000 high-quality registered dairy cows, plus 1,000 acres of productive land on which to grow the forage, plus the necessary milking parlors, bulk cooling equipment, barns, milking machines, silos, and forage choppers.

"But I don't have a million dollars! And I don't want to milk 1,000 cows a day; one hundred cows are enough for me! Is there no hope?" Yes, if you dearly love cows and insist on working 12 to 16 hours a day and have $100,000 in cash or credit to buy 100 cows and other essentials and you are a good manager, you can net $6,000 to $10,000 a year when times are good. But only the rich can afford the luxury of poor management.

QUESTIONS

1 What have been the trends in dairy farming?
2 Why is milk considered to be an excellent food?
3 How have the food habits of the average American changed over the years?
4 Name the six principal dairy breeds.
5 What is the most popular breed of dairy cattle?
6 How would you select foundation stock for starting a dairy herd?
7 Name the mineral elements essential for animals.
8 Why is it important to dairy farming for synthetic urea to be used in mixed feeds?
9 How much green forage will a Jersey cow eat each day?
10 What is "zero grazing"?

ACTIVITIES

1 Obtain samples of whole milk, preferably from different sources. (Small glass bottles with screw caps are good containers.) Shake the samples so that the cream is mixed with the rest of the milk. After several hours examine each sample to see how the cream has separated from other materials in the milk. Note any differences in the amounts of cream in the samples.
2 Select a breed of dairy cattle that you like, and do additional reading to obtain information about the origin, characteristics, and importance of the breed. Make notes so that you can share your information with other mem-

bers of the class. (Ask your teacher about writing to the breed association of the breed that you have chosen).

3 Visit a local dairy farm to observe the breed of cattle, buildings and equipment, hay and pasture crops, grains and other feeds used, care and marketing of milk products, records kept, and the methods used to apply manure to the fields.

4 Make a simple form for keeping daily records of feed costs, milk produced, cost of labor, and receipts from sale of products. Have a class discussion on the importance of records on a dairy farm.

5 Prepare simple score cards and use them in learning to judge dairy cattle.

13 BEEF CATTLE

There were 91.1 million head of beef cattle, with a total value of $20.2 billion, on the nation's farms and ranches in 1970. The record number of beef cattle in the United States was 107.2 million in 1965. The demand for beef continues strong, and the outlook for the beef cattle industry seems to indicate continuing growth.

Beef cattle are adapted to the vast open ranges of the arid West, as well as to the improved pastures of the humid East. The beef animals have the remarkable ability to walk as much as two miles daily for water and salt and to thrive on the range with no protection from the weather. Beef cows produce rugged calves that often find and need no shelter other than sunny valleys, clumps of trees, or warm southern slopes.

Beef cattle will do well with very little care. But like all other livestock, beef animals are healthier and make more money for their owner if they are properly managed.

The raising of beef calves is a very popular activity for people of all ages, but especially for boys and girls. Few experiences provide more satisfaction for a boy or girl than that of starting with a little bawling beef calf and raising it to maturity. Feeding, watering, grooming, and exercising calves for the possibility of winning a blue ribbon at a livestock show is a challenge that has attracted thousands of members of 4-H Clubs, Future Farmers of America, and other boys and girls.

Beef cattle have been raised on our farms and ranches since the first boatload of cattle arrived on the Atlantic coast soon after 1603. Beef animals were a necessity for pioneer families throughout the period of settlement.

BEEF CONSUMPTION PER PERSON

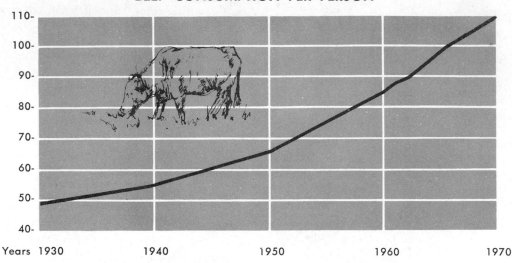

Years 1930   1940   1950   1960   1970

Beef consumption per person has been increasing since 1930 and is expected to continue upward. (*Courtesy* Allied Mills, Inc.)

Beef cattle are now grown on approximately four million farms and ranches in the United States. Since 1900 the consumption of beef in this country has varied from 60 to 80 pounds per person per year and continues to increase. People eat more beef during prosperous periods and less during depressions. This means that people buy all the beef they can afford, and in preference to other meat.

The center of beef-cattle production is in the West-Central states. In their order of beef production, the other regions rank as follows: South-Central, Western, East-North-Central, South-Atlantic, and North-Atlantic.

BREEDS OF BEEF CATTLE

The selection of a breed of beef cattle is largely a matter of personal prefer-

ence. As in the selection of a breed of any kind of livestock, the choice of a breed that is popular in the local community simplifies the problems of obtaining breeding stock and of marketing the animals.

A good beef animal is almost rectangular in shape. The legs and neck are short, giving the animal a compact appearance. The body should be deep, the hindquarter well developed, and the back straight and smooth.

The main beef breeds are: Hereford, Polled Hereford, Shorthorn, Polled Shorthorn, Aberdeen-Angus, Brahman, Galloway, Santa Gertrudis, and Charolais.

Hereford. England is the native land of the Herefords. This breed was introduced into the United States about 1840 and has had a great influence on the beef-cattle industry throughout the nation. Herefords are red with white faces. The throat, chest, underside of the

Left: A descendant of one of the early beef cattle, the historic Texas Longhorn. (*Courtesy* U. S. Fish and Wildlife Service) *Right:* Hereford beef cows produce rugged calves, with no shelter in many parts of our nation. *Courtesy* American Hereford Association)

body, and switch are white. The Herefords are horned.

Polled Hereford. The Polled Herefords are hornless, as the first part of their name indicates. These cattle were developed into a distinct breed as a result of careful selection and breeding which began in Iowa about 1850. The foundation animals were purebred Herefords that had been born without horns. Except for their hornless characteristic, the Polled Herefords look much like the other Herefords.

Shorthorn. This, the largest of the beef breeds, also originated in England. The parent stock had long horns, but many years of careful selection and breeding produced cattle with much shorter horns.

Shorthorns vary in color from red to white, and individual animals may have almost any combination of these

This Aberdeen-Angus cow has all the characteristics of a good beef type. (*Courtesy* American Aberdeen-Angus Breeders' Association)

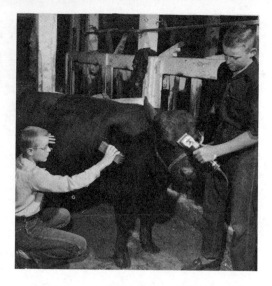

These boys are dreaming of winning a blue ribbon at the fair. (*Courtesy* National Committee on Boys' and Girls' Club Work, Inc.)

colors. The roan color that is characteristic of many Shorthorns results from a mixture of red and white hairs.

Polled Shorthorn. These cattle were developed from naturally hornless Shorthorns. Many Polled Shorthorns that have descended from purebred Shorthorns can be registered as purebreds of both breeds. The appearance of the Polled Shorthorns is essentially the same as that of the standard breed.

Aberdeen-Angus. This breed apparently originated in northern Scotland in the highlands of Aberdeenshire. Aberdeen-Angus cattle were first brought to the United States in 1873 by George Grant of Victoria, Kansas.

The breed is widely distributed and is particularly popular among cattle feeders in the Corn Belt. In recent years there has been a great increase in the

Young Hereford cattle on an improved pasture in Georgia. (*Courtesy* Soil Conservation Service)

number of Angus cattle in the South. These cattle are black and do not have horns. The Angus breed excels in making a carcass with the greatest percentage of edible meat of all other beef cattle.

Brahman. Brahman cattle originated in India. They are the oldest breed of domestic cattle in the world.

The large hump over the shoulders is a unique characteristic of the Brahmans. The hump is neither bone nor muscle: it is gristle. The loose skin on the lower part of the neck is another characteristic of the breed. Various shades of gray are the common colors, although some Brahmans are red. Their resistance to heat, flies, ticks, and mosquitoes makes the Brahmans an outstanding breed for the South.

The Brahmans are used extensively for crossing with Herefords, Angus, and Shorthorns. The crossbred calves make fast-growing and excellent beef. Where crossbreds are raised, Brahman bulls usually are kept in herds of cows of other breeds.

Galloway. Scotland is the native country of the Galloway. These cattle have been raised in the United States for almost a century.

The smallest of the beef breeds, the Galloway is noted for its hardiness in cold weather. Galloway cattle are polled (hornless), black, and have long curly hair. The Galloways have short legs and very compact bodies.

Santa Gertrudis. This breed was developed by crossing Brahmans on other beef breeds, particularly Shorthorns. The strain of Santa Gertrudis developed at the King Ranch in southern Texas is the one that is best known. This strain is a Brahman-Shorthorn cross. The Santa Gertrudis cattle are solid dark red, compact in form, and produce high-quality

Left: **A good specimen of Brahman.** (*Courtesy* American Brahman Breeders' Association) *Right:* **Santa Gertrudis cattle were developed in the semiarid region of South Texas where summer temperatures are very high.** (*Courtesy* Texas Extension Service)

beef. Like the Brahmans, the Santa Gertrudis cattle have great resistance to heat.

Charolais. This breed of beef cattle is becoming very popular in the United States.

Selective breeding of beef animals in France around the eighteenth century developed a big-scale, fast-growing breed that we know today as the Charolais.

The modern Charolais is white or light cream in color. Physically, Charolais are massive, one of the largest breeds of cattle. Purebred bulls weigh from 2,000 to 2,500 pounds or more at maturity. The cows average from 1,250 to 1,500 pounds. The general structure of the Charolais is broad and muscular, giving an appearance of extreme stoutness and ruggedness.

The Charolais breed was developed in France and has become popular in the United States. (*Courtesy* American-International Charolais Association)

The breed is noted for its fast rate of gain and growth; and in the feedlot, for economy of feed conversion. Charolais gain flesh without waste fat and have a very high percentage of tender red meat. They are excellent for crossbreeding with other breeds, and resulting calves have been found to average from 10 to 20 percent heavier than straightbred animals.

MANAGING THE COW HERD

Beef cattle are best adapted to farms that have abundant summer and winter pasture and enough good cropland to produce an adequate supply of forages and other feeds. The cropping system should include corn and legumes if beef cattle are fattened on the farm.

The breeding herd can be wintered entirely on high-quality hay. Corn fodder, oat straw, and ground corncobs are good roughages. The feeding of a little grain is beneficial to cows and calves when the calves are born during the winter.

Sometimes it is advisable to give cows a legume hay or protein concentrate late in the summer when the grass is maturing. Corn silage, grass silage, or stock carrots may be included in the ration of breeding cows when they are not on pasture or when pasture forage is scarce. Every ration should also include some leafy legume hay that has been properly cured. The daily vitamin A requirements of a beef cow can be supplied by approximately 5 pounds of green-colored leafy hay.

A good winter ration for a 1,000-pound beef cow that is to have a calf in late winter or early spring could include 30 pounds of corn silage or sorgo silage, 5 pounds of alfalfa hay, and all the straw the cow will eat. Another good ration includes 40 pounds of sugar-beet pulp, 10 pounds of corn stover or grain hay, and 5 pounds of alfalfa hay.

Beef cattle need plenty of clean, fresh water; and salt should always be available. The average consumption of salt is about 2 pounds per head per month. Calves need a little less than that amount, while mature beef cows and heavy steers require a little more.

A mineral supplement should be provided if cattle are fed mainly on crops produced on soil that is known to be deficient in calcium, phosphorus, or other essential minerals. A mixture of 5 parts by weight of finely ground limestone, 5 parts of sterilized bone meal, and 1 part of salt is usually a satisfactory mineral supplement. Mother cows and cows that are to have calves in a few months sometimes need more minerals than are supplied by the regular feeds. Cows of the dual-purpose breeds (such as Shorthorns) may need extra minerals if they are being fed for heavy milk production.

A farmer may start the breeding herd with purebred cows and a purebred bull, or the foundation stock may be grade animals. The initial investment for breeding stock can be reduced by starting with grade cows and gradually improving the quality of the herd by using purebred sires.

On many farms a few of the best heifer calves are kept as replacements. Bulls usually are purchased from owners of other herds, because it is best to select bulls that are not related to the cows and heifers. The mating of closely related animals, a practice known as *in-breeding*, may perpetuate certain good characteristics, but it also tends to make the weaknesses more conspicuous. Farmers sometimes buy purebred bull calves and raise them to maturity as replacements for older herd sires.

CARE AND MANAGEMENT OF BEEF CALVES

Beef calves generally need little attention when they are with their mothers on good pasture. A few days after birth, however, it is often best to take calves away from their mothers if the cows are to be milked. The newborn calf should have its own mother's milk for 4 or 5 days, after which the calf may drink milk from a bucket or be placed with a cow that has a young calf. The

Branding of cattle is necessary on ranges where there are no fences. Note the brand of "A" on the flank of the cow at the lower right corner of the photograph. (*Courtesy* Charolais Association)

An example of international assistance in improving the breeding, feeding, and care of beef cattle. The man on the left is a member of the Masai tribe in eastern Africa. The Masai depend entirely on beef cattle for a living. (*Courtesy* Food and Agriculture Organization of the United Nations)

practice of having two calves suckled by one cow is known as *double-nursing*.

A calf needs 3 or 4 pounds of whole milk daily for the first day or two after it is weaned. Occasionally a calf will refuse to drink from a bucket; when this happens, the milk should be taken away for several hours. When the calf gets hungry it will be more likely to accept the milk.

Skim milk may be substituted for whole milk after the calf has been weaned about two weeks. Because skim milk is very low in vitamin A, the daily feeding of 2 teaspoonfuls of cod-liver oil or other fish oil is beneficial to the calf.

The use of an oil rich in vitamin A is not necessary after the calf begins to eat hay, silage, or grass. But to be a satisfactory source of vitamin A, the forage must be of high quality and have some green color. If the hay is green in

color, 2 or 3 pounds will provide enough vitamin A for one day.

Calves should have access to good pasture as soon as possible. If pasture is not available when the calves are a month old, a growing crop may be cut and fed to them. A small quantity of silage or carrots may be fed as a substitute for other forage until pasture is available.

Calves restricted to a dry lot (no pasture) should have alfalfa, clover, lespedeza, or other green-colored leafy legume hay. A little freshly cut grass or grain hay of good quality and about one-fourth of a pound of a protein concentrate may be satisfactory if legume hay is not available.

On some farms the beef calves are allowed to stay with their mothers on pasture. This system requires excellent forage to keep the cows and calves in good condition. The calves may be given extra feed in a pen that has an opening large enough to admit the calves but small enough to keep out larger cattle. This system is known as *creep feeding*.

FEEDING CATTLE FOR MARKET

The fattening of cattle for market is a common practice on farms where both beef cattle and corn or grain sorghum are raised. Cattle feeding is a major farm enterprise in the Corn Belt.

In regions where corn and other grain is not available, the cattle may be sent from the ranges to market or to a region in which cattle feeding is a special enterprise. Many cattle from the Southwest go to feed lots in the Corn Belt and in the Northern Great Plains.

The length of the feeding period may vary from two to 12 months, depending on market prices and on the age and condition of the cattle. Steers and other cattle that are more than two years old may be finished quickly if they are in good condition at the beginning of heavy feeding. Calves and other cattle less than two years old may attain a satisfactory degree of finish in four to nine months. Sometimes it is profitable to feed calves nine months or longer.

Feeder cattle should be given a moderate amount of grain at the begin-ning of the fattening period. If full feeding is desired, the amount of grain can be increased gradually until the cattle are getting all they will eat.

Some farmers prefer to use less grain and more high-quality forage over a longer feeding period. Experiments have proven that feeding costs can be reduced by the proper use of pasture and forage crops. Beef animals should have all the water and salt they want while being fattened for market.

QUESTIONS

1 Name the three regions that rank highest in beef-cattle production. Where does your region rank?

2 Briefly describe the pleasure of raising a beef calf.

3 Tell about the beef breed you like best.

4 Name the breeds of beef cattle that may be red.

5 What is meant by the word "polled"?

6 Approximately how much salt is required each month by the average beef animal?

7 What are the ingredients in a satisfactory mineral supplement?

8 What is "inbreeding"? Is it recommended?

9 When skim milk is fed to calves, what should be added to the milk to supply vitamin A?

10 What is meant by "creep feeding"?

ACTIVITIES

1 Study the cattle-market report in a daily newspaper every day for a week. Make notes on the market classes of cattle and the price range for each class. Note which market class commands the best price.

2 Compare pictures of beef cattle and dairy cattle, or if possible observe each type on local farms. Write a brief description of each type, explaining how the two types differ in body form and other characteristics.

3 Examine charts showing the principal cuts of beef. Explain why a knowledge of beef products is useful to a farmer. Find out how the different cuts of beef compare in price at local meat markets.

4 Select members of your class to do research and lead a discussion on new breeds of beef cattle that have recently been introduced into the United States. The principal ones are Limousin, Simental, Black Welsh, German Brown, German Yellow, Maine-Anjou, Murray Grey, and South Devon. What is the purpose in crossbreeding?

5 Have a class demonstration and practice in judging beef cattle. Give specific reasons for the order in which you rank the individual animals. If you have an opportunity to visit a livestock exhibit when beef cattle are being judged, observe the way the judge examines each animal.

14 SWINE

Hog production is a very old farm enterprise. In fact, hogs were being raised for food before agriculture was a well-established industry in North America. Some of the early Spanish explorers brought hogs into the southeastern part of the continent more than 400 years ago. The hogs were slaughtered when deer and other game could not be found or when the Spaniards could not obtain food from friendly Indians. Many of the colonists who came from England, France, and Spain also raised swine for pork to feed their families.

Hogs are raised on almost two-thirds of the farms and ranches in the United States. Farmers in the Corn Belt market more than half of their corn by feeding it to hogs.

Iowa leads all states in both corn and hog production. Illinois, Indiana and Missouri also rank high in hog production. Approximately 75 percent of the hogs marketed in the United States are raised in the Midwest.

Tennessee, Alabama, Mississippi, Louisiana, Georgia, and South Carolina are the leading hog-producing states in the South. Hog raising is a minor farm enterprise in most of the states in the East, Southwest, and West because of the relatively small supply of corn.

In 1970 the total number of hogs in the United States was 67.5 million, 13.4 million more than in 1964.

BREEDS OF HOGS

The breeds of hogs are divided into two main groups, the *lard* type and the *bacon* type.

In recent years the increased use of vegetable oils in cooking and the decline

193

Left: **Hogs are happiest when they have a pasture to graze. (Spotted Poland China breed)** (*Courtesy* U.S.D.A.) *Right:* **An ideal type Duroc boar.** (*Courtesy* United Duroc Record Association)

in the use of animal fats in soap have reduced the demand for lard. This trend has encouraged the production of bacon-type hogs and resulted in the marketing of lard-type hogs at an earlier age and lighter weight. Twenty-five years ago there was a strong demand for hogs weighing 300 pounds. Recently the packing industry has favored weights of 200 pounds.

The main lard-type breeds are the Duroc, Poland China, Spotted Poland China, Chester White, OIC, Berkshire, Hampshire, and Hereford. A large percentage of the hogs produced in the Corn Belt belong in this group. The two bacon-type breeds are the Yorkshire and Tamworth.

Duroc. The popularity of this breed is well deserved. The Durocs probably are more numerous than hogs of any other breed. The color variation of Durocs is from light red to dark red. The sows have large litters and produce an abundance of milk for their pigs. Because of their good dispositions the Durocs are easily managed.

Poland China. This breed was developed in Ohio. The Poland China is black with white feet and white face. The tip of the tail also is white. Well-developed hams are an important characteristic of the breed. Because of their large size and excellent form, the Poland Chinas are excellent for crossing with other breeds.

Spotted Poland China. This very popular breed differs from the Poland China mainly in its color, which is black and white. In size and form it is comparable to the Poland China. The ancestry of the breed also is similar to that of the Poland China. Spotted hogs were crossed with Black Poland Chinas and the Gloucester Old Spots, an English breed.

Chester White. The name of this breed suggests its white hair and skin and its place of origin. The breed was developed in Chester and Delaware counties, Pennsylvania, from three English breeds. Cheshire, Lincolnshire, and English Yorkshire hogs were the foundation stock from which the Chester White was developed. Chester White sows are noted for raising large litters of thrifty pigs.

OIC. Ohio is the native state of the OIC, originally known as the Ohio Improved Chesters. The OIC is much like the Chester White in form and color.

Berkshire. The Berkshire was imported from England and has always been considered one of the best lard-type breeds in the United States. The wide, dished face and short snout are identifying characteristics of the breed. The body is black, with white markings usually occurring on the head, feet, and tail. A very long body is an outstanding characteristic of the Berkshire.

Hampshire. Kentucky is the home state of the Hampshire. Belted hogs and Thin Rinds from New England were the ancestors of the new breed. The Hampshire is easily identified by the white belt that encircles the black body, including the front legs and usually most of the

shoulder area. The erect ears and black head and tail are other common characteristics. The hind legs are black, although some Hampshires have a small amount of white below the hock.

This is one of the smaller lard-type breeds. The Hampshire can be finished to the desired market weight without excessive fat. Because the Hampshires are good rustlers, they are excellent hogs to follow beef cattle that are being fattened for market. The Hampshires also are good grazers and make economical gains on nutritious pastures. Hampshire sows usually have large litters of pigs.

Hereford. The Hereford breed was developed in Missouri at the beginning of this century. Durocs, Chester Whites, and the OIC were the foundation stock. The color of the body is light to dark red. The head is white, and there are white markings on the switch of the tail, the feet, and the underline of the body. Herefords must be at least two-thirds

Left: **A Poland China sow of good type.** (Texas) (*Courtesy The Paris News*) *Right:* **A typical young Hampshire.** (*Courtesy* Hampshire Swine Registry)

red in order to be registered as pure-breds. In addition to having a common name, Hereford hogs and Hereford cattle are very similar in color.

Yorkshire. The long, deep body of the Yorkshire makes it an ideal bacon hog. The Yorkshire originated in England and was introduced into the United States more than a century ago. Hogs of this breed are white but there may be some small black spots on the skin. Large, erect ears are a breed characteristic.

Yorkshires are raised extensively in England, Scotland, and Canada. Most of the Yorkshires in the United States are in the North. The Yorkshire is noted for high-quality bacon and a high dressing percentage.

Tamworth. Much of the recent popularity of the Tamworth is due to the market preference for hogs that do not produce much lard. Light to dark

The Yorkshire breed has a long, deep body and represents the bacon type of hog. (*Courtesy* Encyclopedia Britannica Films, Inc.)

red is the usual color of the Tamworth. The long body is narrow in proportion to its length. The sides are smooth, the back is well arched, the ears are erect, and the snout is long. The Tamworth is a good hog to cross with lard-type breeds to reduce the amount of fat in the carcass of the offspring.

SMALL-SCALE HOG RAISING

Should a family on a small farm raise pigs for meat? It may be profitable to raise one or two pigs for the family meat supply if they can be fed chiefly on surplus garden produce and table scraps. However, if the feed must be purchased at retail prices, a home pork factory may not be a profitable enterprise. For town and suburban residents the question to "raise or not to raise pigs" is answered by a city ordinance; the answer is "No."

A helpful publication on raising hogs is *Raising Livestock on Small Farms,* Farmers Bulletin 2224, United States Department of Agriculture, 1966. Parts of the following paragraphs are adapted from that bulletin.

A good time to buy pigs is in the spring when they are being weaned. The buyer should be certain that the pigs have been raised on clean ground under a strict system of swine sanitation, and that they have been vaccinated for hog cholera. It is best to choose a female pig, or a male pig that has been castrated (a barrow). A male pig that has not been castrated produces meat with an undesirable odor and flavor.

Feeding and Housing. Commercial feed manufacturers have formulated hog rations that some hog scientists consider ideal. The main problem is to select the

Above left: There were 30 hogs in this class project in swine production at the Molokai High School, Molokai County, Hawaii. (*Courtesy* University of Hawaii) *Above right:* Swine of all ages need green grazing, sunshine, and plenty of grain and protein concentrate. (*Courtesy* United Duroc Record Association) *Below:* Antibiotics are sometimes added to hog feed to increase growth. On the left a small amount of antibiotic (Terramycin) added to the feed resulted in 20 percent faster growth as compared with the hogs on the right that received no antibiotic. All four pigs are litter mates and were fed identical rations except for the antibiotic. (*Courtesy* Pfizer International Inc., New York)

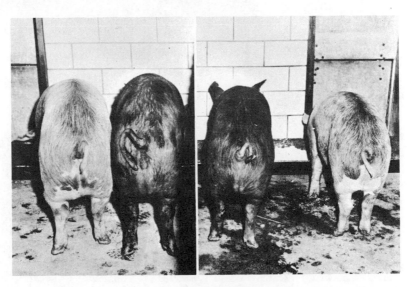

feed dealer and manufacturer with the best reputation among the successful hog raisers in your community. Several recent suggestions may be worth repeating: never try to save money on hog feed by buying the one with the lowest percentage of protein; and if you plan to feed hogs one of the bird-resistant varieties of grain sorghum, have it ground or cracked and use no more than 50 percent of this kind of mixture.

Antibiotics have been used to increase growth but are being banned as hog feed supplement because of hazards to people.

Hogs thrive on less feed if they have access to good pasture. In the northern half of the United States, good hog pastures may include alfalfa, ladino, red clover, alsike, white clover, bluegrass, burclover, timothy, or a combination of

these crops. Annual and biennial pasture plants may be killed if grazed too closely to allow adequate seed formation.

Bermudagrass, carpetgrass, lespedeza, and dallisgrass are used extensively for hog pasture in the South. In this region, as in parts of some of the other regions, temporary pasture may be provided by planting rye, oats, wheat, rape, soybeans, and cowpeas in or near the hog lot.

Without table scraps, garbage, or pasture, a hog will eat about 600 pounds of concentrate ration from the weaning age of 8 weeks until it weighs 200 pounds. A good ration contains grains, a protein supplement, and a mineral supplement. Corn is the standard grain for hogs. Yellow corn is better than

Hampshire hogs are good grazers and make rapid gains on pastures and grain stubble. (North Carolina) (*Courtesy* Hampshire Swine Registry)

white corn when hogs do not have pasture crops. Barley, wheat, sorghum grain, and hog millet may be used as hog feed.

A good mineral mixture should be kept in a self-feeder where it is available at all times. Equal parts of steamed bonemeal, ground limestone or airslaked lime, and common salt are a satisfactory mineral mixture. Hogs should have an ample supply of fresh drinking water at all times.

Hogs get along well with very simple shelter. The main requirement is that it provide protection from drafts, snow, and rain, and give shade in hot weather. A dry floor is essential in a hog shelter. The pen and shelter should be located at least 500 feet from any residence to prevent annoying odors. Strict sanitation also does much to accomplish this objective.

LARGE-SCALE, MODERN PORK PRODUCTION

At one time hog raising consisted of raising three to five hogs on table scraps, moldy feed, skim milk, and faith. Modern hogs are as good as ever at garbage disposal, but this is not the modern way to make a living in the hog business.

Now small pigs walk into one end of an air-conditioned swine "factory" and 5½ months later walk out the other end of the building into a truck that takes them to a market where they soon become first-quality pork. During this 5½ months the pigs never see the sun, never feel the pain of acute hunger, never root into the soft, sod-covered earth in search of tasty roots and worms,

are never allowed to wallow in mud, and are never chased by a dog or a boy.

Concrete slatted floors are clean, automatic waterers always work, and feed augers automatically move scientifically formulated feed to each pen of pigs. In this way the feed is available to each pig for 24 hours a day. Below the slatted floor is a manure pit where the manure is kept in a liquid and is aerated by a large paddle wheel to reduce odors. Once every six months to a year the manure is pumped out automatically. (Ask your State Agricultural College about the recommended spacing between slats in a slatted floor.)

The hogs are kept cool in summer by 24-inch fans, and by insulation in walls and ceiling. In winter the hogs are kept warm by gas heaters. Temperatures are kept between 70° and 75° F.

With such modern methods, one or two men can raise and sell 2,000 to 3,000 hogs a year.

This does not mean that all present-day hog raisers produce hogs in air-conditioned castles. Many farmers still use the following system:

1. Plant a grass-legume mixture on five to 10 acre plots.

2. Place A-shaped hog houses, perhaps 25 per plot, evenly over the area.

3. Into each plot turn 25 bred sows that are about ready to farrow.

4. Let each sow choose an A-shaped house (without much quarreling) to farrow in.

5. The sows eat a large amount of the grass-legume mixture (alfalfa-bromegrass is one of the most popular mixtures), and feed themselves some

scientifically prepared ground grain and protein from a self feeder.

6. When the baby pigs are 10 days old, a creep feeder is placed near them so that baby pig feed is always available. (A creep feeder is one from which a baby pig can eat but its mother cannot.)

7. When weaned, the pigs glean corn fields, and eat more grass-legume forage, ground grain, and protein until large enough to be slaughtered.

QUESTIONS

1 Is there an anti-hog ordinance in your neighborhood?
2 Name the state that ranks highest in both corn and hogs.
3 What are the two main types of hogs?
4 How can garbage be made safe for use as feed?
5 Why has there been a change in demand from lard-type to bacon-type breeds?
6 What breeds appear to be more popular?
7 What does OIC mean?
8 Traditionally, hogs like to wallow in mud. Why, therefore, is it necessary to provide clean, dry hog houses?
9 What can hogs get from nutritious grasses and legumes that is difficult and expensive to obtain from the feed trough?
10 What constitutes a good mineral supplement for hogs?

ACTIVITIES

1 Review the section in Chapter 3 that describes the farm enterprises in the Corn Belt. Then locate on a map of the United States the main centers of hog production.
2 Study the hog-market report in a daily newspaper every day for a week. Learn the main market classes of hogs. Note which class commands the best market price.
3 Use charts and pictures to learn the principal cuts of pork. Explain why a farmer should know the cuts of pork and the market prices of these cuts. Perhaps you or another member of the class can visit or call a local store to obtain information about the current price of ham, bacon, shoulder, salt pork, sausage, ribs, and pork loin.
4 Select the breed of hog you like best and read several references to obtain more information about this breed. Have a class discussion on the breeds of swine.

5 If possible, arrange to demonstrate and practice judging hogs in class. Be prepared to explain the reasons for your ranking of the individual hogs.

6 Visit the meat counters in several food stores to see what kinds of pork products are available. Compare the prices of bacon, sausage, pork chops, and ribs in the different stores.

15 SHEEP AND GOATS

Sheep raising has been an important farm enterprise since Colonial days. The early settlers in New England kept small flocks, and at one time that region ranked first in the United States in the production of sheep and wool. As new agricultural lands were developed in the West, there was a decline in sheep production in the Northeastern states.

In the eastern half of the United States, small farm flocks are kept to supplement other types of farming. The large Eastern cities are good markets for lambs, and farmers near these cities can market the lambs quickly and at considerable saving in transportation costs. Consumers in states north and east of Ohio buy about one-third of the lamb eaten in the United States, while consumers in states east of the Mississippi buy two-thirds of all the lamb produced in this country.

Sheep are raised throughout the Appalachian Mountain Range in Pennsylvania, Maryland, Virginia, West Virginia, Kentucky, Tennessee, and North Carolina. Farmers in this mountain region raise high-quality early lambs for the eastern markets.

Large areas of cheap land in the hilly sections of southern Missouri and northern Arkansas are suitable for sheep. The cost of raising sheep in these areas is relatively low. But predatory wild animals and dogs are a menace to flocks in many of these Ozark communities.

The cutover forest land in the Lake States provides good summer grazing for sheep. The principal disadvantage of this region is the high cost of feeding and sheltering the flocks during the long and severe winters. The mild climate of the South permits the grazing of cut-

In the eastern United States, sheep farming is usually combined with other types of farming. (Kentucky) (*Courtesy* Tennessee Valley Authority)

over land and other pastures all summer and during most of the winter. Because of this advantage, sheep production is becoming more important in the South, particularly in Mississippi and Louisiana.

Sheep raising is profitable in the Corn Belt, a region which has a large acreage of permanent pasture and forage crops. In addition to being a good source of farm income, sheep improve the use of labor and feeds on Corn Belt farms.

Large flocks of sheep are raised on rangelands in the West. On many ranches sheep and cattle are raised together to make maximum use of the range forage. The sheep eat several kinds of range plants that are not acceptable to cattle.

Sheep raising is combined with other enterprises on some irrigated farms in the West. Irrigated pastures provide much of the summer pasture, while al-

falfa and other irrigated crops are excellent winter feeds. The raising of sheep in irrigated sections reduces the cost of marketing grain and hay.

The number of sheep and lambs on farms in 1971 was 19.6 million, the lowest of any year during the past 20 years. One reason for this decline is the competition of synthetic fibers and wool.

RAISING SHEEP ON SMALL ACREAGE

Sheep raising does not require expensive equipment or heavy labor, but it does take sufficient pasture and excellent fences. This reminder is given in *Raising Livestock on Small Farms*, Farmers Bulletin 2224, United States Department of Agriculture, 1966. Additional information in the following paragraphs is adapted from that bulletin.

Above: Many western forests provide good range land for sheep. (*Courtesy* Weyerhaeuser Sales Company) *Below:* Sheep make an ideal farm enterprise on a small acreage for the youth of our nation. (*Courtesy The National Future Farmer*)

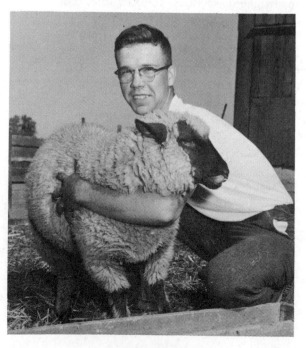

There should be enough pasture to feed the sheep in spring, summer, and fall, and still allow for frequent rotations to clean ground or for separation of lambs from ewes. Rotation helps protect sheep from internal parasites (stomach worms) that can cause serious losses. Two acres of good pasture will provide feed for three to eight ewes for the summer months.

All pastures should be enclosed with a dog-proof fence. Such a fence combines woven and barbed wire and is 56 to 58 inches high. It has a strand of tightly stretched barbed wire close to the ground. Above this is 36-inch woven-wire fencing with a 4-inch mesh, and above this are two strands of barbed wire.

The number of sheep that can be cared for properly on a limited acreage depends largely on how much well-fenced pasture can be provided.

Do not go into sheep raising with the idea that sheep require little or no attention. Their needs are varied, and their habits are different from those of other farm animals. Ewes and lambs may require special care at lambing time.

Selection of sheep. The beginner in sheep raising will probably find it best to choose one of the dual-purpose breeds that satisfactorily combines wool and lamb producing ability. The beginner may want to consult the county agricultural agent or State Extension Service livestock specialist for help in selecting a breed adapted to the locality.

Pure breeding or registry is not always synonymous with quality, merit, or productivity, and the beginner will probably decide that good-quality grade sheep will be satisfactory. One who plans to sell rams for breeding purposes will want to start with pure-bred sheep.

If it is possible to use a ram from a local breeding service, a farmer can start with only ewes. If such a service is not available, it may be best to buy a small flock of ewes and one ram.

Buy young ewes, and breed them to give birth to their first lambs at about two years of age. The gestation period is about 145 days. Ewes usually come in heat in the late summer or fall. It is good to breed them as early as possible so that lambs will be born in the winter months. Cold weather discourages parasites which may afflict lambs born in spring or summer.

Late summer or early fall is the best time to go into sheep raising. Desirable ewes are more likely to be available for purchase then, and they can be put on pasture or late forage for a while before breeding.

Feeding and care. Sheep on good pasture do not need grain. If sheep on pasture lose weight or show other signs of poor condition, it generally means that the pasture is inadequate and that some grain should be added to their diets.

In the winter when no pasture is available, sheep should be fed a ration of good quality legume hay, preferably alfalfa. Three to four pounds of alfalfa hay a day is sufficient for a ewe weighing less than 150 pounds. About a month before ewes are to lamb, supplement the hay with one half to three quarters of a pound of grain a day. Salt and fresh water should be kept available at all times.

Lambs should be taught to eat as soon as possible; they will begin to nibble at feed when 10 to 16 days of age. A small pen or creep can be used to hold the lambs' feeding troughs. Upright slats, about 3 feet high, placed 9 to 12 inches apart will let the lambs in and keep the ewes out.

Lambs can be fed ground grain from a creep as soon as they start to eat and until they are weaned or marketed at about 120 days of age or 70 to 90 pounds in weight.

Sheep need no special kind of housing. A closed barn or shed should be

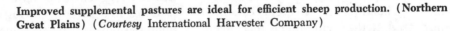

Improved supplemental pastures are ideal for efficient sheep production. (**Northern Great Plains**) (*Courtesy* International Harvester Company)

Orphan lambs enjoying a dry bed and direct warm sunshine. (*Courtesy* Libbey, Owens, Ford Glass Company)

available for bad weather and at lambing time. Small lambs and their mothers should be kept in a closed shelter to separate them from the rest of the flock.

Sheep are sheared in the spring or early summer, either before or after lambing. The owner of a small flock will probably find it best to get a custom shearer if one is available in the locality. If the owner wishes to do the shearing himself, hand-power shearers can be purchased. Other equipment that may be needed includes tools for castrating and docking (removal of lambs' tails, which is done when they are 7 to 14 days old) and scales for weighing lambs and fleeces.

Equipment need not be expensive. Probably the greatest expense to the beginning sheep raiser is the fencing necessary to keep out predatory dogs and to make pasture rotation possible.

Further information on raising sheep can be obtained from county agricul-
tural agents or the United States Department of Agriculture. Also, there are several good books on the subject that may be available in your public library.

RAISING SHEEP ON FARMS

Sheep are best adapted to land that is high and dry, although they may be raised successfully on almost any land that is not too wet. The fine-wool breeds have a strong preference for drier areas. Sheep raising is not a good enterprise in regions that have a combination of high temperatures and high rainfall.

Owing to recent progress in the control of internal parasites, sheep can be raised in sections once considered unsatisfactory.

Pasture and feed. Pasture and forage crops are the best and cheapest feed for sheep. In addition to the forage grasses and legumes, cereals are used as supplemental sheep pasture.

Grazing areas should be changed frequently to keep the sheep healthy. Rotation grazing of pastures helps to control diseases and parasites. This practice requires fences to divide permanent pastures into several grazing areas. Sections of movable fencing may be used to regulate grazing on small areas of forage crops.

Sheep do not need grain if good pasture is available. On many farms the breeding flock is kept in good condition and lambs are raised to market age without supplemental grain feeding. Sometimes lambs are fed grain so that they can be marketed earlier at better prices.

The winter feeding of ewes (mother sheep) may be profitable if the lambs are to be born before spring pasture is available. Keeping the ewes in good condition during the winter helps to produce a larger crop of strong, healthy lambs. Legume hays should make up a large part of the ewes' winter ration. After the lambs are born, grain may be fed to ewes and lambs to supplement available forage.

Grain crops sown in the fall provide pasture for sheep during late fall, winter, and early spring. Wheat, barley, and rye make good winter forage. The sheep are removed from the small-grain pasture in the spring so that a grain crop can be harvested in the summer. Corn silage and grass silage are good winter feeds for sheep, but hay should be available to ewes when silage is fed.

Care of ewes and lambs. When lambing occurs in cold weather, the ewes should be kept in barns or sheds. It is best to keep each ewe in a separate pen. A temporary pen may consist of two light panels connected by a hinge and placed in a corner. This permits the ewe to see the other sheep and keeps her from being nervous or excited.

The lambing pens should be in a well-ventilated building that is free from drafts and as warm as it can be made without artificial heat. A blanket thrown over the lambing pen helps to protect the ewe and lamb in cold weather. A lamb has a much better chance to survive if it is kept warm for several hours after it is born.

A strong, healthy lamb needs little care. A weak lamb should be wrapped in warm clothes or a sheepskin for about two hours. The lamb should be returned to its mother as soon as it is strong enough to nurse. If the lamb cannot nurse, bottle feeding may be necessary as a temporary measure. In this case,

the lamb should have a few teaspoonfuls of milk each hour until it is able to nurse.

Ewes seldom disown their lambs when lambing pens are used. When a ewe refuses to own her lamb the situation can sometimes be corrected by drawing milk from the ewe and rubbing the milk on her nose and on the lamb. Sometimes a heavy-milking ewe can be persuaded to adopt an orphan lamb or a twin lamb of a ewe that does not produce enough milk.

Lambs that are permanently orphaned can be raised by bottle feeding. If possible a newborn lamb should receive milk from a ewe that has a very young lamb. During its first two days the orphan lamb should have 1 ounce of milk every 2 hours. After that the lamb may be given whole cows' or goats' milk. The bottles should be sterilized, and the milk should be warmed to about 100° F.

The following rules for raising sheep were developed at Mississippi's Black Belt Substation:

1. Provide night corral for protection from dogs.
2. Always keep a mixture of 1 part phenothiazine and 9 parts salt before sheep to prevent stomach worms.
3. Drench with phenothiazine when parasites multiply.
4. Place rams with ewes by June 1.
5. Shear rams second time, July 1.
6. Remove rams from flock by Nov. 1.
7. Clip wool from ewes' udders by Nov. 1.
8. Keep ewe and young lamb in pen until lamb is strong.
9. Keep ewes with lambs separate from the other flocks.

10. Place ewes and lambs on pasture when lamb has enough strength.
11. Dock and castrate one day each week until all lambs are worked.
12. Shear ewes as early as possible, from about March 15 to April 1.
13. Creep-feed lambs only when grazing is insufficient.
14. Remove all lambs from ewes by June 1.
15. Keep ewes in drylot with hay and water until well dried up.
16. Sell lambs as early as ready; they should weigh above 60 pounds.
17. Carry over for fall market any lambs grading below medium.
18. Feed oats for 20 to 30 days.
19. Shear all lambs July 1.
20. For older lambs provide plenty of grazing and water all summer.

These recommendations are particularly applicable to farm flocks in the South. On most farms in this region, sheep raising is part of the general farming program rather than a highly specialized enterprise.

THE RANGE SHEEP INDUSTRY

The early Spanish explorers brought sheep into Mexico and the Southwest. As settlement progressed, native sheep were driven to rangelands throughout the West. Due to uncontrolled breeding, most of the herds were inferior in quality. The range-sheep industry was later improved by the introduction of improved types of sheep from the East.

Today more than 60 percent of the sheep in the United States are in the 11 Western states and Texas. The value of these range sheep is approximately 100 million dollars. Several times that amount of money is invested in land, equipment, feed, and labor. Because of the soils, climate, topography, and lack of transportation facilities, large areas of Western rangelands are better adapted to sheep raising than to any other enterprise.

On ranches where early-market lambs are the main objective, many of the lambs are born in sheds during the period from late January to early March. The ewes and lambs graze on the valley and foothill ranges until the lambs are marketed early in the summer. This practice gets the lambs off the ranges while the forage is still green. Lambs born in some of the valley sections in November and December are marketed in April or May. This is a common practice in parts of California and Arizona.

On many Western ranches the general practice is to have the lambs born on the open ranges, usually in May. After the ewes are sheared the flocks are moved toward the mountains. The lambs are finished on the range forage and marketed in September and October.

The raising of sheep to sell for breeding stock is recommended for sections that do not have adequate feed and forage crops for the production of fat lambs. Lambs for feeding and restocking can be raised successfully in Plains areas where water is scarce and the forage is mainly grass. Most of the lambs are born on the range in May and sent to feed lots in September and October. Some of the lambs are finished on Corn Belt farms. In the raising of sheep for breeding, the income from the sale of wool is somewhat more important than when fat lambs are the main product.

Left: More than 60 percent of the sheep in the nation are on western ranges where the soils, climate, and grass are ideal for their production. (*Courtesy* Encyclopaedia Britannica Films, Inc.) *Right:* Summer sheep range in the western mountains where lumber production, watershed protection, and range grazing are integrated uses of the land. (*Courtesy* Weyerhaeuser Sales Company)

The raising of purebred sheep is a highly specialized enterprise on some Western ranches. Purebred rams are raised and sold for the improvement of range flocks. This type of sheep raising requires a relatively large investment for breeding stock, equipment, feed, and labor.

The wintering of flocks is one of the most expensive phases of the range-sheep industry. Hundreds of thousands of sheep graze on winter ranges where forage is abundant and where the sheep can find shelter during storms. Some of the best winter ranges are the desert areas in Nevada, Utah, and southern Wyoming.

Many range flocks are wintered in feed lots. Alfalfa hay is one of the best winter feeds. Hay is hauled from stacks and fed on the ground. The feeding places should be changed frequently to maintain sanitary conditions. Natural windbreaks are the main shelter, and the sheep sometimes have no water except what they get from snow. Sheep that receive only dry feed should have drinking water. Cottonseed cake and other concentrates may be fed to supplement native hay. Grain is not usually fed with alfalfa or other high-quality legume hay.

BREEDS OF SHEEP

The breeds of sheep are classified according to the six types of wool: fine-wool type, medium-wool type, long-wool type, crossbred-wool type, carpet-wool type, and fur type.

Fine-wool breeds. This group includes American Merinos and Rambouillet. The fine-wool breeds are particularly well adapted to Western ranges. The Rambouillet has become more popular in recent years. Ewes of this breed are used extensively as foundation stock in the production of crossbred sheep. The Merinos and Rambouillet are very hardy.

Medium-wool breeds. The medium-wool breeds are well represented in farm flocks. These breeds include Shropshire, Hampshire, Southdown, Oxford, Suffolk, Dorset, Cheviot, Montdale, and Tunis.

Long-wool breeds. As their name implies, these breeds have long wool, and the wool is usually coarse and curly. The group includes Lincoln, Leicester, Cotswold, and Romney.

Crossbred-wool breeds. Each of these breeds was developed from two or more breeds. For example, the Columbia was developed by crossing the Lincoln rams and Rambouillet ewes; and the Corriedale was developed in New Zealand from foundation stock consisting of Lincoln and Leicester rams and Merino ewes. Other breeds are Panama, Romeldale, and Targhee.

Carpet-wool breeds. The long, coarse wool produced by this type of sheep is used in the manufacture of carpets and rugs. One breed of Scottish origin, the Black-faced Highland, is of minor importance in the United States.

Fur sheep. A few Karakul sheep are produced in this country. The breed originated in Russia. The lamb pelts are used for furs, but the wool of mature sheep is inferior in quality. Unlike most of the breeds in the other types, the fur sheep do not produce good mutton.

GOATS

More Angora goats are raised in Texas than in all of the other states combined. Much of the open-brush range in the public domain and the national forests is grazed by goats.

In the Northwest, particularly in the Willamette Valley in Oregon, goats have had a special role in the clearing of land. Some fine orchards have been developed since the numerous herds of goats cleared the brush. Large areas of cutover land in Washington and Oregon are suitable for permanent goat range. Other areas can be grazed by goats a few years and later developed for orchards and farm crops.

Goats are well adapted to the brushland in the Ozarks in southern Missouri. More goats could be raised with winter feeding to supplement the native vegetation. In almost every state goats could be raised for clearing and controlling brush, but this enterprise requires winter feed and shelter.

The mohair which is clipped from Angora goats is used to make car upholstering, rugs, braids, robes, and artificial furs. Some mohair of superior quality is used in the linings of men's suits. Most of the mohair required by American manufacturers is now produced in the United States. Annual production of mohair usually exceeds 20 million pounds. The Edwards Plateau in Texas is the nation's leading mohair-producing section.

In 1969 the number of goats clipped for mohair totaled 3.2 million, the lowest since 1956. Since 1963 the average clip has been 6.6 pounds per goat. In 1969 the total of 21.2 million pounds of mohair was the lowest annual production in ten years.

DAIRY GOATS

The small family may find it more convenient and economical to milk one or two goats than to buy milk or keep a dairy cow. Goat's milk can often be tolerated by infants and invalids who are allergic to other milk. This information and that in the following paragraphs of this section are adapted from *Raising Livestock on Small Farms*, Farmers Bulletin 2224, United States Department of Agriculture, 1966.

A good dairy goat produces at least 2 quarts of milk daily for 8 to 10 months of the year, and can be fed for about one-sixth of the cost of feeding a cow.

A dairy goat costs $35 to $75, depending on her breed and production record. Buyers should be sure that any goats purchased are from a tuberculosis-free and brucellosis-free herd.

Feeding. Goats producing milk should have all the clover, alfalfa, or mixed hay they will eat. Root crops such as turnips, carrots, beets, or parsnips are good feeds for goats. Good-quality silage can be substituted for root crops.

For a milking goat not on pasture, a good winter ration would be: good alfalfa or clover hay, 2 pounds; root crops or silage, 1½ pounds; concentrates, 1 or 2 pounds. Concentrate mixtures should consist of oats, bran, and linseed oil or another protein supplement. Goats on pasture need slightly less grain and fewer concentrates.

Above: Goats thrive on rocky brushlands where a cow would starve. (Pennsylvania) (*Courtesy* Soil Conservation Service) *Below:* These twin goats need milk now, but soon they will browse on leaves, twigs, weeds, and grass. (*Courtesy* Encyclopaedia Britannica Films, Inc.)

Has anyone seen the boss lately? It's milking time. (New Mexico) (*Courtesy* Soil Conservation Service)

Pregnant does should have all the roughage they will eat in fall or early winter. They should also be fed 1 pound of root crops or silage, and ½ or 1 pound of the same grain mixture fed to goats producing milk.

Strongly flavored feeds, such as turnips and silage, should be fed after milking so the milk will not be affected by off-flavors.

Rock salt should be kept before goats at all times, and occasionally a small quantity of fine salt should be added to the grain mixture. No other minerals are necessary if legume hay is fed. Calcium and phosphorus supplements will be needed if nonlegume hay is used. Goats should have plenty of fresh water at all times.

Care of dairy goats. Although goats do not need any special kind of housing, they should be protected from rain, snow, and cold. Goats are natural climbers and, unless tethered, will climb on low buildings and machinery around the farmstead.

Cleanliness is essential in handling and feeding dairy goats. Does kept in sanitary surroundings do not have an objectionable odor. Bucks, the principal offenders against cleanliness, need not be kept if a breeding service is available in the community.

Milking. A milking stand built with a stanchion at one end and a seat for the milker at one side is a real convenience in milking a goat. Such a stand can be constructed at little cost.

Young does usually object to being milked at first. A stanchion and stand help confine them, and a little grain in the box attached to the stanchion helps quiet them. After being milked a few times, the young doe becomes accustomed to milking and will jump on the stand and put her head in the stanchion without assistance. Twice-a-day milking is usually often enough for grade does. Heavy-producing does may need to be milked three times a day for a short time after freshening.

The usual breeding season for does is between September and January. After this time they usually cannot be bred until late in August. They come in heat regularly about every 21 days and stay in heat 1 to 2 days. Gestation averages 149 days. Does usually give birth to two kids, but occasionally there may be three or even four offspring.

If the family needs the does' milk and the farm produces only limited green feed, it may be best to dispose of the young goats. However, they are not hard to raise and can be bottle fed until they learn to drink by themselves.

A good goat will give milk for 8 to 10 months after freshening. One that gives milk for less than 6 months should not be kept.

The United States is helping the developing countries to improve their sheep. Four hundred Rambouillet sheep were donated to India to start a research station to improve fine-wool quality. The United Nations Special Fund supplied the money and the FAO the technicians. (*Courtesy* Food and Agriculture Organization of the United Nations)

QUESTIONS

1 Why is sheep raising important in the South?

2 Why does rotation grazing help to prevent diseases?

3 How would you treat sheep for stomach worms?

4 Where are most of the sheep in the United States raised?

5 Where are some of the best winter ranges for sheep?

6 Name a breed of fine-wool sheep.

7 How does sheep raising on a small acreage differ from sheep raising on a farm?

8 Where are most Angora goats raised?

9 What is mohair used for?

10 For what special purpose is goat milk used?

ACTIVITIES

1 Use encyclopedias, agriculture and geography textbooks, and other references to obtain additional information about sheep, goats, wool, and mohair. For further information consult the national registry associations for sheep and goats (see Appendix A).

2 If any member of the class has owned a pet lamb, have a report and discussion on the care of orphan lambs.

3 Compare samples of wool and cotton. Note the color, length, and texture of the fibers. Compare samples of wool cloth, cotton cloth, and synthetic fabrics.

4 If sheep are raised in your community, obtain as much information as you can about this enterprise. If sheep are not raised locally, explain some of the possible reasons why they are not.

5 Write an essay on this subject: "The Goat is the Only Happy Animal in India" (or in any location not ideal for most other livestock).

16 POULTRY

Chickens are the kind of poultry best known to most of us, but ducks, geese, turkeys, pigeons, and guinea fowls are poultry also. We shall consider all of these groups in this chapter.

Poultry farming was once a sideline business on almost every farm, especially on small family farms. Broiler and egg production are now specialized commercial farm enterprises. Both major kinds of poultry farming have been stimulated by labor-saving equipment and practices, improved feeding and management, and contract production. Under the contract arrangement, a processor, feed dealer, or commercial hatchery usually supplies much of the credit and assumes most of the price risks. Poultry farms in the South produce two-thirds of the broilers marketed. Egg production is widely distributed throughout the nation.

CHICKENS

In 1971 there were 443 million chickens on farms, the greatest number since 1951. The total value of all chickens in 1971 was 537 million dollars.

For many years the average egg production per hen in the United States has increased each year. In 1950 it was 174 eggs per hen, and in 1969 it was 220 eggs. This was an increase of more than 26 percent in 15 years.

Many helpful suggestions on chicken raising are given in *Raising Livestock on Small Farms*, Farmers Bulletin 2224, United States Department of Agriculture, 1966.

Most small farms can provide facilities for a flock of chickens. The beginning poultryman should start with about 50 chicks. This number allows for some

Above: Chicken production in the Hawaiian Islands is concentrated in West Oahu County, not far from Honolulu. This is a typical flock in the 3 million dollar egg and meat poultry business. (*Courtesy* University of Hawaii)

Below: By the year 1980, poultry probably will increase on a percentage basis faster than any other farm animals. (*Source:* Leo Figurski and Kenneth McReynolds, "Dairying on a Business Basis," Kansas State University *Circular* C-399, September 1969)

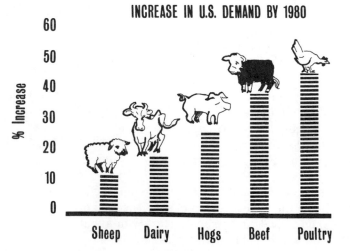

INCREASE IN U.S. DEMAND BY 1980

losses from disease or accident and will supply enough eggs and some meat for a small family. A larger flock may become a source of a little income if there is a demand for fresh eggs.

Several breeds of chickens are suitable for small-farm flocks. Egg-production breeds, such as White Leghorns, are popular on farms where eggs are the main poultry product. If the farm flock is to supply both eggs and meat, a general-purpose breed may be better. Rhode Island Reds, New Hampshires, and Plymouth Rocks are good for meat as well as egg production. Crosses of these breeds also are satisfactory. Chicks of good quality are usually available from local hatcheries.

A flock may be started by buying day-old chicks. Very young chicks require much care and must be kept in a heated brooder house. A beginning poultry raiser may prefer to buy older chicks that are well started, or pullets almost ready to lay.

It is extremely important to buy only from reputable hatcheries or breeders. The buyer should be sure that the chickens have been tested for and are free from pullorum and typhoid diseases. Most hatcheries now have chicks separated by sex, so it is possible to choose mostly pullets for egg production or to include cockerels to be used for meat.

Feeding baby chicks. Newly hatched chicks should be put in the brooder house and immediately provided with a starting mash in chick feeders. There should be plenty of fresh water in drinking fountains. Finely cracked corn can be fed instead of starting mash during the first two days after hatching.

Mash is then usually fed as the entire diet until the chicks are 4 to 6 weeks old, except for some fine grit that can be mixed with the mash or fed separately. Each bird should have at least 1 inch of space at feeders and one-half inch at drinking fountains. The feeding and drinking space must be increased as the chicks grow. Mash and water should be available at all times.

Feeding older chicks. When chicks are 6 to 8 weeks old, the starting mash should be replaced with an all-mash growing diet or a combination of growing mash and grain. Small amounts of grain should be added at first and the proportion of grain increased gradually until the birds are getting equal parts of mash and grain at about 15 weeks of age. Grit must be fed when the diet contains whole grain.

Grain is usually cheaper than mash, but it contains less protein and vitamin value than mash. This difference is not very important; as the birds grow older they have less need for protein and vitamins. It is now possible to produce a pound of chicken with about 2 pounds of feed.

Corn, wheat, oats, and barley are satisfactory grains for poultry. Most poultrymen now use commercially prepared feeds. The quality of these feeds is carefully controlled. It is important to follow the manufacturer's directions exactly in feeding these formulas.

Farmers who want to mix their own feed from home grown grain can get feed formulas and directions for mixing from the county agent or State Extension Service office. These formulas have been tested, and it is very important to follow directions carefully in using them. It is especially important to follow directions in adding minute quantities of vitamins and other additives to large

Egg production per hen has increased until now the average hen lays an egg more often than every second day. (Ohio) (*Courtesy* Libbey, Owens, Ford Glass Company)

Chickens are especially responsive to good feeding practices. Both chickens are 6 weeks old. *Left:* No vitamin B-12, weight ½ pound. *Right:* Received vitamin B-12, weight 1¾ pounds. (*Courtesy* The Squibb Institute)

batches of feed. These additives must be stirred into the feed sufficiently to distribute them evenly throughout.

Chicks can be put on range by the time they are 6 weeks old if space is available and weather is favorable. Range gives the chicks exercise and sunshine. Movable shelters, feeders, and drinking fountains are sometimes used to keep chickens on clean ground.

Egg-production breeds usually start to lay at 20 to 24 weeks of age; general purpose breeds lay at 22 to 26 weeks. About 2 weeks before pullets are expected to start laying, the growing mash should be gradually replaced with an all-mash laying diet or laying mash with grain.

Feeding layers. Laying mash or a mixture of laying mash and grain should make up most of the diet of laying hens.

Pullets enjoy a clean, dry poultry house. A lot of sunshine is obtained here with the use of large panes of insulated glass. (Pennsylvania) (*Courtesy* Libbey, Owens, Ford Glass Company)

When grain is fed, grit and oyster shell should be added to supply the necessary calcium for normal eggshells. If mash alone is fed, the manufacturer's directions should be checked to see whether limestone or oystershell should be added. Mashes usually contain sufficient calcium.

Feed is the main cost in egg production. Laying hens of the light breeds eat an average of 85 to 90 pounds of feed a year. Heavier all-purpose hens eat 95 to 115 pounds.

Hens from flocks that have been developed for egg production should produce 200 to 240 eggs each per year. Some commercial flocks maintain a higher average.

Care of eggs. Laying hens which are properly fed and cared for produce high-quality eggs for family use or for the market. Eggs should be gathered from the nests twice daily. The eggs should then be cleaned and cooled. Eggs should be held at a temperature between 45° and 55° F.

Undersized or thin-shelled eggs should be removed from those to be marketed. The size and shell of eggs affect price. Eggs of uniform size have the best appearance.

Brooder houses and equipment. Day-old chicks require a well-built, draft-free brooder house that contains 1½ square feet of floor space for every two chicks. The brooder stove may be heated by coal, oil, gas, or electricity. Electric brooders are satisfactory and are less of a fire risk than other brooders. Some poultrymen use homemade brooders.

For additional or more specific information on raising chickens, consult your county agent or write to your state agricultural college or the United States Department of Agriculture, Washington, D.C. 20250 (see Appendix A).

TURKEYS

In 1970 there were 6.7 million turkeys in the United States, and their total value was 36.8 million dollars. That was the highest total value since 1967. The record number of turkeys was 7.8 million in 1967.

Turkeys can be raised on small farms if proper care and equipment is provided. Young turkeys must be kept warm and dry. Turkeys should not be allowed to run with chickens and should not be put in buildings that have housed chickens in the past 3 months. Young turkeys should not be kept with older ones. Land used for chickens or turkeys should not be used as range for another flock until at least 3 years have passed. These precautions are necessary to keep turkeys from getting blackhead or other serious diseases.

Two common market classes of turkeys are mature roasters and fryer-roasters. The latter are also known as broilers. Small mature roasting turkeys of both sexes are ready for market at 22 to 24 weeks of age; large, at 24 to 28 weeks. Large hens can often be marketed at 20 weeks. Small white turkeys make excellent fryer-roasters at 16 weeks. Large white females make satisfactory fryer-roasters when marketed at about 13 weeks.

Turkey raising on a small scale may start with the purchase of about 100 day-old poults from breeding flocks tested for and free from pullorum, typhoid, typhimurium, and sinusitis. The

Checking the weight of turkeys to determine if they are making satisfactory gains. (*Courtesy* W. T. Spanton)

poults should be fed and watered as soon as possible after they are brought home.

Feeds and feeding. Very small poults need a starting mash containing 28 percent protein. After their first 8 weeks, the poults should be fed a growing mash with grain. The mash may be loose or pelleted and should have a protein content of 20 to 22 percent. The little turkeys should have free choice of both grain and mash. Feeds may be commercial, or they may be home mixed according to formulas recommended by state agricultural colleges or the United States Department of Agriculture.

Any common grain or combination of grains may be used with a growing mash. Corn should be cracked until the turkeys are 16 weeks old. Confined turkeys which are not given supplementary green feed should have a well-balanced growing mash. When green feed and direct sunlight are available, turkeys may grow well on a less expensive mash without vitamin supplements.

Houses and range. Poults require a well-built, artificially heated brooder house until they are 8 weeks old. There should be 1 to 1½ square feet of floor space per bird.

Sand is best for litter during the first 2 weeks, after which wheat straw or splinter-free shavings may be added. Instead of using litter, the little turkeys may be started on a floor with narrow slats three-fourths of an inch apart. Another type of floor which is satisfactory is one covered with No. 2½ hardware cloth nailed to removable frames.

If older poults and adult turkeys are kept in confinement, they require a well-ventilated building with a dry floor and tight roof. All openings should be screened with heavy wire to keep out small birds and predators. The floor should be littered with straw, hay, or splinter-free shavings. The litter should be replaced as required for sanitation. Until market age and beyond, debeaked turkeys require about 5 square feet of floor space per bird. If not debeaked they need 7 to 8 square feet of floor space each.

Turkeys raised on range should be moved to clean ground every 2 to 4 weeks. When the weather is mild, poults can be started on range when they are about 8 weeks old. Under less favorable conditions, they should not be put on range until they are 10 to 12 weeks old.

A grass or legume pasture that is well drained and fenced is a satisfactory range. Shade and roosts should be available on the range. A portable range shelter on skids generally is needed. The shelter should have wire walls which are strong and close enough to keep out such predatory animals as dogs, foxes, skunks, weasels, and raccoons. The shelter door should be latched each night after all birds are inside.

DUCKS AND GEESE

Ducks and geese offer opportunities under special conditions.

Ducks. Successful duck raising usually requires a running stream to provide a sanitary supply of water.

Each year approximately 10 million ducks are raised in the United States. States ranking high in duck production are New York, California, Illinois, Massachusetts, Michigan, Pennsylvania, and Wisconsin.

The most popular breed for market is the Pekin. Most ducks are raised for the specialty trade in the larger cities where "roast duck" is a delicacy.

Geese. Geese are extremely hardy. They require almost no housing and very little extra feed if an ample supply of green grass or clover is available. Geese are actually poultry, but they graze almost like sheep.

Each year more than a million geese are raised in the nation. The leading states are Minnesota, Iowa, Wisconsin, Illinois, and South Dakota. The most popular breeds are Emden and Toulouse.

Artificial ponds or natural bodies of water are desirable, but water is not as essential for geese as for ducks.

In some places, such as in young strawberry beds, geese are good "hoe hands." They will keep the grass cleaned out but will not bother the strawberries.

The larger cities provide good markets for the geese, where they are cooked like chicken.

SQUABS

Squabs are young pigeons 25 to 30 days old. Squabs for the family table or market can be raised successfully on small farms not suited to chicken raising.

Anyone planning to market squabs should first investigate the local market. Squabs usually bring good prices, but the demand for them is more limited than the demand for chickens and eggs.

Pigeons can be raised in simple, inexpensive houses or in an unused part of a barn or shed. Adult birds feed their young on a substance called pigeon's milk, which is produced in the adult bird's crop.

Each pair of breeders will produce 10 to 14 squabs in a year.

Breeds recommended for producing early-maturing squabs of high market value are King, Carneau, Mondaine, and Giant Homer. Squabs of these breeds should weigh 14 to 24 ounces, a desirable weight and size for an individual serving.

THE GUINEA FOWL

There are approximately one million guineas on farms in the United States. Guineas are "watch dogs" for the farm, since they are easily excited and let out a harsh cry when a hawk, owl, fox, dog, or strange person enters the farmyard.

Left: Geese have kept the weeds out of this strawberry patch. (*Courtesy* New Hampshire Extension Service) *Right:* Successful duck raising requires a running stream. (*Courtesy* National Committee on Boys' and Girls' Work, Inc.)

Guineas are also good in destroying insects such as grasshoppers. And since they do not scratch, as do chickens, they are not destructive in the garden, which makes them easy to keep on small farms.

The main source of income from guineas is in the sale of young ones to be eaten like young chicken. The meat is dark and has a flavor similar to that of wild game.

QUESTIONS

1 Why has the average American hen become more efficient?
2 What breed of chicken do you like best? Why?
3 Why should hens be supplied with grit and ground limestone?
4 Why is a running stream necessary for successful duck raising?
5 Contrast the housing needs of geese and of chickens.
6 What states rank high in turkey production?
7 Describe the essentials of modern turkey raising.
8 Can a person make a profit raising squab?
9 How does the management of turkeys differ from the management of pigeons?
10 Why are guineas not destructive to gardens?

ACTIVITIES

1 Examine an eggshell and observe whether it is brittle or very flexible. Crush a dry eggshell thoroughly. Pulverize some oystershell in the same way. Explain why laying hens need oystershell or other sources of lime.
2 Use bulletins and other references for additional reading on the breeds of poultry. Note which breeds are best for egg production, which are best for meat, and which are considered dual-purpose breeds.
3 Plan a balanced ration for laying hens, using local feed crops as much as possible. Indicate the kinds of grasses and legumes that could be used for a poultry pasture.
4 Arrange a class demonstration and practice in culling hens. If this is not possible, have a class discussion on the purpose and methods of culling.
5 Collect information from a local poultry producer on the disposal of poultry manure. Is it dumped in a gully, applied on a garden or field crops, or given to anyone who will haul it away? How much is poultry manure worth per ton as a fertilizer?

17 SMALL ANIMALS

In this chapter we shall consider the raising of rabbits, fur-bearing animals, and bees. Most boys and girls at some time want to raise some kind of small animal. Many adults are tempted by advertisements describing the possibilities in fur farming. Most of us are fascinated by the social life of bees.

RABBITS

Anyone who is interested in raising rabbits can find helpful information in *Raising Livestock on Small Farms,* Farmers Bulletin 2224, United States Department of Agriculture, 1966. This section of the chapter is adapted from that publication.

Domestic rabbits can often be raised by boys or girls on small farms. Rabbits can be raised in towns if there are no community regulations that restrict or prohibit such an enterprise. Only a modest cash outlay for stock, housing, and equipment is required.

Rabbit meat is all-white, fine grained, and high in protein. Only about 20 percent of the dressed carcass is bone. This is a high meat yield compared with that of many meat animals.

In addition to providing some of the home meat supply, rabbits can be sold for meat or as laboratory animals if there is a local market. Rabbit skins have some market value, especially white rabbit skins that can be dyed any color.

Selecting stock. Rabbit raising may begin with the purchase of young rabbits just weaned, or with a few animals ready for breeding. Young rabbits cost less, but there is a waiting period of five or six months until the does are ready to breed.

Field mice like this one (called meadow vole) are now helping Michigan State University scientists and others who are waging war on world hunger problems. The meadow vole (*Microtus pennsylvanicus*) is used in evaluation of small quantities of high-protein food grains that are being developed by plant breeders in Canada, the United States, and Mexico. (*Courtesy* Fred Elliott, Michigan State University)

To be successful in rabbit raising requires a love of the animals and attention to many timely details. (*Courtesy* Julian Donahue)

Medium and heavy breeds of rabbits are best suited for home and commercial production of meat. Some of the popular breeds are New Zealand, American, Bevern, Champagne d'Argent, Chinchilla, and Flemish Giants.

Stock should be purchased only from a reliable dealer who will guarantee the rabbits to be healthy and productive.

The gestation period of rabbits is only 31 days. A good doe usually raises 6 to 8 young in a litter, and it is possible for her to produce four or five litters a year. A doe can be bred when the young rabbits are 5 or 6 weeks old.

Young rabbits of the medium breeds should weigh about 4 pounds when weaned at 2 months. At this age and weight, they are ready to eat or market. On this basis, a good doe can be expected to produce more than 100 pounds of marketable rabbits each year.

Feeding practices. Does suckling young should have a ration which is 20 percent protein supplement, 39.5 percent grain, 40 percent roughage, and 0.5 percent salt.

A good ration for dry does, bucks, and young rabbits might contain 8 percent supplement, 31.5 percent grain, 60 percent roughage, and 0.5 percent salt. Dry does, bucks, and young rabbits may be maintained on alfalfa hay alone, or hay plus a few ounces of grain daily.

A number of ingredients or combinations of ingredients can be used to meet the nutritive requirements of rabbits. Linseed meal, soybean meal, or peanut meal can be used for the protein. Corn, oats, barley, wheat, or milo can make up the grain portion of the diet. Alfalfa or clover, or other good-quality hay is recommended for roughage. Salt may be added to other feeds, or a small piece of

Guinea pigs are good laboratory animals because they are sensitive to differences in diet. *Left:* regular diet. *Right:* regular diet plus liver. (*Courtesy* The Squibb Institute)

compressed salt may be provided for rabbits to lick. Many rabbit growers prefer to buy a pellet feed, especially prepared for rabbits, that provides all necessary nutrients and saves the labor of mixing feeds.

Whether the grower uses a pellet feed or mixes the ration himself, it is important to regulate the amount of feed to keep the rabbits from becoming too fat. About 400 pounds of grain and other concentrates are required to feed an average-size doe and her four litters to the age of 8 weeks. Small amounts of green feed may be added to the diet. Freshly cut grass, clover, or garden crops are good green feeds. Rabbits should not be given garden crops and grasses that have been sprayed with or exposed to insecticides.

Housing. Rabbit hutches usually are about 2 feet high, no more than 2½ feet deep, and 3 or 4 feet long. Inexpensive hutches can be made at home. All-wire quonset type hutches may be used inside buildings. Partially enclosed hutches are often used outdoors.

Several kinds of flooring can be used in hutches. Wire mesh flooring is used extensively where self-cleaning hutches are desirable. Solid and flat flooring, or a combination of solid flooring at the front and a strip of wire mesh at the back, can also be used. In areas of mild climate, hutches can be placed outdoors in the shade of trees or buildings. Hutches may also be placed under superstructures for protection from sun and rain. Rabbits do not require sunlight. During hot weather some cooling measures must be provided in addition to shade. All buildings which house rabbits should be adequately ventilated.

THE GUINEA PIG

Since 1870 the tame guinea pig has been used extensively for a wide variety of experimental purposes, including inoculation with pathogenic (disease-producing) organisms for studying the symptoms of a disease; the standardization of vaccines, serums, and anti-toxins;

These rats are litter-mates. *Left:* regular diet—weight 7 ounces. *Right:* regular diet plus vitamin A—weight 9 ounces. Note the weak eyes of the rat without vitamin A. Low vitamin A in *our* diets weakens our eyes too, and makes it difficult for us to see at night. (*Courtesy* The Squibb Institute)

genetic (inheritance) experiments; and vitamin studies. Many food-manufacturing companies maintain colonies of guinea pigs for standardizing the vitamin content of products. The guinea pig is the only rodent requiring vitamin C in its diet.

Guinea pigs weighing 250 grams, or about one-half pound, are desired for many experimental purposes. If fed properly, guinea pigs reach this weight when they are four to six weeks old.

Many institutions using guinea pigs now maintain their own guinea pig colonies or establish contracts with people who raise large numbers that have been developed for specific tests. Many others are supplied by agencies that obtain guinea pigs from numerous small producers. Anyone who plans to raise guinea pigs on a commercial scale should first arrange to sell surplus stock to a hospital, laboratory, or manufacturer of biological products. On a small scale, guinea pigs are ideal pets, and surplus stock can be sold or given away for this purpose. They are among the cleanest of all animals used as pets.

MICE AND RATS

Because of its fertility, convenient size, and inexpensive maintenance, and its resistance or susceptibility to certain diseases, the tame white mouse has become a favorite research animal. Its variability has made it a valuable animal for genetic research, and it has been more intensively studied in this connection than any other mammal. It is widely used in medical research, especially in work on cancer, and in fertility studies. Thus the mouse as a subject of study has contributed much to the biological sciences.

The tame white rat shares with the mouse a reputation as a useful animal for research. It has been used more extensively in nutritional investigations than the mouse. The tame white rat maintains an important place in the study of hormones and in the testing and standardization of drugs. It is also a favorite animal for use in animal psychology experiments because its learning ability is high compared with that of other small laboratory animals.

Wild gray mice and rats are the "bad guys." Rats eat 33 million tons of the world's food grains a year. They are in every country, city, village, and farm. Rats also transmit certain diseases to people (they have caused epidemics) and kill many beneficial birds.

Rats cannot be eliminated, but they can be controlled by traps, poison baits, poison gas, sanitation, and rat-proofing of buildings. All control work can be done effectively on a community cooperative basis. For more information on how to kill rats, contact your county agricultural agent or city health department.

HAMSTERS

The golden hamster is a rodent that has been gradually increasing in favor as a laboratory animal. It is smaller than the rat and guinea pig and reproduces more rapidly. In general, hamsters may be used in the laboratory for the same purposes as rats, mice, or guinea pigs.

The nutritional requirements of hamsters have been extensively determined. Since they are especially susceptible to the absence of vitamins D and K, without which they die, hamsters are especially valuable in the study of these vitamins. Hamsters are less valuable than guinea pigs in the study of vitamin C. Calcium and phosphorus deficiency produces in hamsters the symptoms of rickets, as it does in man.

In zoos, hamsters are sometimes given to certain birds and other animals that require live food.

Besides being popular as pets, hamsters also are useful in high school and college classes for studies on reproduction. Hamsters have a short gestation period of 16 days and mature 60 days after birth.

The main advantage of the golden hamster over other laboratory animals is its ability to reproduce rapidly. This saves time and space and cuts the maintenance cost per animal. For some experiments the hamster cannot replace other species of laboratory animals.

Neither the Department of Agriculture nor any unit of the United States Government recommends raising hamsters as a source of large profits. It is not advisable to raise many hamsters without first making arrangements for marketing the surplus animals to some hospital, laboratory, or biological supply company. It is largely up to the individual to develop his own market, as the Federal Government does not maintain any list of purchasers.

CAUTION: Breeders of hamsters are cautioned to prevent the escape of any of these animals. Such release under favorable conditions might establish the hamster in the wild and thereby create a serious rodent problem, as they destroy growing crops, gardens, and other agricultural enterprises. Purchasers should be aware of the danger of escapes and should make every effort to prevent the establishment of a wild colony.

FUR FARMING

The scarcity and the high value of certain wild fur animals have led to efforts to raise fur animals in pens. Not all of these ventures have been profitable, but by selecting the best animals for mating, the fur farmer can improve the color, length, density, quality, and other attributes of the animals' fur, increasing the value of the pelts for use in fine clothing.

In general, colder climates are more desirable for fur farming, though good pelts have been produced in pens in moderate climates and at sea level. Marginal land can often be used satisfactorily.

An abundant supply of cheap feed must be available, suitable for the type of animals to be raised. Refrigeration is essential if the mink, the fox, or another meat- or fish-eating animal is being considered. Electric power for mixing feeds is a necessity.

Certain states and localities have licensing requirements or other restrictions, so the proper authorities should be consulted before starting a fur farm.

Getting a start. After thorough investigation of the possibilities of the particular type of fur farming in which he is interested, the fur farmer should buy equipment and animals on a modest scale and, if possible, from local sources.

Good equipment is important: it lasts longer, reduces labor, and provides better sanitary conditions. Also important is protection against marauding animals and thieves. Of course, the equipment must also be adequate to prevent the escape of the fur animals.

Excellent breeding stock, though costing more, usually returns more profit and certainly affords the owner more pleasure in his work. The highest priced stock is not necessarily the best. It pays to buy only from reputable breeders.

The United States Department of Agriculture maintains a list of fur-farm associations, but not a list of breeders. It is left to the associations and the fur journals to give out information on prices and on places to buy breeding stock. The Department does not vouch for the integrity or the financial standing of any individual or organization on its list.

Silver foxes. Not more than one litter of silver foxes is produced each year. A few litters of 9 or 10 are obtained, but an average production of 3 or 4 young for each vixen (female fox) is considered satisfactory, and, of course, some vixens do not produce each year. Some ranchers with large numbers of animals leave the pups with the parents until September; others wean them after about 8 weeks, when they weigh about 4 pounds.

Young foxes attain the weight of mature animals when they are about 7 months old. The pelts are taken in November, when the animal is 9 to 10 months old, or in December—depending on climatic conditions.

At the larger ranches, the breeding pairs of foxes are provided with ground-floor pens, 40 feet wide and from 20 to 40 feet deep, with a 2-foot wire overhang to prevent escape. At other ranches the males and females are kept in separate pens 6 to 8 feet wide, from 16 to 30 feet long, and 4 to 6 feet high. A wire netting covers the entire top of the pen. The raised floors are of wire, for sanitation and for control of parasites.

As many as 350,000 silver fox skins have been produced annually in the United States, but the number is rapidly decreasing. Average prices of from $12 to $16 for furs that cost more than $30 to produce send many fox farmers out of business and cause others to reduce their operations.

Though indications are that there will be a limited demand for choice fox skins, it is not advisable for beginners to engage in fox farming until the demand for these furs has increased.

Minks. Though minks were raised in captivity for their fur as early as 1866, mink skins were not produced in quantity until the 1930's. In 1948 there was a tremendous increase to 2 million skins. About 40 percent of these were of the dark standard type; the rest were hybrids or mutations of various kinds. The softer, more delicate shades are most popular. Through selective matings the standard dark mink became very dark or almost black.

On the average, a mature female mink weighs from 1½ to 2 pounds; males weigh 2½ to 3 and even up to 4 pounds.

Breeding time is about the first week in March. Litters of 9 or 10 may be produced, but an overall ranch average of 4 for each female is considered satisfactory. Young minks are weaned at 7 or 8 weeks, when they should weigh about 0.9 pound, the females being slightly heavier than the males. They are ready for pelting about the last of November of the year they are born.

Minks may be kept in individual pens having 10 to 20 square feet of floor space, the size depending on whether the animals are to be pelted or kept for breeding. Large colony houses with smaller pens having raised wire floors, with part of the pen projecting beyond the roof, are satisfactory. Small detachable nest boxes facilitate care of the animals.

Feed for minks is similar to that for foxes except that more raw meats are required and larger quantities of fish may be fed with safety. A mature mink consumes about a quarter-pound of feed at its single daily feeding. Females suckling litters naturally eat more and are fed twice daily. Minks to be pelted should be fed all they will eat. A ration composed entirely of dry feeds has not proven satisfactory for minks.

Anyone interested in raising minks should study all phases carefully before starting. Unless he has exceptional ability, it is questionable whether he should enter this business. Bulletin 229, *Fundamentals of Mink Ranching*, issued by Michigan State University, gives further details.

Chinchillas. The chinchilla is a small rodent native to South America. A few animals were first imported into the United States in 1923.

Most chinchillas are sold at high prices as breeding stock. Pelt values have not yet been established on prime pelts from pen-raised animals, but the ultimate object of a sound industry should be the production of pelts for commercial use. Prices of breeding stock will probably be materially lower when this has been accomplished.

Chinchillas weigh about 1½ pounds at maturity, the females weighing slightly more than the males. They look very much like squirrels. Their fur is extremely fine and silky, blue-gray in color, with characteristic white bands around the guard hairs.

Each year a pair may produce two litters of from one to three young. The gestation period is about 111 days. The young are fully furred and have their eyes open at birth. They are weaned at about 2 months of age and will breed at 6 to 8 months.

Many chinchillas are kept in small metal cages in basements. Larger producers have special buildings. A peculiarity of the species is the regularity with which these small animals like to dust

Above: **Apple orchards and beehives are meant for each other. The apple trees supply shade and nectar for making honey. The bees cross-pollinate the apple blossoms and increase the number of apples.** (*Courtesy* The A. I. Root Company) *Below:* **A worker bee helping to pollinate an apple blossom. Note full basket of pollen on her left hind leg.** (*Courtesy American Bee Journal*)

themselves by rolling. CAUTION: There is a hazard that the bite of any warm-blooded animal except the opossum can cause rabies.

BEES

Bees are important to our well-being primarily for two reasons. They produce honey and they are essential to the pollination of many plants.

The mastering of any trade or profession takes time, work, and study; beekeeping is no exception. One should start with some good bulletins and only one or two colonies. Then he may increase his operations as he learns, and as the bees pay their way. If, after a year or two, he decides he is no beekeeper after all, or that his location is not suitable, he may dispose of his colonies and equipment with little or no loss of time or money, but with some valuable experience gained.

Pollination of plants. Apples, cherries, plums, pears, and certain varieties of peaches, blackberries, and strawberries are some of the fruits honeybees aid in pollination. Bees are also important in the pollination of watermelons, muskmelons, cantaloupes, and squash, as well as other truck crops. The presence of bees is necessary to produce satisfactory crops of seed of alfalfa, white (Dutch) clover, alsike clover, and red clover. Repeated observations have shown clearly that the services of bees result in greater crops of seed of hubam clover, white sweet clover, and yellow sweet clover, as well as many other crops.

The colony. Bees are not domesticated insects. There is no difference between the behavior of a colony in a hollow tree and the behavior of one in a modern apiary. But if you know the bees' biology and instincts, just about whatever you want can be done with them. All beekeeping manipulations are based on such knowledge. The first step, therefore, is to learn something of the colony and its way of life.

A colony of bees consists of a queen, a few drones, and a large number of workers. There may be less than 10,000 workers in early spring, but exceptionally strong colonies may grow to as many as 100,000 in the fall.

The nest consists of a number of combs composed of hexagonal (6-sided) cells. The combs are made of wax secreted by young bees. Experiments have shown that bees must eat an average of almost ten pounds of honey to produce one pound of comb. Therefore, beekeepers should extract the honey and return the comb to the hive.

Eggs are deposited singly in the cells by the queen. The egg hatches into the larva, or grub, which later changes into a pupa, from which the adult bee emerges.

The queen. A mother bee, the queen, is found in every normal colony. The queen differs from the other bees in the colony in having a longer and more tapering body. Her body has no color bands. Normally only one queen is found in a colony, but on rare occasions an old queen and a daughter live together in harmony for several weeks; this is not, however, a satisfactory arrangement.

The sole duty of the queen is to lay eggs. She has nothing to do with the governing of the colony; this duty apparently belongs to the workers. At the height of the brood-rearing season in the spring, the queen may lay 2,000 or more

eggs in a single day—a weight of eggs greater than the normal weight of her own body. However, this rapid rate of egg production is continued only for a short time. Eggs laid in worker or queen cells are fertile and develop into females; those deposited in drone cells are not fertilized and these eggs develop into males.

The drone. Drones are male bees, developing from infertile eggs in drone cells. They are even larger and stouter than the queens, but their bodies are not as long. Drones have no stingers and are physically disqualified for any work. Their sole purpose in nature is to fertilize young queens. Since much honey is consumed in their rearing and upkeep, and as only a few are necessary for mating purposes, the modern beekeeper tries to eliminate the production of drones as much as possible. This is

The queen bee is the large bee in the center. She lays eggs for the entire hive. (*Courtesy American Bee Journal*)

best done by using full sheets of wax foundation with impressions only for worker cells so the bees build mostly worker-size cells and construct only a few of the larger drone cells.

When the swarming season is over and the honey flow has drawn to a close, the worker bees will not tolerate the presence of these boarders any longer. Drones are starved and driven out of the colonies to die.

The worker. Workers are the smallest and most numerous occupants of the hive. They develop in worker-size cells from fertile eggs. Since their development occurs in small cells and the larvae are not fed the same rich nutritious food throughout their larval lives as the larval queens, they mature as sexually undeveloped females. Three weeks are required for development from egg to adult.

The length of the adult worker's life depends upon the amount of work the bee does. It may live from fall to spring, but during the honey-gathering season it spends its energy and dies in about six weeks.

When a queen dies the workers can raise a new queen by feeding royal jelly to any young larva hatched from a fertile egg. Should a colony become queenless, certain individuals will assume the duty of egg laying.

A field bee usually visits many flowers to gather one load of nectar amounting to about 30 milligrams and consisting of about one-half excess water which must be evaporated in the ripening process. Therefore, the bee actually delivers about 15 milligrams of water-free honey from one field trip. Since there are 453,592 milligrams in a pound, about 30,000 trips must be made to collect enough nectar to produce a pound of honey. Assuming each trip covers a distance of two miles, this would mean that bees travel a distance of more than twice the circumference of the earth at the equator and visit hundreds of thousands of flowers to make one pound of honey.

The beekeeper should provide an adequate supply of water for bees near the hives. They use water especially in the preparation of larval food, and much water is consumed during the height of the brood-rearing season.

When collecting pollen and nectar, bees show a remarkable fidelity in working only one kind of a plant at a time. This greatly increases their value as pollinating agents. Bees do not overcrowd the plants they are working; only enough are present to take care of the available nectar or pollen. The bees arrange this through use of a sign language.

Additional information on bee management may be obtained from *Selecting and Operating Beekeeping Equipment*, Farmers Bulletin 2204, United States Department of Agriculture.

QUESTIONS

1 How can the farmer increase the value of the pelts he produces?

2 What are some of the things the fur farmer must do when starting a fur farm?

3   What is the average production of young in each silver fox litter?

4   What is the average weight of mature male and female minks?

5   What kind of feed does a mink require?

6   What is the ultimate object of a sound chinchilla industry?

7   To which plants is bee pollination important?

8   What does a typical bee colony consist of?

9   What is the function of the bee queen?

10 How many trips must a worker bee make to produce one pound of honey?

ACTIVITIES

1   Use biology textbooks, encyclopedias, and other references to obtain more information about the way in which bees aid in the pollination of apple blossoms, white clover, sweet clover, and many other plants.

2   Perhaps some members of the class have rabbits, hamsters, guinea pigs, or other small animals. If so, have them make reports on the care and feeding of the animals.

3   If there is a pet store in your community, perhaps some members of your class could visit the store to obtain information on the kinds of animals sold there, the prices of different animals, the equipment needed to raise each kind, and how to care for the animals.

4   In spring and summer, watch bees moving about among flowering plants. Notice how long a bee stays on a flower. Do additional reading to obtain information about the making of honey.

5   Arrange for the class to see a film, film strips, or a set of slides on a topic related to this chapter.

18 MECHANIZED FARMING

We live in a hungry world. Half of the 3.6 billion people who inhabit the earth are on starvation diets, and two in ten do not have balanced diets. This leaves only three persons in ten who are well fed. Do *you* get enough of the right kind of food?

Although many people in our country do not have proper diets, the cause is more likely to be a poor selection of food than a lack of food. We are a well fed nation. When crop failures do occur in parts of the country, surplus food and feed from other areas are quickly supplied.

What do we have that most other nations do not have? Is it more land per person? No—the density of population here is about average. Do we have higher yields per acre? Here again the answer is no. In fact, our crop and livestock yields are less than those in many European countries and in Japan.

What we do have is more freedom, more fertilizer, better seed, more grass, and more farm equipment. Our laws favor individual initiative far more than do the laws of any other nation. Fertilizers are plentiful and reasonable in price. The better seed, such as that of hybrid corn, is readily available and produces high yields per acre. The United States has a billion acres of grass. Also, our farmers are blessed with an abundance of farm equipment with which to produce more per man. Our laws have encouraged our nation's creative minds to put science to work for us.

In colonial days it took 85 percent of the population to produce enough food for itself and for the other 15 percent. Then nearly everyone was a farmer, including George Washington and Thomas Jefferson. Now the situation

"Modern" tillage implements around the world in the decade of the 70's. (A) More than 90 percent of all tillage in Dahomey, western Africa today is done with the village-made hand hoe. (*Courtesy* United Nations/FAO) (B) More than 90 percent of all tillage in India today is done with bullocks and a village-made wooden plow with a 2-inch metal tip (Andhra Pradesh, India). (*Courtesy* Roy L. Donahue) (C) Camels are used to pull the plow in northern Africa and the arid Middle East (Tunisia). (*Courtesy* United Nations/FAO) (D) Most plowing in the United States is done with a tractor similar to this. (*Courtesy* John Deere)

is completely reversed; 5 percent of our people produce enough food and fiber for the other 95 percent, and have enough in reserve to feed people in other countries the produce from one acre in five.

The production efficiency of each farm worker is increasing so rapidly that by the time you complete the reading of this sentence farm production will be greater. In 1948 one farmer produced enough food and fiber to feed and clothe himself and 15 other persons, compared with 44 today.

During the 15-year period 1951 to 1965, the United States farmer produced 35 percent more on 11 percent fewer acres and with 45 percent less farm labor, but with 48 percent more farm production supplies, than in the preceding comparable period. A large part of the increase in farm production supplies was for farm mechanization.

Rice is the primary food for 60 percent of the world's people. In the United States, rice production is so highly mechanized that it requires only 7.5 man-hours per acre. In most parts of the world the production of one acre of rice requires 750 man-hours.

TRACTORS

United States farmers own about 5 million tractors that have taken the place of 22 million work animals that required the feed produced on 76 million acres.

A large share of the credit for increasing the farmer's efficiency goes to the farm tractor.

A recent study was made by the United States Department of Agriculture to find out how much time a farmer really saves when he uses a tractor in-

Above: Modern power harvesting equipment permits the rapid harvesting of large acreages in a short time when weather and crop conditions are favorable. (*Courtesy* the Ford Company) *Below:* Land planing (smoothing) is a practice for which the farmer needs technical help. Land planing makes the operation of power equipment easier and is a necessary step before establishing either an irrigation or a drainage system. (*Courtesy* North Carolina State University)

One farm worker can now do the hay baling that would have required several workers a few years ago. (*Courtesy* the Ford Company)

stead of horses as farm power. The study showed that each farm tractor saved 85 working days a year. This amounts to almost one-fourth of a year. In other words, a farmer who changes from horses to a tractor can go fishing every fourth day and still maintain the same rate of production.

ANTIQUE AND MODERN TRACTORS

In 1860 an Illinois farm paper reported the first successful use of a steam engine for plowing. The steam engine was a great labor-saving machine at the time, but by modern standards its first accomplishment seems almost comical.

The first steam engine plowed for 23 minutes, stopped 6 minutes to refuel with wood; ran 13 minutes, stopped 8 minutes for water; then plowed one more minute before stopping for repairs. The crew consisted of a man and team to supply fuel and water, a fireman, two men to operate the plows, and the engineer who invented the machine and who stood by in case of a breakdown.

The first gasoline tractor, built in 1901, weighed 10 tons. Today a modern tractor capable of doing the same amount of work weighs only 1½ tons.

In 1932, rubber tires appeared for the first time on some tractors. These permitted faster speeds in moving from one field to the next and allowed the tractor to cross asphalt and other paved roads. Previously, the steel lugs on the wheels would have ruined an asphalt road. Rubber tires also allowed the tractor to roll so easily that they saved 25 percent on tractor fuel.

Each year farmers have a wider choice of tractors. The new models are designed for greater safety, more uses, and easier handling. There are special tractors for use in home gardens, in small commercial gardens, and in nurseries.

NEW WAYS TO PREPARE THE SOIL

The first plow was a forked stick that was used to scratch the earth. This stick was pulled by slaves or women. Oxen were later hitched to a larger

A new family of moldboard plows includes a two-way model (top) and a high-clearance model (bottom). When matched to tractors with from 50 to 131 horsepower, these plows are an excellent choice for farmers with large acreage, heavy trash conditions, and stony fields. (*Courtesy* the Ford Company)

forked stick, and in this way the soil was turned up in preparation for planting seeds. Drawings in stone, made as early as 6,000 years before Christ, showed a Y-shaped stick being used to till the soil.

For some unknown reason, progress in the design of a better implement than the forked stick was extremely slow. There was almost no improvement until about 1800 A.D. Even in many foreign countries today the same kind of forked stick is in use.

In 1788 Thomas Jefferson used principles of mathematics to improve the design of the plow. The cast-iron plow was patented in 1797, but it was not until the 1830's that an all-steel plow was made to turn the tough sod and black soil of the Illinois prairies.

With early wooden plows, eight oxen and two men took a day to plow an acre. Now a man on a modern tractor with a two-bottom plow can turn more than ten acres a day.

After the land is plowed it needs smoothing and packing before it is ready to be planted. This job is done with some type of harrow.

The first harrow was a tree branch which was dragged over the plowed ground. Even today, in some parts of our nation, a log or a railroad iron pulled like a drag serves as a smoothing harrow. A spike-tooth harrow or a disk harrow is more desirable equipment.

Other equipment sometimes used in preparing the soil for planting crops includes a lister, spring-tooth harrow, disk plow, disk harrow, field cultivator, and chisel.

A lister (middle buster) turns a furrow both ways at once. It looks like two moldboard plows mounted together. In

Decisions in farm management must be made on whether to terrace specific fields. Technical assistance in making such a decision can be obtained from the Soil Conservation Service in each state. (*Courtesy* Allis-Chalmers)

dry areas, crops are planted in the bottom of the furrow; and in wet land, crops may be planted on the lister ridges.

A spring-tooth harrow is designed to smooth land and to tear out grasses and weeds, exposing their roots to the sun. This harrow is an ideal tool on stony land or where many tree roots are present.

The disk plow was patented in 1847 but did not become popular until after 1900. It consists of one or more large concave disks set at an angle so that they rotate while being pulled through the soil. Disk plows are popular in hard or sticky soils where the conventional moldboard plow has trouble in turning a furrow slice.

The disk harrow is an implement for smoothing soil. It is made with 20 or

This self-propelled irrigation system is being used on newly planted brome and alfalfa following corn cut for silage. (Nebraska) (*Courtesy* U. S. D. A., Soil Conservation Service)

Airplane seeding rice into flooded fields in Sacramento Valley. (*Courtesy* University of California Agricultural Extension Service)

more disks set upright, much like wheels, or at a slight angle.

Field cultivators—cultivator shovels mounted behind a tractor—are often used as a substitute for a plow.

Chisels are strong, narrow points or shovels mounted on heavy beams. They penetrate the soil deeper than do plows, disks, or field cultivators. Chisels are used to break up hardpans in the soil. The hardpans may be natural, or they may be caused by packing the soil with heavy machinery or with a plow that is used at the same depth year after year.

PLANTING MACHINES

The Chinese are credited with having invented the first seeding machine as early as 2800 B.C. Mounted on a wheelbarrow, it had three spouts through which seed was sown.

In the United States the first seeding machine was patented in 1799. Then in 1840 the machine was further improved: a series of gears was added to regulate the flow of seed down each spout.

The improved grain drills sowed the seed more uniformly, and thereby accomplished a twofold purpose. First, less seed was needed per acre. Second, the seeders were responsible for a more uniform stand and, as a consequence, higher yields per acre.

It is reported that during the American Civil War, grain drill salesmen offered to take as their pay for the drill the increase in grain yields from 40 acres that resulted from using the drills instead of sowing the grain by hand. At this time two bushels of seed planted with a drill were reported to plant the

This machine can spread a liquid complete fertilizer (N + P + K) in variable ratios as the valves are adjusted. (Illinois) (*Courtesy* Farmall)

same acreage as three bushels would plant by hand.

The time saved by a modern seeding machine is tremendous. A man can sow 10 acres per day by hand, but with a 12-foot tractor-drawn drill he can plant 50 acres. Only one-fifth of the time once required for seeding is necessary now.

Just as corn is a native American crop, so also is the corn planter an American invention. Writing of the early American Indians, Captain John Smith said, "The greatest labor they take is in planting corn."

The first corn planter was patented in 1828. It was mounted on a wheelbarrow pushed by hand and planted six rows at a time. Improvements in 1890 permitted corn to be planted in check rows, like a checkerboard, allowing cultivation in two directions.

Mechanized planters for cotton, potatoes, and truck crops were developed

several years later. Fertilizer attachments for planters began appearing in 1880, when one was developed for the potato planter. Similar attachments to spread fertilizer at the same time that seed is planted are available now for almost all planters, including grain drills.

CULTIVATORS

The horse-hoe, which first appeared about 1820, was an early device to reduce the amount of hard hoeing that was necessary. A generation later the horse-hoe was mounted on wheels to make it draw more easily and do a better job of cultivating.

The next logical development was the two-row, horse-drawn cultivator. Then came tractor-drawn equipment. Now it is common practice over the nation for farmers to use 6- to 12-row cul-

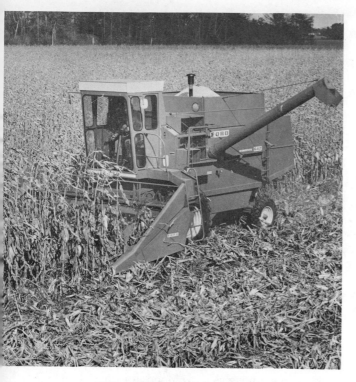

This self-propelled harvester sends clean grain into a special compartment while the plant residue is deposited on the soil to be turned under or chopped into the soil to add organic matter. (*Courtesy* the Ford Company)

tivators and do a job better, easier, and 100 times faster than hand hoeing could do.

MECHANIZATION NOT MAGIC

Tractors and mechanized equipment increase the efficiency of the farmer, permit more timely operations, reduce drudgery, and increase the number of acres that can be farmed. Mechanized farming may increase the gross farm income if used according to recommendations. Net income may also be increased

if decisions are based on the farm and farm machinery record book rather than the glamour of the newest models with excess horsepower.

When a farmer must decide on tractor power replacement, he has these options: (1) repairing the old tractor, (2) buying a new tractor, (3) buying a new tractor in cooperation with a neighbor, (4) renting or leasing a tractor, or (5) hiring the work done by a neighbor or by a custom operator.

Decisions may be made on impulse, effective advertising, or glamour unless an analysis sheet is used to list facts. The sheet given in this chapter on page 245, is from United States Department of Agriculture *Leaflet* No. 427. Your County Extension Director can inform you if a better form exists for your county. He can also help you estimate costs for all items indicated on the sheet.

Many farmers do not realize that depreciation on a 60-horsepower tractor will usually amount to 3 dollars a day, and operating costs 10 to 15 dollars a day. Wise farmers buy a tractor with the minimum horsepower rating to do the work required, care for it as they do their automobiles, and use it for as many productive hours in the year as possible.

MINIMUM TILLAGE

One way farmers can increase net income with the use of tractors and farm equipment is to operate more acres with the same farm power by practicing minimum tillage. This means using the least farm power and equipment possible. For example, one pass over a field with a tractor and specially designed imple-

Desirable soil structure in the seed and plantbed can be maintained fairly well by minimum tillage practices, but destroyed by continuous tillage. (A) Virgin grass prairie soil with ideal soil structure. (B) Minimum tillage in a rotation of corn, oats, clover, and wheat, resulting in good soil structure. (C) Continuous corn with continuous tillage, resulting in poor soil structure. (*Courtesy* University of Illinois)

Left: This first successful cotton picker was used in 1941. A new model two-row cotton picker is shown here in operation. (*Courtesy* John Deere and Company) *Right:* The cleanest cotton that brings the highest market price is obtained when a defoliant is sprayed on the cotton plants just before harvest to make all the leaves drop. (*Courtesy* Bell Helicopter Company)

ments can chop the previous year's crop stubble into the soil, plant the new crop, spread fertilizer, and apply weedicides, insecticides, and fungicides all at the same time.

Minimum tillage has the following advantages: (1) less labor of man and machines, (2) lower production costs, (3) conservation of soil moisture, (4) maintains or increases yields, (5) makes double cropping possible, (6) furnishes winter cover for wildlife, (7) reduces wind and water erosion, thus reducing air and water pollution, (8) increases soil moisture in areas of winter snow by trapping snow in crop residues, and (9) increases desirable soil structure in the plantbed (see figures on soil structure, Chapter 20).

Behind nearly all efficient mechanized agriculture is a multi-million dollar complex of fertilizer ore deposits, giant mechanized operations, and television and push-button controls to ensure efficiency and uniformity of low-cost fertilizer. *Above:* This giant dragline picks up 82.5 tons of rock phosphate ore near Bonnie, Florida. *Below:* Closed-circuit television sets and many push-buttons are used to guide the production of the world's largest plant making phosphoric acid for use primarily as a farm fertilizer. (*Courtesy* International Minerals and Chemical Corporation)

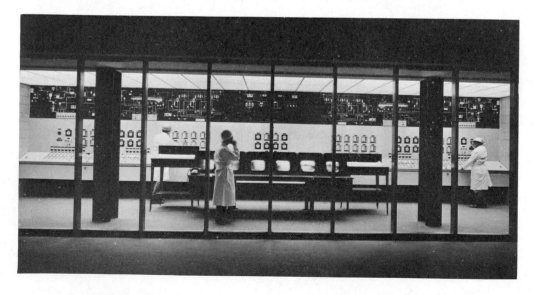

ANALYZE YOUR MACHINERY NEEDS WITH THIS WORKSHEET

ITEM	MODEL YEAR	YEAR ACQUIRED	COST	PRESENT VALUE	REMAIN-ING YEARS	REPLACEMENT		TRADE-IN VALUE	CASH OR CREDIT NEEDED
						YEAR	COST		

HOW TO USE THIS WORKSHEET:

List each of your machines in column 1. Then write in the model year, the year you bought it, and its cost.

Next, estimate the current value of each machine and the number of years you expect it will last.

Put down the year you hope to replace each machine. Consider the physical condition of each machine, probable repair costs to keep each one running, your need for dependability and efficiency, and how soon you might be able to get an improved model.

Next, estimate the probable cost of each machine. Use current prices even though the price may have changed by the time you are ready to buy.

Record also the probable trade-in value of each machine at the time when you are ready to replace it.

Finally, enter the extra amount you would need in cash or credit to buy each new machine or piece of equipment.

You will now have all the data needed to help you decide more intelligently how to replace your farm equipment.

QUESTIONS

1   What percent of the people of the world are well fed?

2   What do we have in the United States that other nations do not have, that contributes to a high standard of living?

3   What are some of the probable new developments in mechanization?

4   Give figures to show that farming is more efficient now than it was a few decades ago.

5   How many working days a year does the tractor save for the average farmer?

6   Describe the first successful use of a steam engine for plowing.

7   What is a lister?

8   Give an example of the greater efficiency of modern methods of seeding.

9   In what year was the cast-iron moldboard plow introduced?

10   When was the one-man baler developed?

ACTIVITIES

1   Use encyclopedias, agriculture textbooks, and other references for additional reading about the development of tractors, harvesting machines, plows, and other farm implements.

2   Observe the kinds of mechanized farming operations that are common in your community. Make notes for use in class discussions.

3   Visit a local implement dealer to obtain information about the cost of tractors, combines, grain drills, plows, forage harvesters, mowing machines, and other equipment used in your community.

4   Select a farming operation such as harvesting grain or cultivating corn. Write a comparison of this operation 100 years ago and today.

5   Read descriptions of agriculture in other countries. Compare the farming methods with those in the United States.

6   Examine several farm magazines to find articles on mechanization of farming.

7   Find out all you can about tractor-pulling contests, and if possible attend a local farm equipment exhibit.

19 WATER FOR A QUALITY ENVIRONMENT

Water is everywhere. All forms of life had their beginning in water, and water is still necessary for all living things.

Our bodies are about three-fourths water. As sturdy as we think we are, water makes up 80 percent of our muscles and 90 percent of our blood. If it is surprising to learn that human flesh and blood are mostly water, remember that the jellyfish is 99.8 percent water.

Water means different things in different situations. A timely rain will save a crop from drought; yet the same rain can ruin picnic plans for thousands of people. A drainage ditch helps remove excess water so that farm crops will grow better, but at the same time the habitat for muskrats and waterfowl is destroyed when the water table is lowered. Water provides the cheapest transportation known, but claims the lives of many people. Yes, water may be a blessing or a curse, depending on how intelligently it is managed.

Our childhood memories of water may be focused most clearly on fishing or on the old swimming hole. We grow up to be more conscious that water is vital to us every day. Without water there would be no crops to give us food, no wildlife for recreation, and no beautiful streams or lakes to enjoy. Without water there would be no forests to provide hundreds of useful wood products.

The modern home uses water in ever-increasing quantities. As our standard of living rises, there is a corresponding upward trend in our per-capita consumption of water. So great is the use of water by industry that some manufacturing plants measure their daily water supply in millions of gallons.

A billion raindrops can be used in a thousand ways. On their journey from

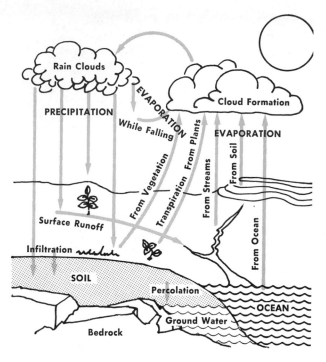

The never-ending water cycle can be managed to improve our environment.

Most large cities exist only where there is an adequate supply of water close by. (New York City) (*Courtesy* U.S.D.A.)

land to sea, the same raindrops can help trees to grow, supply spring and well water for man to drink, and flow in creeks that provide water for livestock. Water generates electricity to make our work easier, gives us places for swimming, boating, and fishing, and provides routes of cheap navigation. Water also helps protect cities from fire and carries away sewage. After performing these and hundreds of other tasks, some of the water comes to rest in the oceans and seas—but not for long.

Evaporation begins immediately, sending the water back into the atmosphere. Clouds form, and another billion raindrops soon are ready to start their journey to land, to sea, and back to clouds again. Water is always in action. The continuous movement of water from air to earth and back again is called the *water cycle*.

In some arid regions, eight of every ten drops of water falling from the clouds are sucked right back into the atmosphere again by evaporation. Only two of every ten drops stay on earth long enough to be useful.

Some of the factors that affect absorption and runoff are the texture and structure of the soil, the organic matter content, and the slope. The vegetative cover, the rate and intensity of precipitation, and the amount of moisture in the soil are other important factors.

Sandy soils admit water readily, while most clay soils are less permeable. The latter, however, have greater water-holding capacity. Soils rich in organic matter absorb and retain water better than soils in which the matter has been depleted by burning or excess tillage.

Level land allows more time for water to soak into the ground than does steep land. Steep slopes tend to have less absorption and more runoff.

Grass and other vegetative covers break the fall of the raindrop, slow the runoff, and allow more water to soak into the ground. Live roots provide channels that carry water into the soil. When roots decay, they serve as organic wicks that hasten the downward movement of water. Dead roots increase the capacity of the soil to hold water.

Soil, like air, has a saturation point. When soil contains all the water it can hold, any additional water that falls moves downward to replenish the water table. When rain comes faster than the soil can absorb it, the balance becomes runoff and may cause erosion. Soil and water must be managed together.

WANTED: MORE WATER

Rainfall records of one type or another have been made in the United States since about 1870. These data show that although wet years and drought years occur, there has been no long-time trend in either direction. The average annual precipitation for the nation over the past 80 years has been about 30 inches.

The pressure of an increasing population has pushed hard against this constant water supply. In the past century our population has increased sixfold while rainfall has remained unchanged. This is sad news but does not begin to tell the whole story. Centers of population have moved toward the arid west and more people have become concentrated in cities. The movement to cities has been so rapid that half of the people in the nation now live on 2 per-

Forests provide the best type of protective cover for the soil. Under forests, more water goes into the soil and less flows off. Mississippi) (*Courtesy* Soil Conservation Service)

cent of the land area. Such a concentration of people puts an added burden on water facilities.

Water as a prerequisite to settlement. The settlement of the West was a struggle from one water hole to the next, and fighting for control of the water supply was a common occurrence. By 1850 most of the forested East had been settled, and the bolder men and women moved into the treeless West. In the East the trees that had to be felled, piled, and burned had been a barrier to the production of food for families and grass for the cattle. As people settled the grasslands, the problem changed from too many trees to too little water.

Springs, so common in forest areas, were left behind as the settlers approached the grasslands. Grass was as

Above: When railroads were built across the West, water was always a critical problem. (*Courtesy* Paramount Pictures) *Below:* Successful irrigation requires agreements on the building of dams and irrigation canals, and the amount and cost of the water to each farmer. (*Courtesy* U. S. D. A.)

effective as trees in the protection of soil, but there was not as much rainfall on the grassland as in the forest.

No springs, few permanent streams, and ground water at a depth of 100 feet or more were some of the real barriers to the settlement of the West. People settled at the mouths of canyons where water poured out and at the bases of mountains. Water was "the boss."

Agriculture. Ranching and dry farming are practiced in arid regions, but irrigation dominates all intensive agriculture.

Modern irrigation agriculture was begun by the Mormons in 1847. Today about half of the farms in 11 Western states are irrigated. The value of irrigated land is well over half the value of all farm land in the United States. In Utah each acre of irrigated land is entirely dependent for its productiveness on the water from seven acres of forest-

and-range-watershed in the mountains.

Water runoff in Utah. The mountains of Utah receive fairly high rainfall and considerable snow, but the arid lowlands must be irrigated to produce satisfactory field crops. The problem is to build dams to collect the mountain runoff for use as irrigation water. The size of the dam and the amount of water to be stored must be calculated to determine the number of acres of land that can be irrigated.

Special Report Number 18 from the Utah State University indicates that runoff water can be estimated fairly accurately from the annual precipitation (rainfall plus the rainfall equivalent of snowfall). For example, when the mean annual precipitation is 35 inches, the mean annual runoff is 17 inches.

Irrigation agriculture seems ideal because moisture is fully controlled. But it cannot be successful without disciplined cooperation among the water users. Water conservation involves regulation of all users of the watershed area to prevent erosion, floods, and the movement of silt into reservoirs. Much land can be ruined by heavy concentrations of salts carried in the irrigation water unless management practices are sound.

Settlement in an irrigated area must be carefully planned. It is easy to bring in more people than a given irrigated area will support or to plant more cropland than the water supply justifies.

Some 30 inches of rain, snow, and other precipitation fall on the United States in an average year. This is a total of approximately 5 billion acre-feet of water. Two-thirds of this water never reaches streams, but sustains vast acreages of nonirrigated crops, pasture, rangeland, and forests. The other third finds its way to streams and is available for navigation, recreation, power generation, and "withdrawal uses" for which water is diverted from streams or pumped from underground.

Water that is withdrawn and then returned to streams may be used again as long as the quality of the water is maintained. Only water that is used and not returned to streams reduces the total water supply. Of the water actually used, or consumed, and not returned to streams, 9 out of every 10 gallons are for agriculture.

The use of water will continue to grow as our nation grows. By 1980 farmers will withdraw about 360 million acre-feet. Water use by cities and industries will grow even faster than farm uses of water. By 1980 the water available for reuse will be about 85 percent of the total withdrawn.

Each person living in a modern home uses about 50 gallons of water each day. A mature hog requires 3 gallons a day, a steer about 12 gallons, and a milk cow about 16 gallons.

Relation of mean annual runoff (inches) to mean annual precipitation (inches) in Utah. (*Source:* Utah State University)

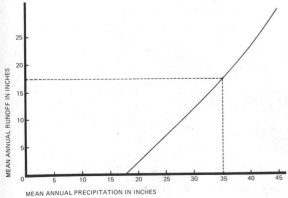

MEAN ANNUAL RUNOFF IN INCHES

MEAN ANNUAL PRECIPITATION IN INCHES

Source: Utah State University

WATER AND ITS FUNCTIONS

Water is a great moderator of climate when it occurs in such large quantities as the Great Lakes, the Gulf of Mexico, and the Pacific and Atlantic oceans—especially when winds move from water toward land. Water heats slowly, but once it is warm it stays warm longer than does land. This means that land near large bodies of water is cooler in summer and warmer in winter than land a long way from water.

Water and plant growth. Water and temperature largely determine the type of vegetation that grows naturally in any part of the world. In general, the relationship between natural vegetation is shown in the table below.

Water relationships divide plants into three broad groups. The camels of the plant world are the xerophytes such as cactus. These plants have remarkable adaptations to drought. At the other extreme are the hydrophytes, the ducks of the plant world. Rice is practically a hydrophyte. In between are the mesophytes, the average citizens of the plant world. Mesophytes include practically all agricultural plants.

Actively growing plant parts, not including woody parts, usually contain 75 to 90 percent water. All green plants use water in the process of photosynthesis (the manufacture of food with the aid of sunlight and chlorophyll). In addition, relatively large amounts of water are constantly passing through the plant to be transpired by the leaves. Evaporation occurs as moisture that is released through the tiny openings in the leaf surfaces (stomata). The root systems that supply this steady flow of water are far more extensive than most people realize. For example, grass and alfalfa roots often extend to a depth of 16 feet. Tree roots may extend deeper if the soil permits.

Good soil structure is more important than anything else from a moisture standpoint. A clay soil with the right structure may hold enough available water to mature a crop even if there is no rain throughout the entire growth period.

Plants as well as soils differ in moisture efficiency. In one study, a variety of alfalfa used 963 pounds of water to build a pound of dry matter, while another variety used only 651 pounds; one variety of millet required 444 pounds of water and another variety needed only 261. This is a fundamental factor in the adaptation of agricultural plants to soils and climates.

Water as a universal standard. Water is so common and so uniform that it is used as a standard in accurate scientific weight measurements. Under standard conditions, the weight of 1

Climatic Region	Inches of Average Annual Precipitation	Vegetation Type
Superhumid	more than 40	Rain forest
Humid	30–40	Forest
Subhumid	20–30	Tall-grass prairie
Semiarid	10 to 20	Short-grass plains
Arid	less than 10	Desert

Note: These precipitation and vegetation relationships are true for the Temperate Region; for the Tropical Regions, where evaporation and transpiration are higher, precipitation should be about 50 percent higher.

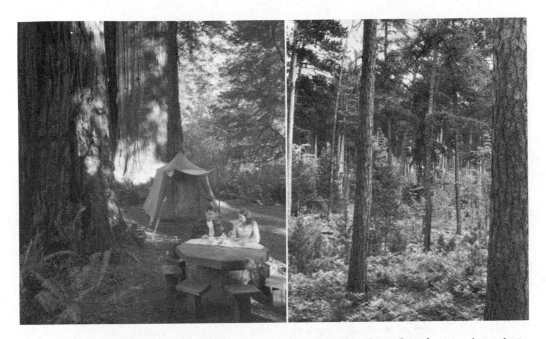

Above left: With high rainfall and high humidity the redwoods grow into giants. (*Courtesy* Redwood Empire Association) *Above right:* Thirty to 40 inches of precipitation in the Lake States produces Norway pine like these. (Michigan) (*Courtesy* Michigan Department of Natural Resources) *Below left:* Short-grass plains in Arizona receive 10 to 20 inches of precipitation. (*Courtesy* U. S. Forest Service) *Below right:* Tall grass grows under 30 inches of annual precipitation in Oklahoma. (*Courtesy* Harley Daniels, Red Plains Experiment Station)

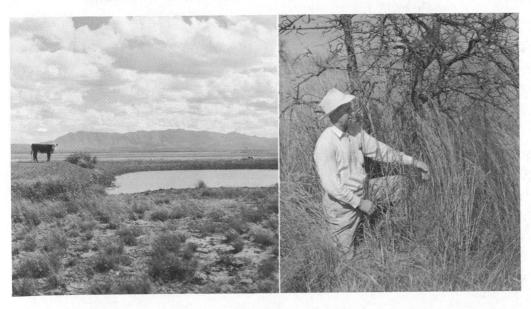

cubic centimeter of water is given as 1.0. Any substance that weighs twice as much as this standard volume of water is said to have a density, or *specific gravity*, of 2.0 (rock salt). Materials that weigh only half as much as the same volume of water are said to have a specific gravity of 0.5 (certain woods).

Efficient use of water. To serve man's needs adequately, water must be cheap and available in the right amounts at all times. Also, it must be chemically, physically, and bacteriologically satisfactory for the uses intended.

In most cities of the United States, purified water is delivered to customers at about 5 cents a ton. This is cheaper

This blast furnace uses a tremendous amount of water for cooling, but most of the water can be used over again. (*Courtesy* Bethlehem Steel Company)

than dirt. However, water is a major item of expense to a manufacturing plant because so much is needed. For example, a paper mill uses 20 to 40 million gallons of water a day. A typical iron smelter requires approximately 17 million gallons of water a day.

An attempt to "stretch" the supply of water for industry and for people in cities consists of purifying and treating polluted water so that it can be used safely and more easily. A pilot project to treat and reuse more water is now being conducted at Lebanon, Ohio. The processes consist of forcing water through a fine screen, filtering, chemical coagulation, and absorption with powdered carbon. Early results indicate success.

It is probable that one future source of more usable water will be sea water from which salt has been removed by evaporation with heat from atomic energy. This technique works, and several plants are in operation in the United States and around the world. The main difficulty is cost. The plants now in operation are supplying water for municipal use, but the cost of producing water to irrigate farm crops is prohibitive at present.

STORAGE OF WATER

Water may be stored in the upper layers of the soil, at lower depths as ground water, or as open surface water. The water stored in shallow layers of soil is used mainly by growing plants. Both ground water and surface water are used for household and industrial purposes and for irrigating farm crops.

About three-fourths of the world's fresh water is locked up in ice which covers one-tenth of the world. Some

Above: The dam under construction will provide more usable water for drinking, industry, and sanitation. (Arkansas) (*Courtesy* Soil Conservation Service)

novel proposals for increasing the amount of nonsalty water available to towns along the Atlantic Ocean, Pacific Ocean, and Gulf of Mexico may now seem impossible but may become a reality in the future. One such proposal is to use tug boats to move floating icebergs to destinations where water is needed. The fresh water could be pumped from a plastic bag around the iceberg into the city water supply. The flattened shelf ice of Antarctica (toward the South Pole) is considered more suitable than the Arctic floating icebergs because the flatter pieces would not turn over in transit. Such theoretical solutions to water supply may be prohibitive in cost for many years to come.

Pollution of water may be caused by biological agents, chemical agents, and physical agents. Biologically, water may be polluted by bacteria, protozoa, and viruses. Some of these organisms may contaminate drinking water and cause severe diarrhea, typhoid fever, or hepatitis.

The chemical pollutants may be or-

ganic or inorganic. The organic pollutants include such substances as animal and human manures, and wastes from canneries, textile factories, paper mills, and vegetable processing plants. These organic pollutants are especially harmful because they increase the reproduction of bacteria which deplete the oxygen in the water below the level required for fish. Inorganic pollutants include nitrates, phosphates, mercury, lead, and radioactive isotopes from atmospheric explosions of atomic bombs or from leaks from nuclear generating plants.

Physical pollutants include sediments from soil erosion, sawdust, and discoloration of water by discharge of soaps and certain chemicals from factories.

A detailed research study was made of the nitrate content in shallow wells and deep wells in the Chino-Corona milkshed area 24 miles east of Los Angeles, California. With 45 parts per million of nitrates as the critical level, the waters from deep wells averaged 27; waters from shallow wells averaged 221.

Lakes provide for storage of runoff water for many uses, such as swimming. (California) (*Courtesy* Redwood Empire Association)

Water in the soil. For each foot of depth, soil in good condition can hold 1 to 3 inches of rainfall. Variations in storage capacity of soils are caused by differences in organic-matter content; root channels; kinds and proportions of sand, silt, and clay; stoniness of the soil; and the insect or animal burrows present.

Although a soil may be holding all the water it is capable of holding against the pull of gravity, if it has been well managed there are still channels or pore spaces through the soil that can carry additional water downward to the water table. Much of this water reappears in springs and wells. The ability of soil to conduct water downward makes for the more uniform supply of water that is so essential for our day-to-day living.

Clemson University reported that a study was made in South Carolina of the available-water-storage capacity of 43 different types of soil. The first 3 feet of soil were used for this study. Results showed that four of these were fine-textured soils that retained between 6 and 8 inches of water in the first 3 feet; six retained between 5 and 6 inches; twenty retained between 4 and 5 inches; eleven between 3 and 4 inches; and two soils were sandy and held only 2 to 3 inches of water. These data give a good indication of the capacity of soils to hold water for normal plant growth.

Ground water. When mineral resources are mentioned, you may think of coal, petroleum, or iron. Within the earth is another great mineral resource: ground water. This is one of our greatest resources. On its use and management depends much of the agricultural and industrial development of the nation.

This underground reservoir of ground water is a reliable source of water between rains. The flow of streams in dry weather depends upon ground water. Conservation of ground water is necessary for the conservation and management of fish and other aquatic (water) animals.

You know what happens if you keep taking money out of the bank without putting any back. The water-bearing beds of the earth are our "water banks." If the supply of ground water is to last forever, there must be a balance between the quantity of water that is taken out of the earth and the amount that gets back.

Recent use of radioactive methods to test the age of water has shown that our ground water supplies are being used faster than they are being replenished. From a 900-foot well at Ft. Davis, Texas, and from a 300-foot well near Champaign, Illinois, the water currently

being used was determined to have ac-
cumulated there more than 50 years ago.

When water is withdrawn faster
than it is replaced in a water-bearing
formation, the water table falls. If water
seeps into a formation more rapidly than
it is withdrawn, the table rises. Ground
water acts this way only in a permeable
formation, such as sand and gravel, that
lets water pass through rather easily.

In many communities there is a
grave concern over the lowering of the
water table. New wells in certain areas
must be deeper than wells developed
several years ago, while older wells
must be deepened so that they will not
fail during dry periods. Some artesian
wells no longer flow. Such conditions
are a matter for serious thought, although
geologists assure us that the supply of
ground water is still very large.

One solution to the problem of
water supply seems to be in the proper
use and distribution of wells. Certain
wells are capable of yielding more water
than at present. Other areas are being
depleted by overpumping.

Development and conservation of
the nation's ground-water resources is a
long-range program. The United States
Geological Survey, state geological surveys, state boards of water engineers,
water districts, and cities are working
together in the management of ground-
water resources. Records are kept on
about 7,000 observation wells throughout the nation.

When the water table is held down
by a formation of rock or clay that it
cannot penetrate, pressure develops.
Artesian wells may flow from the pressure of water at a higher elevation in a
formation several miles distant. Such
wells may flow through natural open-

Water is being used from this Texas well
faster than it is being replenished by rainfall.
(*Courtesy* Roy L. Donahue)

ings in the earth's surface, or they may
be drilled through the water-tight layers
above water-bearing formations. Some
artesian wells have enough pressure to
force water high into the air.

The drilling of numerous wells in
an artesian formation has a tendency to
reduce the pressure and flow of water.
Uncontrolled flow of open wells causes
a tremendous waste of water, and eventually many of the wells cease to flow.
Checking the flow of artesian wells is
one of the most important water-con-
servation practices.

Another way to increase the amount
of ground water is to recharge it by
diverting surplus surface water downward through sandy and gravelly soil,
known as a "water-bearing formation."

Fresh water from salty sea water at a reasonable cost is now a reality. Since 1952 the Office of Saline Water in the U. S. Department of the Interior has been establishing desalting plants around the borders of the United States, using energy from the sun, a membrane process, and a process of distillation with atomic energy as the source of heat. This atomic energy desalting plant at San Diego, California, has a capacity of 2.5 million gallons per day. (*Courtesy* Office of Saline Water, U. S. Department of the Interior)

Surplus surface water can often be diverted downward into the same sand-gravel water-bearing formation that serves as the medium through which the ground water flows to supply water for wells.

Surface water. Surface water is stored mainly in rivers and lakes.

Most of the water in rivers comes from runoff, but a large amount also comes from ground waters that seep out along river channels. Seeps like this are called *springs*.

About one-third of all rainfall flows over the land, finding its way into streams and rivers. Sometimes more than one-third is runoff, partly because man has reduced the absorbing capacity of the soil. For example, Kentucky, with 45 inches of precipitation, has an average of 18 inches (40 percent) of runoff.

Where rainfall is below 10 inches almost no water is available to feed the rivers and lakes. This is the reason for the dry stream beds in many western states where rainfall is low. During heavy rains there is some runoff, and floods may occur. However, there is not enough regular runoff to insure permanent stream flow.

Lakes, both artificial and natural, provide easy storage of readily available water. Lakes add to the beauty as well as the usefulness of the landscape.

By the end of 1969, farmers and ranchers had built an estimated 1,948,340 ponds, with an average capacity of two acre-feet. (One acre-foot is equal to 43,560 cubic feet.)

A watershed. All the land and water area having common drainage into a stream, lake, or ocean is a watershed.

Everyone lives in such an area. It may cover only a few acres draining into a brook. It may include thousands of square miles draining into a large river and finally into a lake or ocean.

In its broadest meaning, a watershed is a regional river basin with a wide range of farmland, many streams, lakes, towns, cities, and other subdivisions all draining into an ocean.

A watershed provides for maximum water management in a specific area. People with varied interests can work together on common problems in such a management unit. A watershed is also a workable unit with which public agencies can cooperate in water management activities.

FLOOD CONTROL

Proper use of the land can do much to reduce flood hazards. Planting the right vegetation increases the amount of water that enters the soil, by slowing surface runoff. Mechanical operations such as terracing, strip cropping, and contour cultivation also reduce the danger of floods. Most of the time a combination of mechanical and vegetative protection is necessary in flood control.

Prevention of floods. A dense cover of trees, grasses, or legumes provides a protective blanket that intercepts part of the rain and snow. A portion of this precipitation is evaporated back into the air without reaching the ground. This

Approximately a third of all rainfall eventually gets into a river. Does the appearance of the stream indicate that it is the wet season or dry season? (North Carolina) (*Courtesy* Soil Conservation Service)

same canopy also breaks the impact of the rain on the ground, thereby retarding soil dispersion and preventing the clogging of pores in the soil.

Stems and fallen leaves of plants obstruct the overland flow of water and in combination with roots furnish an important means of soil stabilization. Furthermore, between storms, vegetation transpires water and thus tends to deplete the moisture in the soil. This increases the capacity of the soil to absorb storm water that otherwise would contribute to rapid stream flow.

Terraces are ridges of earth that are built across the slope for the purpose of gently guiding water downhill at a nonerosive velocity. On most kinds of soil, terraces perform this function successfully. On the Texas Blacklands, terraces without sod-forming crops are not recommended.

When row crops are grown on terraces, the rows should be parallel to the terraces.

Tillage along the contour on slopes helps to hold storm water on the land. Continued cultivation has the disadvantage of lowering infiltration rates by destroying good soil structure. Loss of soil structure can be counteracted to some degree by the addition of organic matter. Any loosening of the surface soil generally makes the soil more susceptible to erosion when surface runoff occurs. Intensive cultivation is also conducive to the clogging of soil pores. "Minimum tillage" encourages more water to enter the soil. The term means running a tractor and tillage implements over a field the least possible number of times. This kind of tillage reduces soil compaction and destruction of soil structure. The spaces between the rows are left as

Left: **Floods are always a hazard, but most of them can be prevented. (***Courtesy*** Soil Conservation Service)** *Right:* **In one time over the field, this tractor implement breaks the wheat stubble, applies fertilizer, and plants corn seed. This kind of minimum tillage reduces soil compaction so that the soil will absorb more rainfall. (***Courtesy*** Allis-Chalmers Corporation)**

"rough" as possible, and only the soil along the plant rows is stirred intensively. The rough soil between the rows absorbs rainfall very quickly and does not seal over with the beating of the raindrops.

The Tennessee Valley Authority (TVA) is an example of how floods are controlled on a coordinated basis. Here in the valley of the Tennessee River, touching seven states, the Authority has built dams for flood control, navigation, and hydroelectric power. In addition, the TVA has made, and encouraged people to use, high-analysis phosphate fertilizer to grow more legumes, grasses, and other sod-forming crops. Also, the TVA has made nitrogen fertilizer and promoted its use, especially on winter pastures. The TVA has helped landowners to plant more trees and to improve the wildlife habitat. While one result is better flood control, the ultimate result is a higher standard of living for all the people.

When entire watersheds are treated properly, the soil is open, porous, and

Above left: Chiseling on the contour helps to increase water infiltration on western range lands. *Above right:* Terracing helps control the flow of water from fields in Nebraska.

Improved tillage implements such as this helical digger for shattering tillage pans also leave the surface rough and absorptive like a sponge. (*Courtesy* David Brown Tractors)

absorptive, like a sponge. Good water-shed management involves planting trees, cutting forests selectively, terracing cultivated fields, reducing the acreage in row crops, and increasing the acreage of grass and other sod-forming crops. Growing more legumes for soil building, using proper fertilizers, turning under green manure crops, keeping the proper number of livestock so as not to overgraze grasslands, and building dams are other recommended watershed management practices.

The worst flood damages occur along the Gulf coast of Texas, in the Northwest, in the Northern and Southern Great Plains, and in the Ohio and Mississippi River valleys.

A tropical hurricane, accompanied by high winds and heavy rains, caused serious damage to the Southeastern and Middle Atlantic states as late as 1972.

QUESTIONS

1 What percent of the weight of our bodies is water?

2 Where does your community obtain its water supply?

3 In Utah how many acres of watershed are necessary to supply irrigation water for one acre of cropland?

4 How deep may grass roots extend for water?

5 What diseases may be transmitted by polluted water?

6 Trace the journey of a single raindrop.

7 Where is water stored?

8 How can the storage capacity of the soil be increased?

9 Are ground water supplies being recharged as fast as they are being used?

10 How has the TVA helped to make proper use of water resources?

ACTIVITIES

1 If you live in a rural community, observe the sources of water for livestock and household use. Find out whether water is scarce or abundant in the community.

2 If your school is in or near a city that has a water department, have one member of the class call or write the department to obtain information on the source of water and the average amount of water used daily in the community.

3 Keep a diary of the precipitation that occurs in your community during a period of two weeks. Indicate approximate amounts of rain or snow, and record the occurrence of dew, frost, hail, or sleet.

4 After a rain obtain several samples of muddy runoff water from bare land and several samples of runoff water from woodland and good pasture. Allow the soil to settle to the bottom of each container. Explain how vegetation affects runoff.

5 A stream flowing through an area offers an interesting study for those interested in land, water, wildlife, and recreation. The stream furnishes drinking water for animals and favorable living conditions for many kinds of wildlife. Besides being the natural drainageway for the land, it may offer opportunities to fish, trap, swim, skate, and for other recreation. During periods of high water it may even be dangerous and cause losses from flooding. Is the stream beautiful, clean and clear? Find a stream nearby. Study it and learn about it. Cooperate with a group of interested youths to clean up a stream or lake.

20 SOIL AND ITS CONSERVATION

The soil is the godmother of all living things. From the soil come our food, clothing, and shelter. Bread, the staff of life, started as a grain of wheat planted in the good earth. The clothes on your back may have been picked from the open bolls of cotton plants growing in moist, fertile soil. This same soil also grew the forest trees that were cut to build our homes.

We have become so accustomed to concrete and steel, automobiles and airplanes, and the comforts and luxuries of modern life that we are inclined to forget our dependence on the soil.

There was a time in man's history when he, like the animals of the forests and plains, had to spend most of his time seeking food and a means of protecting himself from the weather. As recently as the time of Columbus, 19 of every 20 persons were engaged in the production of food, clothing, and shelter. Today, in the United States, because of the development of our scientific agriculture and because of our bountiful soil, one farmer is able to produce enough food for his family and for 44 other people. This means that fewer people make a living directly from the soil. But all who eat are just as dependent as ever on the land.

There is at least a little love of the soil and of growing things in all of us, and there is a lot of love of the soil in some of us. It is truly astonishing how much we are all gardeners. From the window box of the apartment dweller to the farm, and from the suburban garden to the great ranch, flowers, fruits, and vegetables sprout and grow. Nor are these masters of growing things always country people. On the contrary, the plants may be tended by hands accustomed to grease and wrenches, or lovingly cared for by mothers of the next generation. This interest in growing plants is very heartening. It makes one people of us all.

The failure to maintain an average crop yield over a period of years is usually blamed on the soil. The truth is that a crop of corn or cotton is a product of the joint efforts of man and soil. Mostly the responsibility for failure is man's —for he is the intelligent and directive head of the partnership.

The farmer's failure to maintain crop yields may result from the selection of a soil unsuited to corn or cotton. Or, having selected a proper soil, he may have failed to manage the soil intelligently. However, we must admit that on some soils it is almost impossible to make a living.

Sometimes the soil is referred to as a tool with which man accomplishes something, but it is not a simple tool such as a saw or a plane. Soil is as highly complex a tool as an engine or an animal.

Soils differ in their character and in their capacity to perform service, just as factories, engines, and animals differ. Different types require different care and treatment. It is upon the recognition and use of these differences in capacity and treatment that success depends. Through wise use man can make the soil productive; by misuse he may destroy its productivity.

The history of the 21 civilizations that have come and gone is dominated by man's desire "to own a piece of land." Traditionally, land means security. When social and economic laws discourage land ownership by every citizen, civilization declines.

The love of growing things may express itself in a productive garden. (*Courtesy* National Committee on Boys' and Girls' Work, Inc.)

To some people "a piece of land" means a 10,000-acre ranch. "All the land I want is what touches mine" is the expressed desire of greedy people. Other people are content to make a modest living from a section (640 acres) of rangeland. Still others want a quarter-section (160 acres) farm with good soils and a diversified farming system. Most people are happy just to own the city lot where their home is located. Vegetable gardens, flower gardens, beautiful lawns, and well-cared-for shrubs help to satisfy man's craving for a sense of "soil and security."

Left: Soils that have developed in humid regions from limestone rock are usually dark in color, clay in texture, and very fertile. Note that a grassed waterway (**G**) has been left. (*Courtesy* Allis-Chalmers Corporation) *Right:* A class of boys in agriculture study the soil in a national land judging contest in Oklahoma. (*Courtesy* Oklahoma Extension Service)

Many people live in such crowded quarters that they cannot even afford the luxury of growing flowers outdoors. These people get tremendous satisfaction from raising and loving house plants that can be grown in flower pots in the windows. This feeling of security is better than no security at all.

CHARACTERISTICS OF SOIL

Some people regard soil as a material to support buildings, bridges, and highways. Others consider soil mainly as part of the study of parent rocks from which it developed. Some people look upon soil as a storehouse of nutrients for crops, trees, grasses, and legumes.

The soil serves these and many other purposes. It nourishes our food crops and provides a means of livelihood for those who choose farming as a vocation. Soil is both a cause of and cure for pollution. It causes pollution by choking rivers and lakes with erosion sediments and sending clouds of dust into the air; it cures pollution because of its capacity to purify what man discards, such as garbage and other human waste products. Soil provides recreation and supplementary food for the backyard gardener. It is a very interesting—yet baffling—material for scientific study; and finally, all living things return to the soil when their life is finished.

Many people express interest in the soil only when it makes the headlines. It is spectacular news when dust storms fill the sky with small grains of soil that obscure the sun, when concrete highways are made impassable by soil subgrade failure, or when soil and rock trap miners in a coal tunnel.

Is it not just as interesting to marvel at the way a farmer raises a good crop of corn on a rich, mellow silt loam; or to feel sad when we see a tenant family cultivate a worn-out, eroded clay hillside that supports only stunted cotton plants?

Do you ever wonder how soil comes into being? The true story of the dynamic birth of soil gives us an insight into man's greatest resource.

The surface of the soil is the common meeting place of the climate above and the bedrock beneath. The surface soil is also the place where plant and animal life, which are organic matter, meet the mineral matter from the rock below.

The hills and valleys and plains of Mother Earth are acted upon for centuries by Father Time, with the help of the sun, rain, plants, and animals. As a result of this action a soil is born.

Now what is soil? *Soil is a natural medium for plant growth.*

Perhaps you have seen water hyacinths growing in water. Is the water a "soil," too? Yes, very broadly speaking, this is a water soil or *hydrosol.*

Lichens grow on rocks! Are rocks called "soils"? Rocks are soil also; they are *lithosols.*

Plants such as Spanish moss grow in the air on telephone wires. Could the air be called a "soil"? Yes, the air is a natural medium for plant growth, so these soils are *episols.*

Most soils with which we are familiar have developed from upland unconsolidated (soft) materials. These are mixtures of organic matter, sand, silt, clay, gravel, and rocks. Any such soil material exposed to the climatic elements will eventually grow plants. When it does, it is a soil.

Parent material and soil formation. Mother Nature left sandstone in some places, and in other places she left limestone or granite. In a few locations, shale, schist, gneiss, and lava occur as the native bedrock.

Whatever bedrock is present, the natural forces of weathering by rainwater and the action of plants have been working for centuries to wear it away. All the time that the rocks have been decomposing, other forces in nature have been building up a soil from the decomposing rock products. New chemicals have been formed in the soil from the decay products of the rocks and plants.

Humus has been built mostly from the plant residues. Soils that develop from granite and sandstone are sandy (coarse-textured) and infertile. Soils from shale have the texture of clay and are not very fertile. By contrast, soils that develop from limestone or lava bedrock in humid regions are usually fine-textured and very fertile.

After rocks are weathered into finer particles by climatic and biological forces, they may be moved before coming to rest to make soil. This means that a limestone or sandstone bedrock may be covered by soil material moved there by nature's transporting agencies: water, wind, gravity, and glaciers. When this

Many soils in the arid West contain so much sodium that they will not grow anything until some of the sodium is replaced by calcium in calcium sulfate (gypsum), as shown here. The crop on the right is barley. (*Courtesy* L. E. Dunn, University of Nevada)

soil material is 10 feet or more in thickness the soil that develops may escape being influenced by the bedrock. Parent materials deposited by glaciers over the northern part of the United States, and the marine-deposited sediments along the Gulf and Atlantic coasts, are examples of soils that have developed without influence from the deep bedrock.

Climate and soil formation. Suppose that a limestone boulder as big as a city block were to appear above the waves in the Atlantic Ocean near the Statue of Liberty. After exposure to the rain and the sun, the surface of the rock would start to get soft and "dust off." The changes in temperature from day to night and from season to season would break the rock into small pieces. Rainwater would dissolve some of the limestone and soften other parts of the stone. Wetting and drying would cause expansion and contraction of some of the loosened particles, and more destructive weathering or "rotting" of the stone would result.

Life and soil formation. While the climatic weathering was taking place, small plants would start growing on the thin layers of rock flour. After many years of the growth of these tiny plants, other plants of a higher form such as the true mosses would become established. Generations of lichens and mosses would grow and die until the organic matter from their decaying tissues helped to make a better place for higher plants to start growing. Many annual weeds might migrate into the area, followed in a few years by grasses. Perhaps grasses would occupy the area for several centuries. Then finally a forest might gradually dominate the original rock.

Topography and soil formation. If the limestone boulder in the Atlantic or Pacific Ocean or the Gulf of Mexico were tilted slightly, permitting rainwater to drain rapidly, the time required to develop a hardwood forest would be longer. With less water staying on the boulder after each rain, plants would grow more slowly, and as a result less organic matter would be returned to help the next generation of plants to grow. On the other hand, if the boulder had a slight depression in it that would hold more water, the process of plant succession would be faster. Under these conditions soil would develop more rapidly. In like manner, when the land is hilly the steep slopes develop soil slowly, while the more level areas make soil faster.

Age and soil formation. One way to determine the age of a soil is to start from the time the rock was first pushed out of the water and record the number of years to the present. This is *geologic age*. The time required for the development of a mature soil from this rock may vary from 30 to several thousand years.

The second way to determine soil age is by the distinctness of the soil horizons (layers) in the profile (cutaway view). This is *pedologic age*.

Soils develop faster from sandy glacial material than from granite rock. A mature soil may develop from soft glacial material in 300 years, but it may take 6,000 years to develop a mature soil from limestone bedrock.

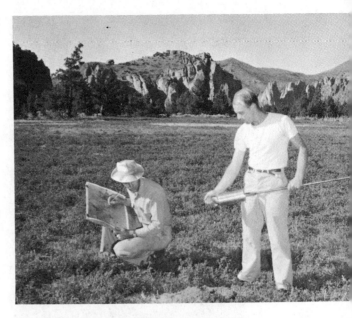

The soil in the foreground formed from rocks. Rocks in the background will be soil some day. The men are mapping soils. (Oregon) (*Courtesy* Soil Conservation Service)

Air and water do not pass readily through this silt loam (Uniontown silt loam in Kentucky). (*Courtesy* Soil Conservation Service)

There is not enough lime in these Minnesota soils to supply high-lime forage for the cows. As a consequence they chew on bones, bark, or whatever they can find to try to satisfy their hunger. (*Courtesy* Minnesota Extension Service)

THE PARTS OF A SOIL

Soil contains minerals, water, air, and both living and dead organic matter.

Minerals in the soil may vary in size from clay to sand to boulders. Salts in solution are too small to be seen but they too are among the soil's minerals.

Air and water in the soil are in the openings (pores). The pores are made by plant roots, animals such as earthworms, and by cracks in the soil that result from wetting and drying.

Organic matter supplies energy for all life in the soil. The dead remains of plants and animals are filled with small living bacteria and fungi, while plant roots, earthworms, and many other organisms are part of the living organic matter in a soil.

Physical parts of a soil. A cubic foot of surface soil in good condition for plant growth contains about half air space and half solid matter. The solids consist mostly of minerals and include about 3 percent active organic matter. When the soil supports plant growth, approximately half of the pore space is filled with air; the other half is water.

Chemical parts of a soil. The organic and mineral parts of the soil are sometimes taken apart by soil scientists. They find that almost all of the 100 or more elements that occur in nature are found in the surface soil, but that most of the elements occur only as traces.

Almost three-fourths of the soil is silicon, the material from which glass is made. (In fact, sandy soils are used in the manufacture of glass products.) Aluminum is the next most abundant mineral element in the soil. The most common aluminum ore, bauxite, is a type of clay. Over 80 percent of most soils is made up of these two elements. Third in amount is iron, which averages approximately 3 percent in most soils.

The quantity of fertilizer and lime elements in virgin surface soils is relatively small. Total nitrogen and total phosphorus (P_2O_5) average approximately 0.15 percent each, total potash (K_2O) about 2.0 percent, and lime (CaO) 0.5 percent.

SOIL CONSERVATION

Soils are always changing. Century by century, year by year, even day by day, soils change.

Soils have always eroded to pollute the environment, but with wrong use they are now eroding faster. *Left:* Water erosion in South Dakota. *Right:* Wind erosion in Oklahoma. (*Courtesy* Soil Conservation Service)

Before men started using the soil, the change was very slow. Centuries were required for marshy soils to become drier, and for some dry soils to be moistened by a rising water table. Many arid soils became saltier, while soils in high-rainfall areas lost salts by leaching.

Soils in the path of great ice sheets were buried beneath glacial debris. Soils on sloping lands lost topsoil each year, and soils in valleys received new layers of topsoil. Soil formation and development was influenced by differences in rainfall and in temperature, by fires set by lightning, and by the grazing of the vegetation by wild animals.

Then along came man. He plowed the soil and planted crops, while domestic animals in increasing numbers grazed the native grasses. Through man's influence, soils changed faster.

No longer do marshes take centuries to dry; man drains them in a few days.

Some dry-land areas are changed into marshlands overnight by irrigation. Salty soils often result from this practice. Salty soils with open subsoils may be flooded with irrigation water and in a few day's time may leach toxic quantities of salt from their surfaces.

Man has opened ways for erosion, too. At one time, centuries were required to erode one inch of soil; now the same erosion may result from a single rain.

Yes, soils are always changing, and man now changes them more rapidly. At first glance it might seem that all of man's influence on the soil is bad. "Not so," say the majority of landowners. For with proper use of the same tools that bring destruction, man now can build soil faster than ever before. With fertilizers and lime so cheap, fertility can be restored. Soil-building legumes and grasses can be planted, fertilized, and limed in a single machine operation.

Never before in history has the land-owner been given so much help in building soil—and never before have these practices paid so handsomely.

MAN AND LAND OUT OF HARMONY

By far the majority of soils never were ideal for the growth of farm crops.

In the beginning, soils were formed largely from limestone rocks, sandstone rocks, and granite. In some places the bedrock was shale. Other soils developed on geologic materials that had never hardened into rock.

The soil inherited its fertility from whatever kind of material was present. Wherever parent material was low in lime, the soil also was low in lime.

Man inherits these contrasting soils and demands that they grow satisfactory yields of corn or cotton or pasture grasses and clovers. Some soils *will* grow good crops of corn year after year without much care; other soils never had sufficient nutrients to do so.

Now, suppose that man wants to raise corn on a hillside soil formed from sandstone. In its virgin state that land would not yield even one satisfactory crop of corn. So the soil must be "doctored" with soil-building grasses and

Man and land out of harmony—lower productivity, more pollution. (*Courtesy* Soil Conservation Service)

legumes, properly limed, and fertilized. In this way man overcomes the natural disharmony between the land he inherits and the kind of crops he wants that land to produce for him.

When man misuses the land. But not all men who inherit land and farm it take good care of it. In fact, the same love-of-land that hastened the settlement of our own country also brought about the misuse of the land.

Men came to America for land, free land. They left their native countries for political, economic, or religious reasons; they selected America as their new home because they knew that land was here for them. Land meant freedom.

In a little more than 300 years, Americans hewed and dug and plowed and built until they set up a nation richer by far than any other recorded in history. America became known the world over as the land of opportunity. Such it was, and such it still remains. Nevertheless, somewhere along the line of progress many people forgot that the bounty of the land was the basis of their prosperity.

Then came dust storms and floods, the results of misuse of the land. There was nothing new about dust storms or floods, for they have occurred since man can remember, but none such as these had occurred before. Dust from the west rose high in the air and blotted out the sun over Washington, D. C. River flood crests rose many feet higher than ever before. The cost in lives and property was enormous.

With continuous cropping, organic matter is gradually decreased because of constant stirring of the soil. Rain does not soak into the ground as fast as it once did. More water flows over the surface, carrying away topsoil. Crop removal also takes away fertility. These factors team up to deplete the soil.

People on the march. Poor soil cannot provide comfortable homes, good schools, and opportunity for recreation. Many homes that were splendid when the land was fertile degenerated as the soil was depleted. The owners of many homes on poor land have moved away, leaving their land in care of tenants. But even the most ambitious tenants cannot make a good living or a prosperous community where the land has washed or blown away.

Short-term tenant leases hasten soil depletion. A "cropper" has little incentive to save the soil when he knows he must leave for another place at the end of the season. When tenants have long-term leases, which may lead to ownership, it is to their advantage to plant more soil-building crops and to follow a system of land use that conserves the soil and increases its fertility. Over a period of years the fertility and the care that are put into the soil return to the farmer in the form of pride and profits.

Irrigation that may spoil the land. In spite of the farmers' many years of experience in irrigation, more than one million acres of irrigated land had to be abandoned in the 11 Western states. The magnitude of this problem is apparent when we realize that the million acres came from a total of only 20 million acres under irrigation in these states. Abandonment of the land was due in large part to the accumulation of salts in the soil.

The salt is either already present in the soil and moves up into the root zone when irrigation water is applied, or it is brought in by the irrigation water.

When man and land are in harmony the land is more useful and beautiful. (*Courtesy* Soil Conservation Service)

In many cases salt-laden water is the only water available. Careful and controlled use of water and the development of adequate drainage are the means by which this type of soil deterioration can be prevented.

Natural erosion. There has always been erosion; there always will be. It is a natural geologic process that man did not cause and that man cannot stop. Geologic erosion is a fundamental process of land sculpture.

Man-made erosion. Although man did not create the wind or the rain, some of man's land-use practices invite greater destruction of the soil by both wind and rain.

For example, the Great Plains were once covered with grass that sheltered the soil. Breaking this sod and planting crops, mainly wheat, started a chain reaction that led from a decrease in organic matter and a crusting of the soil to dust storms and flowing mud. "Wrong side up," a Sioux Indian called it in 1836 as he watched a North Dakota farmer plow under virgin sod.

Many native grasslands felt the pressure of livestock. Overgrazing by too many livestock made some perennial grasses disappear. Short-season annual grasses, weeds, poisonous plants, and brush took the place of long-lived palatable grasses. The rain did not soak into the soil as far or as fast, while deep channels for water to enter the soil, once provided by long-lived grasses, became fewer. Spongy organic matter, once abundant and soft, decomposed into nothingness.

MAN AND LAND STRIVING FOR HARMONY

One little bee, working alone, can make a pound of honey in 64 years. That is, one bee could do this if he could whistle to keep up his spirits and live that long. Fifty thousand bees in one hive, all working together, can make a pound of honey in an hour. This proves that teamwork is necessary to get the job done.

However, in farm life, teamwork alone is not enough. The teamwork must be tempered with the will of the farm people to help themselves, with the intelligence of the agricultural workers who help them, and with the willingness of the public and politicians to understand and assist with the farmers' problems.

Laws to aid soil conservation. In 1928 Congress adopted an agricultural

Left: Soil Conservation Service technicians map soils for Soil Conservation District members to improve environment and increase income. *Right:* Beating raindrops strike bare soil until flowing mud bleeds the life from the land to pollute the environment. (*Courtesy* Soil Conservation Service)

bill with an amendment that provided for the establishment of ten soil-erosion experiment stations. These stations, which were to gather facts about soil erosion on important types of agricultural land in the United States, were located in Washington, Kansas, Iowa, Missouri, Wisconsin, Ohio, Oklahoma, Texas, and North Carolina. Most of the experimental work was done between 1930 and 1940. The results of research proved that destruction of the land could be halted and that damage could be repaired.

In 1933, while these experiments were going on, the Soil Erosion Service was formally established in the United States Department of the Interior. Soil and water conservation projects were set up with the cooperation of farmers and ranchers in areas throughout the country where erosion problems were most severe.

In 1935 the national erosion-control program was placed under the United States Department of Agriculture, and the Soil Conservation Service took the place of the Soil Erosion Service.

Using more than 50 years of previous experience and experiments, the Soil Conservation Service made a survey of the soil conditions throughout the nation. The following conclusions were reached:

1. Soil erosion by wind and water is severe over most of the country.
2. Yields of agricultural products decrease because of continuing soil erosion.
3. Soils are different, each kind is better adapted to some use than to another, and cultivation should not be tried on some soils at all.

4. No single treatment for "sick" soil will solve the problem any more than one medicine will cure all the ailments of sick people.
5. The problem is not only ecological and environmental; it is also economic, social, and political.

Meanwhile, the farmers and ranchers found that any person working alone to control erosion on his land faces an almost impossible task. Wind and water recognize no fences and no boundaries except those of nature. The problem of erosion extends not only from acre to acre but from farm to farm, from ranch to ranch, and beyond. If soil erosion is to be controlled, it must be done by cooperation among farmers and ranchers and among the people in general.

Soil conservation districts. In 1937 the United States Congress authorized the organization of soil conservation districts. The main purpose of this legislation was to encourage soil conservation at the local level. During the first year of this program, soil conservation districts were organized in North Carolina, South Carolina, Georgia, Arkansas, South Dakota, Utah, and Nevada. Now there are 3,026 soil conservation districts throughout the United States and they include 99 percent of the farms and ranches.

Each state has a State Soil Conservation Committee to administer the soil conservation district law. Soil conservation districts are created and operated by the landowners. Once created, the district may call upon Federal and state agencies for assistance. The district is able to undertake projects too large for one landowner.

This Colton Loamy sand in Maine is ideal to reduce pollution from cattle feed lots, sanitary landfills, and septic tank filter fields. The main hazard is that if large amounts of polluted water are released into it the water may move downward so rapidly as to possibly pollute water in shallow wells. (*Courtesy* Soil Conservation Service)

Each soil conservation district is governed by an elected group. Some states call the governing bodies board of supervisors; others, commissioners or directors.

Action of newly-created soil conservation districts begins with a study of existing soil conditions and preparation of plans of action to improve these conditions. The district then calls upon the Soil Conservation Service, the land-grant universities, and others for help. A local work unit is formed to permit farm-to-farm cooperation in such activities as drainage or terracing.

The soil conservation district cooperators work in unison to obtain more effective action to control soil erosion,

build better terraces and farm ponds, establish irrigation systems, and improve the rural environment. Some of the practices to improve the environment qualify for partial payment through the Rural Environmental Assistance Program.

Improvement in the quality of the environment of the farm can be accomplished in hundreds of ways including the timely spreading of animal manure, the proper location of cattle feeder lots, and the sanitary disposal of human wastes.

Manure should be spread every day that the soil is porous enough to permit water to move into it. This reduces the amount of water that runs off the field, carrying manure to pollute the streams

and lakes.

Cattle feeder lots, sanitary land-fills, and septic tank filter fields should be located on soils that permit water to move into and through them readily. This reduces runoff water and holds pollution of surface waters to the minimum.

Hugh H. Bennett, the father of soil conservation and first director of the Soil Conservation Service, estimated in 1939 that 3 billion tons of soil were lost in the United States by water and wind erosion. Twenty-eight years later the director of the same agency estimated that not only had no progress been made, but that 4 billion tons of soil are being lost annually.

Above: Manure should be spread on the fields every day that the soil permits water to move into the soil. An open soil absorbs rainwater instead of allowing it to flow over the surface to carry manure into streams to pollute them. (*Courtesy* New Idea Machinery Company) *Below:* Irrigation need not spoil the soil by increasing toxic soluble salts at the surface. This field on the High Plains of Texas is being irrigated properly with water that is low in soluble salts. Water is pumped from deep wells into the pipe in the foreground. Water enters each furrow from openings (gates) in the large pipe and soaks into the soil along the 700-foot row. (*Courtesy* Texas Agricultural Experiment Station, Lubbock, Texas)

QUESTIONS

1   Why are soils changing faster now than formerly?
2   Are all soils the same? Why?
3   Explain: "Land meant freedom."
4   What is the relationship between impoverished soils and unstable communities?
5   Name two hazards of irrigation.
6   Differentiate between natural and man-made erosion.
7   There are 3,026 soil conservation districts in the nation. Do you live in one?
8   What is "soil"?
9   What is the relationship between rocks and soil?
10   Describe a soil near the place where you live.

ACTIVITIES

1   Examine a road cut or the bank of a gully, observing the layers (horizons) of the soil profile. Note the color and thickness of the top layer, but remember that this may not be the original topsoil.
2   With the help of your teacher, your county agricultural agent, or a Soil Conservation Service technician, obtain soil samples from farms in the community and send samples to your land-grant university for testing.
3   Obtain samples of topsoil rich in organic matter. Spread the soil on paper and look for earthworms and other living organisms. Explain how organic matter affects life in the soil.
4   Arrange a class field trip to a farm on which good soil and water conservation practices are being followed. Make notes for later use in a class discussion.
5   Prepare your own land judging score card, and practice judging land as a class activity.

21 FORESTS

Forests consist of trees that provide lumber, homes and food for wildlife, natural beauty, and soil protection. Trees also temper the weather, screen dust from the air, suppress loud noises, dissipate unpleasant odors, reduce pollution, and increase our supply of oxygen.

A tree is a marvelous factory. Its leaves manufacture food from water and minerals that are pumped from the roots and from carbon dioxide absorbed from the air. Energy from the sun is then used by the leaves to manufacture sugars, starches, fats, proteins, and cellulose.

TREES AND PIONEERS

Trees were a nuisance to the early pioneers. The fear of hostile Indians behind each tree ran uppermost in their minds. And with forests everywhere, not enough grass could grow for the livestock. Neither would corn and tobacco grow under the shade of trees.

Trees had to give way to clearings for the production of food and of materials for clothing. Trees had to be slashed and piled, and burned or rolled into ravines. Can we blame our forefathers for wanting to live and to eat? They could not eat wood. What would you have done under similar circumstances?

But trees also had their uses to the early pioneers. From the widest selection available to any pioneers in the world, our forefathers cut durable white oak for log homes, forts, and stockades. Nearby forests provided sugar maple for excellent sugar and syrup, walnuts, butternuts and hickory nuts to eat, and a dye from the butternut tree.

Squirrel potpie made from acorn-fat squirrels was quite a treat. Acorn-fat

Above left: Sap from the sugar maple tree runs into a vat where it is boiled to make maple syrup and maple sugar. (*Courtesy* University of Massachusetts) *Above right:* The longleaf-slash pine forest occurs on the Gulf Coasts of Alabama, Louisiana, Mississippi, Georgia, South Carolina, and over most of Florida. (*Courtesy* U. S. Forest Service) *Below:* Wild elephants and forests belong together in humid southern India, Ceylon, and Africa. This "wild" elephant has been trained to do the heavy work in loading logs. (*Courtesy* FAO of the United Nations)

This forest of white pine, hemlock, sugar maple, white oaks, and red oaks is almost exactly like the forests which the early American colonists found around them. (*Courtesy* Maine Department of Development)

turkeys were "just around the corner." Hides of deer, foxes, and wolves that lived in the forest made fine robes, clothing, and rugs. The raccoon-skin cap was in fashion.

By the time the pioneers had developed several generations of "hand-me-down" skills in dealing with trees, many people were migrating westward. A new foe, shorter and more numerous, became the barrier. The new barrier was grass.

Grass prairies and plains required new skills. What had happened to the springs that had been so plentiful in the forested East? It was farther to the next stream, and sometimes only dry channels were found.

What kind of land was this? What would they burn to cook their food? How would they heat their homes? Trees were no longer just a few steps away.

To get water and a few trees for fuel and shelter, pioneers stayed close to the few streams as they crossed the plains. When trees were scarce, buffalo chips (dried manure) were burned for fuel.

As the pioneers reached the Rocky Mountains, trees again welcomed them. These trees were not familiar to the people who had left the Appalachian Mountains. Rocky Mountain trees were nearly all cone-bearing. White oak, chestnut, red oak, hickory, and black walnut were no more. In their place grew some pines that bore edible nuts. Lodgepole pine, ponderosa pine, and Douglas-fir grew in abundance. Incense-cedar, white cedar, and several spruces grew at different elevations. Settlers soon learned to use these newly-found species, many of which were giants compared with those they had left behind.

Half of the journey from the Atlantic to the Pacific had been made through forests. Forests covered 900 million acres, and the nation's population at the time of the westward push across the plains was almost 900,000. This was an average of 1,000 acres of forest land for each person.

Now there are 630 million forest acres and over 206 million people. This is more than three acres of forest land per person. With wise conservation, these three acres of forest should supply all of our needs for forest products such as lumber for homes and paper like that on which these words are printed.

PRESENT AND FUTURE
TIMBER SUPPLIES

There is some kind of forest cover on one-third of the nation's land area. Two-thirds of the forested land is suitable for timber production. In other words, almost one-fourth of our land area is in commercial forests, or land which could be managed for that purpose. Softwood sawtimber is the leading forest product. Approximately 80 percent of the sawtimber in the United States is softwood.

The land area east of the Rocky Mountains comprises 73 percent of the commercial forests. In this portion of the United States, the Southeast has more commercial forest area than any other region. It is anticipated that softwood production will shift southeastward as old growth in the West is harvested and young stands of timber in the East are developing.

A major factor in growing timber as a crop is the relatively long time from seedling to commercial tree. The small trees growing now will not provide sawtimber for another 40 to 50 years. The production of timber is a complex, time consuming enterprise. Some of the factors which affect forest management are tree species, size of trees in timber stands, density of stands, growth rates, and damage by fire, insects, and diseases. The value of trees, especially hardwoods, varies greatly from trees worth a few dollars each to high-grade veneer timber worth several hundred dollars a tree.

More than half of the commercial forest land in the United States, exclusive of Alaska and Hawaii, is in small ownership units which average about

The South has most of the fast-growing timber in the United States. Here a southern pine tree is being harvested for lumber. (*Courtesy* McCulloch Corporation)

49 acres in size. One-fifth of all farm acreage is forested. These small farm forests contain approximately one-third of all forest land in the United States. The East has about 97 percent of the small privately owned forests.

The management practices of individual forest landowners and the policies of government agencies responsible for public forest land will determine our future timber supply. Millions of acres of forest land need restocking, improved management, better fire protection, and more efficient harvesting methods.

THE REDWOODS

For almost 50 years the "Save-the-Redwoods League" has been promoting

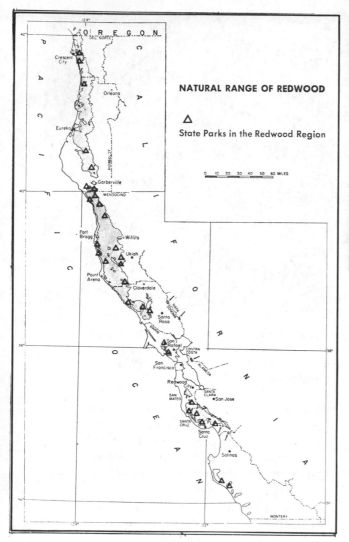

NATURAL RANGE OF REDWOOD

△

State Parks in the Redwood Region

0 10 20 30 40 50 60 MILES

Left: The triangles indicate location of the 30 redwood-type state parks in the natural range of the redwood. (*Courtesy Journal of Forestry*) *Right:* How old is this tree? (*Texas*) (*Courtesy* Soil Conservation Service)

their beginning before the birth of Christ and inspire a feeling of reverence for the wonderful laws of nature.

Let's look at the problem of the redwoods from a scientific viewpoint tempered with a concern for posterity. Redwood lumber is beautiful and durable. The redwoods are among the fastest growing sawtimber trees in existence. For example, in Del Norte County, California, a stand of redwood trees that regenerated naturally from seed and stump sprouts was 120 years old, 60 to 80 inches in diameter, and more than 200 feet tall.

Forest scientists look at a redwood forest as a community of living trees capable of regeneration and requiring much the same kind of management that is needed for a deer herd. If specified male deer were "harvested" by hunters,

the preservation of the redwood tree, the tallest of all living things.

Redwood trees on the Pacific Coast can live more than 3,000 years. See if you can find a record of anything with a longer life span. Some of the redwoods attain a height of 350 feet and a diameter of 15 feet. Such giants deserve preservation for the benefit of future generations. Some stands of redwoods had and if the old, nonbreeding does could be easily identified at a distance and also harvested, the total population of deer could be healthier and the numbers kept in balance with the food supply.

So it is with all trees, including redwood trees. When an individual tree is ready for harvest it should be harvested. Sprouts from the stump and seeds will produce other redwood trees to take its place. If a redwood forest were treated like a petrified forest and no harvesting were permitted, the trees would supply no useful lumber and would eventually become overripe and die. Other species of trees such as the Douglas fir and the tanoaks would finally dominate the over-ripe redwoods. In time the bay and lowland white fir, which are more tolerant, would replace the Douglas fir and the tanoaks.

Redwoods are a renewable resource and can be used forever as a source of lumber and scenic beauty. One hundred and fifty square miles of redwood forests over a 500-mile range have been set aside in 30 California state parks. This is about 5 percent of the forests in which redwoods are dominant. Forest scientists say that these protected redwood areas should be sufficient for sentiment and public enjoyment and that the other 95 percent of the redwood forests should be managed for lumber production. Foresters can be trusted to harvest only the ripe redwood trees and thereby to improve and preserve the redwood forests.

The oldest known redwood (*Sequoia gigantea*) is believed to be 3,212 years old. The life span of this species has been estimated at 6,000 years. In comparison, the life span of the bristlecone pine (*Pinus aristata*) has been estimated

at 5,000 years. The oldest known bristlecone pine is believed to be 4,600 years old. It is in the White Mountains of California.

In 1967, a redwood tree 369.2 feet tall was considered the tallest tree in the United States. The tree is in the Humboldt Redwoods State Park in California. The alluvial soil in which the tree is growing was deposited by Bull Creek.

The tree believed to have the largest diameter is a *Sequoia gigantea*, "General Sherman." The diameter is 34 feet, and the weight of the tree is estimated at 1,320 tons. It is also estimated that the tree contains enough wood to build 35 five-bedroom houses.

In 1967 the tree marked "T" was believed to be the tallest tree in the United States. It is a redwood tree 369.2 feet tall in the Humboldt Redwoods State Park in California. (*Courtesy University of California*)

FORESTS AND THEIR USES

Upon first thought, a forest is simply a large group of trees. Perhaps this is true, but there is a deeper and more meaningful definition and description of a forest.

A lone shade tree with plenty of growing space must struggle to absorb enough water and plant nutrients for proper growth. The same species growing close to other trees in a forest must struggle even harder for survival. Interlocking root systems and overlapping

Forests are more than just trees. Forested areas provide lumber, watershed protection, grazing, recreation, wildlife, and many other things of value.

TIMBER PRODUCTION

WATERSHED PROTECTION

GRAZING

RECREATION

OTHER USES—BERRY PICKING, ETC.

WILD LIFE

branches compete for water, plant nutrients, light, heat, and growing space. But not all closeness is bad; some is good.

When trees grow closely they grow taller, straighter, and with fewer limbs to make knots in the lumber. And seedlings of some trees start growing only in the dense shade that closely growing trees provide.

Trees growing in a forest help each other to protect the soil against loss of plant nutrients by erosion and leaching. Erosion losses are less in a forest because of the mutual protection each tree gives the other trees. Trees intercept much of the rainfall and cushion the force of the falling raindrops. Leaf litter on the forest floor keeps the surface open and porous, thus increasing the absorption of rainfall. Tree roots make channels through which rainwater moves down into deeper soil layers. All

Above: The North Dakota windbreak at the right protects a crop of corn. *Below left:* The principal use of forests is to produce lumber for building homes. (Pennsylvania) *Below right:* A pure spruce stand in this Maine forest can be used to make paper like that in this book. (*Courtesy* Soil Conservation Service)

these protective forces add up to less run-off water and less erosion from forested areas and a better quality environment.

More water entering the soil under a forest should mean more losses of plant nutrients downward to the water table and out of reach of growing trees. This is true to some extent. It is also true that tree roots are constantly "pumping" soil nutrients to the leaves. Leaves then return these same nutrients to the surface of the soil when the leaves fall to the ground and decompose.

Forests are more than trees. Forests provide shade, nesting places for birds, and food for squirrels and earthworms. Forests are the best known protectors of our watersheds. In addition, forests provide many useful products. Forests even invite man's meditation and inspire his thankfulness to God. Forests and man are inseparable.

Forests may be used for many purposes at the same time. Growing trees produce lumber and other wood products. While growing, these same trees protect watersheds, increasing the amount of water available for cities, industries, wildlife, and recreation. Many forest lands can be grazed, and openings in the forest produce wild berries for man and wild animals.

Of the 630 million acres of forest land in the United States, nearly all is available for wildlife food and cover. Much of this huge area is used also for recreation such as fishing, hunting, camping, boating, and swimming.

Nearly 465 million acres of this total forest land offers the best possible watershed protection. Four hundred sixty-two million of these acres produce timber of commercial value. A total of 342 million acres, some of which overlap the acreage of protection forests, are grazed by range livestock.

Noncommercial forest land is land on which forest trees do not get large enough to sell for lumber or pulp. The United States has about 168 million acres of noncommercial forests.

With the proper application of science, our forests may be more widely used to serve our environmental needs.

HOW CLIMATE INFLUENCES FORESTS

Climate controls the plant and animal world. The sun and the wind and the rain determine whether trees and squirrels or grass and antelope will occupy a given area. Of the climatic factors, rainfall and temperature largely determine what vegetation will grow. Cold weather over the years favors conifers such as the spruces and firs.

Rainfall exerts a tremendous influence over the kind of vegetation that is found, as well as how fast the vegetation grows. Annual rainfall in excess of 60 inches a year gives rise in temperate regions to a rain-forest of Douglas-fir, Sitka spruce, cedar, hemlock, and, in certain places, redwood and giant sequoia trees. The same rainfall in tropical regions produces a jungle of hardwoods. Regions receiving this much water are termed *superhumid*. Average July humidity (moisture) for superhumid areas is about 75 percent. This is a very damp air.

Humid regions are those that receive from 30 to 60 inches of rainfall annually. These areas favor the growth of such hardwood-conifer mixtures as sugar maple, beech, yellow birch, hemlock, balsam-fir, and white spruce. Sixty per-

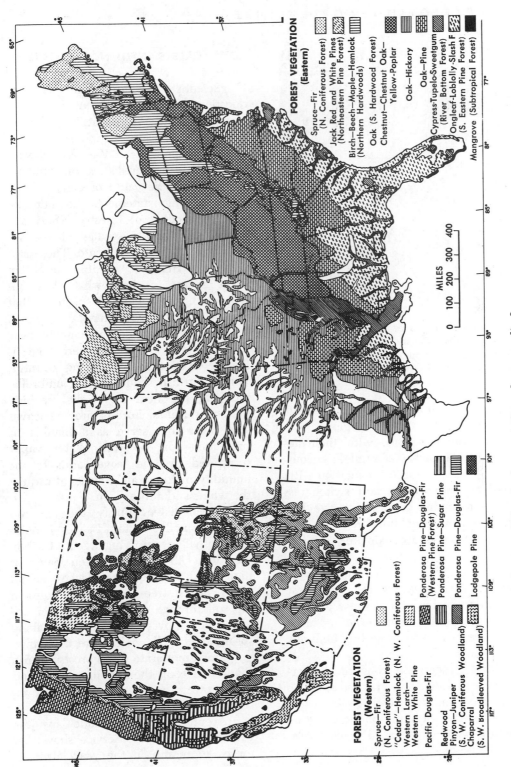

FOREST VEGETATION (Western)

Spruce—Fir
(N. Coniferous Forest)

"Cedar"—Hemlock (N. W. Coniferous Forest)

Western Larch—
Western White Pine

Pacific Douglas-Fir

Redwood

Pinyon—Juniper
(S. W. Coniferous Woodland)

Chaparral
(S. W. Broadleaved Woodland)

Ponderosa Pine—Douglas-Fir
(Western Pine Forest)

Ponderosa Pine—Sugar Pine

Ponderosa Pine—Douglas-Fir

Lodgepole Pine

FOREST VEGETATION (Eastern)

Spruce—Fir
(N. Coniferous Forest)

Jack Red and White Pines
(Northeastern Pine Forest)

Birch—Beech—Maple—Hemlock
(Northern Hardwoods)

Oak (S. Hardwood Forest)

Chestnut—Chestnut Oak—
Yellow-Poplar

Oak—Hickory

Oak—Pine

Cypress-Tupelo-Sweetgum
(River Bottom Forest)

Ongleaf-Loblolly-Slash F
(S. Eastern Pine Forest)

Mangrove (Subtropical Forest)

MILES

0   100   200   300   400

Forest vegetation of the United States. What is the vegetation where you live? (Areas not in forests support tall-grass prairies, short-grass plains, or desert vegetation, depending on rainfall.) (*Courtesy* U. S. Forest Service)

Planted eucalyptus trees are growing rapidly in West Oahu County, Hawaii. Forests in Hawaii occupy approximately one million acres. (*Courtesy* University of Hawaii)

cent relative humidity is the average July reading for a humid region.

Subhumid regions, with their annual 20 to 30 inches of rainfall, encourage tall-grass prairie vegetation. Big bluestem, Indiangrass, and switchgrass are the main tall grasses. Average July humidity in the subhumid region of the United States is approximately 50 percent.

Ten to 20 inches of rainfall characterizes the *semiarid* region of the short-grass plains. Buffalograss and the grama grasses are common here. Thirty-five percent is the average July humidity in this area.

Desert vegetation, consisting of cactus, yucca, and a few grasses, grows in areas receiving less than 10 inches of rainfall. Here the atmospheric moisture for July averages about 25 percent relative humidity.

HOW FORESTS INFLUENCE CLIMATE

Scientists have argued for years the question of whether or not forests influence climate. Arguments run the entire scale from: "Forests increase rainfall and stop floods," to "We have always had floods, and forests are only the result and not the cause of climate."

Sweeping statements do not advance the cause of forestry. Many scientists have made studies of the influence of forests on climate. These studies have made it possible to obtain exact figures to show what close observers have always suspected, that forests *do* influence climate.

Forests influence the climate over the area they occupy and exert some influence over the surrounding country. Forests furnish shade like an umbrella, reducing the heat and light reaching the soil surfaces. Forests provide leaves and twigs that serve as a blanket of protection for the soil. Water vapor transpired by tree leaves adds to the humidity in forest air. Forests also quiet the winds and add oxygen.

Green leaves purify the air and make it smell sweeter. Leaves contain chlorophyll, a purifying agent. Carbon dioxide is liberated by green leaves at night and by decaying leaves at all times. Forest tree roots punch holes in the soil through which more water enters and moves downward, making springs and wells yield more uniformly.

More moderate air. Forests furnish shade that cools the hot summer air and that protects our eyes by subduing intense sunlight.

Accurate records made by the United States Forest Service show that for every month in the year, air tem-

Above: Forests cool the hot summer air and warm the cold wintry gales. (Minnesota) (*Courtesy* U.S. Forest Service) *Below:* In winter the soil is warmer in the forest than in a pasture or a field. (*Courtesy* Wisconsin Conservation Department)

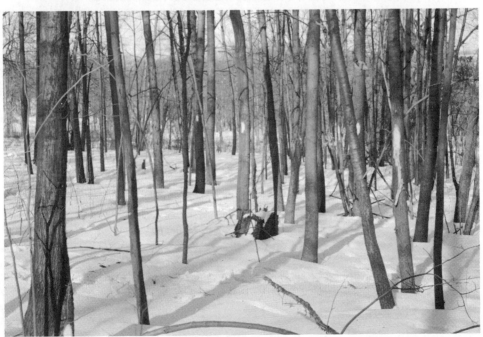

peratures in the forest fluctuate less than do those in open areas. Forests reduce the air temperature when it means the most to our comfort. For example, the hotter the day, the greater is the difference between temperatures in the forest and in the open. On hot days it is often 5° F. cooler in the forest. On the coldest days the temperatures average 2 degrees warmer in the forest.

Year in and year out, forests lower the air temperature an average of 1° F. This figure is for forests growing at low elevations. At elevations of 3,000 feet, the air temperature in a forest is 2° F.

lower than the air temperature in nearby open areas.

More even soil temperature. Partly as a result of the shade and partly because of the protective blanket of forest litter, soil temperatures in forests vary less than do temperatures in the open, thus favoring all forms of desirable life in the soil.

For example, a hemlock-hardwood forest in New York greatly moderated the soil temperature to a depth of 18 inches. At the 6-inch depth in the forest the soil averaged 11° F. cooler on the hottest days than did nearby open land

A model plan adapted to the Northern Great Plains for an 8-row shelterbelt planting of trees to temper the hot or cold winds, provide beauty and shade, serve as a snow fence, and reduce soil erosion by wind. For maximum protection the tree belts should be planted on the north and west sides of the farmstead. (*Source: Farmers Bulletin* No. 2109, U. S. Department of Agriculture, 1966, page 5.)

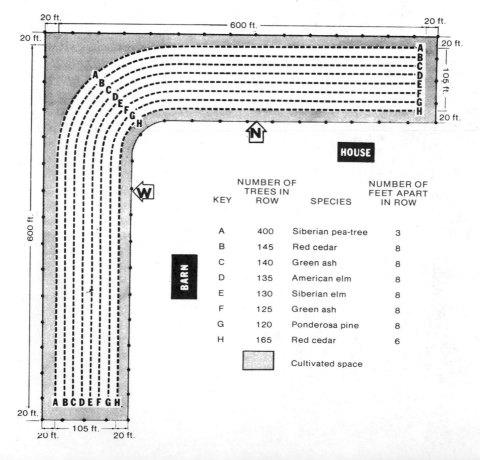

KEY	NUMBER OF TREES IN ROW	SPECIES	NUMBER OF FEET APART IN ROW
A	400	Siberian pea-tree	3
B	145	Red cedar	8
C	140	Green ash	8
D	135	American elm	8
E	130	Siberian elm	8
F	125	Green ash	8
G	120	Ponderosa pine	8
H	165	Red cedar	6
		Cultivated space	

at the same depth. At the same 6-inch depth, the forest soil was 3° F. warmer on the coldest days than was soil in the open. At a depth of 18 inches, the forest soil was 6° F. warmer in winter and 6° F. cooler in summer as compared with open ground. Some temperature differences were noted to a depth of 4 feet.

Forests protect the soil from freezing. Soils not frozen are capable of absorbing much more water than are frozen soils. In this way, forests aid in reducing the surface runoff that causes soil erosion and floods. Even frozen forest soils are more permeable than frozen soil in pasture or cropland.

Quieter winds. On a windy day no place outdoors affords greater protection for man or beast than does a clump of trees. Here is ample proof: In Idaho, winds were reduced from 1.5 miles per hour in the open to 0.2 miles per hour in the forest. This was in a western white pine stand. Nebraska reports a comparison of a 7.5 miles-per-hour wind in the open with a 2 miles-per-hour wind in a jackpine forest. Aspen trees in Colorado reduced winds to only one-third of their open-field velocity.

Shelterbelts are rows of trees planted at right angles to the prevailing winds to reduce wind speeds. Their effectiveness in accomplishing this purpose may be seen from the following examples.

A blue gum shelterbelt in California reduced wind velocities from 21 to 7 miles per hour. Also there was some windspeed reduction on the leeward side of the shelterbelt as far as 3 times the height of the shelterbelt. Four rows of dense Norway spruce in Indiana were effective on the leeward side for twice the height of the shelterbelt.

More rainfall. Can forests make it rain? This is still a much-disputed question. Only one study seems to offer convincing proof that forests can increase rainfall.

The pictures on page 294 show what a twentieth century private company did to right the wrongs of nineteenth century indifference to soil and its protective cover. In 1843 copper ore with a high sulfur content was discovered in Copper Basin in southeastern Tennessee. As the ore was heated in large open piles, using nearby trees as fuel, sulfur dioxide gas escaped and killed trees, shrubs, and grasses over a 50-square-mile area. The 55 inches of annual rainfall caused severe gully erosion (rainfall had been 53 inches in the grass strip and 57 inches in the surrounding forest). In 1899 the Tennessee Copper Company, now a subsidiary of Cities Service Company, took over the operation of the copper mines. The toxic sulfur dioxide was reclaimed as a saleable product, sulfuric acid. In 1899 the company started a long-time revegetation program with loblolly pine trees and kudzu (a leguminous vine). Seven million pine seedlings have been planted, with present plans to plant one-half million more each year. The damaged area has been reduced from 50 square miles to 15, and the effort continues.

A United States Department of Agriculture publication on "Some Plant-Soil-Water Relationships in Water Management" contains information on the relationship between the depth that trees root and the amount of runoff water. In arid regions where trees are planted as shelterbelts and where the objective is to encourage as much runoff as possible for irrigation reservoirs, shallow-rooted

Top: Sulfur dioxide gas killed the trees on this land (Tennessee). *Center:* Almost 50 years later a long-time planting program began (note the use of planting bars). *Bottom:* Pine trees and kudzu (a legume) are restoring the productivity of the eroded land. (*Courtesy* Cities Service Company)

trees are preferred to deep-rooted ones. The following trees are classified as shallow-rooted and root from 1 to 5 feet: jack pine, Scotch pine, Norway spruce, white willow, cottonwood and catalpa.

FORESTS IN SOIL AND WATER CONSERVATION

In cold weather the leaf-litter blanket keeps the soil warmer. When the litter and soil do freeze, they tend to form honeycomb ice that permits rain and water from melting snow to penetrate rapidly in the forest. On the other hand, open land freezes sooner, deeper, and harder than soils in forests.

Forests greatly increase infiltration, thereby reducing runoff and erosion. For example, in a Wisconsin forest, 97 percent of the rainfall and snowfall was held where it fell. By contrast, only 74 percent of the total precipitation was held by the soil in an adjoining cornfield.

Oklahoma reports that 99.99 percent of the rainfall was retained on the land under unburned post-oak forest cover. A similar area recently burned retained only 97.5 percent of the rainfall.

One of the most striking examples of the relation of forests to infiltration was reported in Mississippi. A natural-oak forest with broomsedge in the openings encouraged 99 percent of the rainfall to "stay home." By comparison, the soil in a nearby cotton field kept only 53 percent of the rain where it fell. The other 47 percent of the rainfall flowed off the field, carrying valuable topsoil with it. Fifty inches of water penetrated the woods soil, but only 26 inches moved into the cotton soil.

In some instances forests may actually encourage too much water to enter the soil. This frequently is true in arid regions where surface runoff water is caught behind dams and used for irrigation.

In Colorado in an area above 9,000 feet in elevation, cutting all of the forest trees increased the runoff by an amount equal to 1 inch of rainfall. But the damage to the watershed as a result of cutting the trees must be weighed against the value for irrigation of the increased runoff water.

Floods are related to forests. For example, in New York a total of 6.4 inches of rain fell between March 10 and March 19. Peak flow in nearby streams occurred on March 18. Soils in the open were deeply frozen and from them water moved rapidly to the streams. All snow, too, was melted from open fields by March 18, and water from the snow joined the rainwater to make flood crests higher.

By contrast, the soil under the hardwood forest never froze, and therefore water from melting snow moved downward. Nor did the snow melt as rapidly. On the same day of the flood crest, March 18, there was still enough snow left on the forest floor to be equivalent to 3 inches of precipitation.

During critical spring flood periods, then, forest-covered slopes release less water because of deeper seepage, and

Left: Eastern hemlock is a common cone-bearing tree of the North and Northeast. *Right:* Longleaf pine is one of the four principal southern pines. (*Courtesy* U. S. Forest Service)

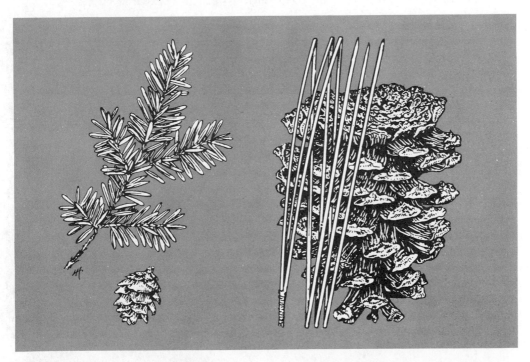

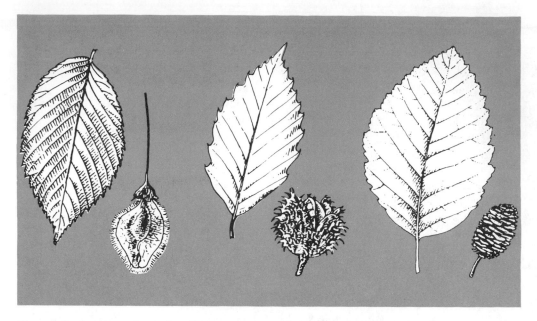

Above left: American elm is found growing over most of the humid region. *Above center:* American beech is most easily identified by its gray, smooth bark. The leaves are thin, smooth and tough. *Above right:* Red alder is one of the few western hardwood trees. *Below left:* Douglas-fir is a very common, large tree, found over a wide area throughout the West. *Below right:* Ponderosa pine is an excellent lumber tree of the West. (*Courtesy* U.S. Forest Service)

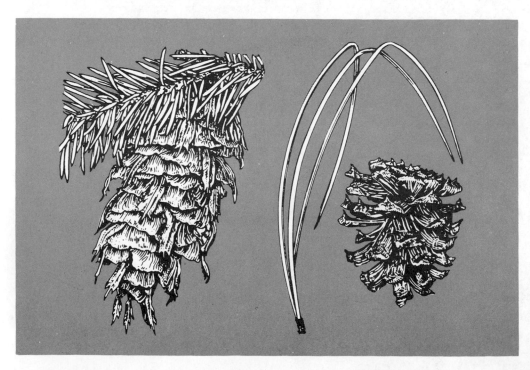

also retain water in the form of un-melted snow. In this way trees and forests help to reduce floods.

IMPORTANT TREES

There are approximately 845 different kinds of trees and shrubs growing wild in the United States. This does not include the hybrids and horticultural varieties such as Lombardy poplar and weeping willow. Of these 845 species of woody plants, 150 belong to the hawthorn group. There are 57 species of oaks, 34 species of pines, and 33 species of willows.

Not all trees are of equal value or importance, but most have some use. Trees of very little value are known as "weed trees" and are usually killed by poisoning to make more light, water, and plant nutrients available for the more valuable trees.

Eastern cone-bearing trees. With the exception of the bald cypress, which sheds its leaves every fall, all the eastern conifers are evergreens.

Eastern broad-leaved trees. The broad-leaved trees are deciduous; that is, they shed their leaves in autumn. Sometimes the broad-leaved trees are called *hardwoods*, although the wood of some broad-leaved trees is softer than the wood of some cone-bearing trees.

Western cone-bearing trees. Cone-bearing trees dominate the forests in the West, while broad-leaved trees (hardwoods) are more numerous in the East.

Western broad-leaved trees. The principal broad-leaved tree in the West is the red alder, a relative of the shrub alders in the East.

1. Slash pine is one of the more popular southern pines for planting. 2. Sugar pine, the largest of the western pines, grows from Oregon to lower California. 3. The shagbark hickory is a large tree of the eastern half of the United States. 4. The redwood is the world's tallest tree. (*Courtesy* U. S. Forest Service)

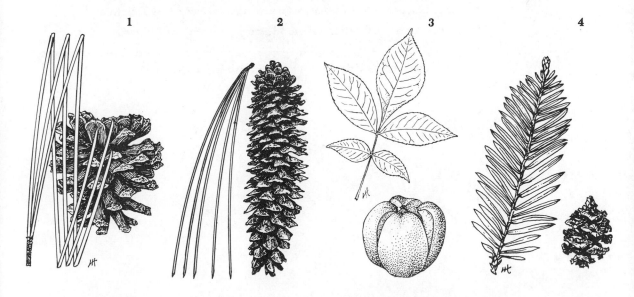

1   2   3   4

QUESTIONS

1   Name the points at which a tree grows.
2   Are trees ever considered a nuisance today as they were in pioneer days?
3   How do trees help to control soil erosion?
4   How are forests beneficial to wildlife?
5   What is the native forest vegetation in your community?
6   What is the relationship between forests and soil temperature?
7   How do forests influence climate and weather?
8   How do western trees differ from eastern trees?
9   What is the native vegetation in areas receiving 20 to 30 inches of precipitation annually?
10 What is being done to preserve the redwoods?

ACTIVITIES

1   See how many trees you can identify near your home and school. With the help of your teacher, learn to identify other trees in your community.
2   Count the growth rings on a log or stump. The number of rings indicates the approximate age of the tree in years. Explain possible reasons why certain rings are narrower than others.
3   Ask your teacher or school librarian for bulletins and leaflets published by the state forestry department and by the U. S. Forest Service. Have a class discussion on forest resources and forestry in your state.
4   Make a class collection of wood samples and tree leaves.
5   Plan a field activity to include tree identification, wood identification, compass use and pacing, and timber cruising.

22 WILDLIFE

Most people define the term *wildlife* according to their own interests and experiences. Sportsmen value game animals for recreation and food. Farm boys sometimes make extra money by trapping furbearers. Wildlife biologists are interested by the fact that animals, plants, soil, and water are interdependent and must be managed together.

For our purpose it is sufficient to consider a few groups of wild animals that are directly related to agriculture. Some are beneficial, others are harmful, and certain species are almost neutral in relation to man.

Agriculture has completely changed the wildlife picture in this country. When the first explorers pushed into the central part of North America, the forests and grasslands supported a large population of big game animals. There may have been more than 150 million bison, elk, deer, antelope, bear, mountain goats, and bighorn sheep. The latter three groups are now restricted to remote mountain areas, and almost all the remaining bison are in parks and game refuges. Elk and antelope have lost large areas of their original range because man had other uses for the land. The white-tailed deer and the mule deer have fared better than the other big game animals.

Wildlife was an important source of food in Colonial days and throughout the period of settlement. When the railroads were being built across the Great Plains, hunters were employed to provide game for the construction crews. As the livestock industry developed, people became less dependent upon wildlife.

The westward expansion of agriculture was accompanied by much needless destruction of big game animals and

A young raccoon in an old maple "den tree."
(*Courtesy* U.S. Forest Service)

upland game birds on livestock ranges. On the other hand, many farmers and ranchers probably were not aware of the steady shrinking of the wildlife habitat. Intensive use of all arable land was the natural result of the growing domestic and foreign markets for grain, cotton, and livestock products. How do you think *we* would have used the land if we had lived 50 to 100 years ago?

IMPROVING THE WILDLIFE HABITAT

Every species of wildlife can reproduce and survive if it has the right kind of habitat. Man's destruction of the natural habitat has been the main cause of the decline of most wildlife species; and the improvement of habitat is the backbone of wildlife conservation.

The three essentials of a good wildlife habitat are food, cover, and water. These must be available within the area which an animal normally covers in its daily activities.

Many species have small home ranges. A tree squirrel likes to stay within 200 yards of its home tree. Bobwhite quail usually occupy a home range less than one mile in radius. The home range of a cottontail rabbit may be 20 to 160 acres.

Food. Each species of wildlife prefers a particular class of food. The choice of foods among those that are acceptable depends upon the seasonal abundance or scarcity. Winter is the season of low food supply for most species.

Tree squirrels eat pecans, walnuts, acorns, hickory nuts, and dry corn in winter. Tender buds and leaves are eaten as soon as they are available in the spring. Then the summer months bring berries, green corn, and other foods.

The summer foods of the white-tailed deer include the twigs of shrubs and trees, grasses, weeds, and berries. Any available palatable plants are eaten in winter. A browse line can often be seen where deer eat the branches and evergreen foliage as high as they can reach.

Several species of wildlife that prefer different foods can share the same habitat. For example, jack rabbits thrive on weeds and herbs; cottontail rabbits tend to concentrate on grasses.

The mule deer and the Columbian black-tailed deer both like grasses, tree and shrub leaves, acorns, and lichens. But competition between the two species is reduced by their preference for different kinds of habitat. The Columbian black-tailed deer likes dense stands of trees and shrubs in forested areas; the favorite habitat of the mule deer is open forest and brushland.

There is, however, some competition between mule deer and the Columbian black-tailed deer when they share the same valley ranges in winter. And white-tailed deer compete with mule deer where the ranges of the two species overlap.

The competition between livestock and grazing or browsing wildlife should be considered in stocking grasslands. Seven or eight mature deer require as much forage as a 1,000-pound cow. Nine or ten pronghorned antelope are also equivalent to one cow.

Many conservation practices increase the food supply of wildlife. The improvement of soil fertility increases both the quality and quantity of wildlife foods; the plants have a higher content of essential minerals and vitamins and are tastier. Crop rotations make

several kinds of food available at different seasons.

In selecting plants to provide cover for the soil it is often possible to include trees, shrubs, legumes, and grasses that provide food and cover for wildlife. Winter cover crops help animals survive in winter when the food supply usually is low. The planting of crops in contour strips makes several feeding areas available to wildlife and provides covered travel routes between these areas. Trees and grasses planted to stabilize gullies also increase the cover.

Emergency feeding of wildlife is sometimes necessary to preserve breeding stock. During severe winters the mortality of deer, antelope, and elk may be very high unless supplementary

Left: **The planting of multiflora roses provides excellent shelter and food for wildlife, and beauty for us. (Delaware)** (*Courtesy* Soil Conservation Service) *Right:* **A snowshoe rabbit in its native habitat.** (*Courtesy* U. S. Forest Service)

Above: **An excellent habitat for these migratory geese. (California)** (*Courtesy* U.S. Fish and Wildlife Service) *Below:* **A nest of a mottled duck in a rice field on the Gulf Coast of Texas.** (*Courtesy* Texas Fish and Game Commission)

forage is provided. Farmers in the eastern half of the United States sometimes scatter grain to help coveys of bobwhite quail through the weeks when snow and ice cover the natural foods. Emergency feeding usually is limited to small areas and short periods of time.

Grassland agriculture includes several practices that improve both food and cover for wildlife. Deferred grazing and rotation grazing give the better grasses a chance to make a satisfactory growth and produce seed. The use of lime and fertilizers on grasslands and croplands improves both the quantity and quality of vegetation. Reseeding of rangelands, removal of brush, control of fire, and the improvement of pastures are generally beneficial to the wildlife habitat.

Some species of wildlife make good use of plants that are poisonous to people or livestock. Quail, doves, and rodents eat croton seed, although the oil of croton is poisonous to humans. Certain rodents are very fond of croton. In the eastern part of the United States cattle and sheep are sometimes poisoned by laurel and rhododendron, but deer are not injured by these plants.

Every farm conservation plan should be based upon the capability of the land. A land-capability map is a good guide in determining how some fields and field borders can be improved for wildlife without interfering with other uses. On many farms and ranches there are areas that are more suitable for wildlife food and cover than for any other purpose. The development of such land for wildlife is good conservation.

Cover. The different families and species of wildlife need different kinds and amounts of cover.

The jackrabbit needs only the sparse vegetation of the open plains, but the cottontail stays close to a dense cover of briars or brush. The pronghorned antelope depends upon protective coloration, keen sight, and speed rather than cover. Mule deer like open forest and brushland, while the Columbian black-tailed deer prefer the dense stands of shrubs and trees in forested areas.

Fowl-like birds use several kinds of cover for nesting, feeding, resting, roosting, and for moving from one part of the home range to another. Waterfowl need nesting and feeding cover in their northern breeding grounds. During migrations and on the southern wintering grounds, waterfowl use cover mainly for feeding and resting. The impounding of large lakes along the main flyways is very beneficial to migratory waterfowl.

The vegetation on a good watershed may also provide cover for wildlife. The best plants for this dual purpose are those that have dense foliage throughout the year and the kind of root systems that make them resistant to fire, grazing, and drought. Plants for wildlife cover and erosion control should be of a kind easy to establish. Vigorous growth is desirable to provide complete cover for the soil as soon as possible.

Certain plants have undesirable characterisitcs that keep them from being popular as wildlife cover. Johnsongrass, kudzu, some of the wild roses, and vine honeysuckle, although good wildlife cover, are so aggressive that they may spread to areas where they are not welcome. The common barberry is a host to the stem rust of wheat; poison ivy is poisonous to people; and several species of crotalaria are poisonous to livestock.

The prevention of fires in grasslands and woodlands is one of the most important and difficult problems in wildlife conservation. If pastures, ranges, and forests are burned every year or two,

Left: This forest fire is destroying more than just trees—it is ruining the habitat for wildlife and polluting our environment. (California) *Right:* Prevent forest fires—break your match—with one hand. (*Courtesy* U. S. Forest Service)

the nutritious grasses and legumes are succeeded by nonforage plants, many of which have little or no value either for wildlife or for livestock.

Rotation grazing and deferred grazing protect the nesting cover of upland game birds and nongame birds that nest in pastures and rangelands. When pastures are mowed to control weeds, more cover will be left if only part of the area is mowed each year. It is best to leave irregular strips and patches that form covered travel routes to all parts of the pasture. In harvesting hay crops it is often possible to leave strips along field borders to provide food and protected trails. Rotations of grasses, legumes, and grain crops improve the seasonal availability of wildlife foods, and the vegetation along field borders provides wildlife with continuous cover between feeding areas.

Water. The water requirements of wildlife vary from the complex aquatic environment of fish and waterfowl to the desert habitat preferred by some birds, mammals, and reptiles. Doves and wild turkeys must have drinking water, but the bobwhite quail gets enough moisture from food and dew. The white-tailed deer requires a habitat in which surface water is readily available, while the mule deer can get along without water for days or weeks.

The quality and amount of runoff from a watershed depends mainly upon the condition of the land. A watershed that is protected by healthy vegetation yields a regular flow of clear, clean water. Excessive runoff during rainy weather and dry streams during periods of drought are caused by overgrazed grassland, overplowed cropland, and overcut forests. The conservation practices that provide the best protection for watersheds also provide a dependable supply of water for wildlife.

Regularity of stream flow is one of the essentials of a satisfactory habitat for aquatic plants and animals. Fish and waterfowl need shoreline plants for food and cover. Frequent and sudden changes in the water level keep such vegetation from becoming well established.

The practices that produce the maximum yield of nutritious forage on pastures, ranges, and haylands help to keep farm ponds and lakes at a fairly uniform level. The control of runoff and erosion on cropland serves the same purpose. The use of lime and fertilizers to improve the fertility of a watershed increases the capacity of streams, ponds, and lakes to support fish and wildlife.

On the other hand, extensive drainage and reclamation projects destroy the habitat of waterfowl and aquatic furbearers. Perhaps the only solution to such conflicts is the intensive development of more nonagricultural land for wildlife.

Pollution is one of the most difficult problems in the management of water resources. The principal sources of pollution are sewage, industrial waste, and the silt that runoff carries into stream channels.

Sedimentation is one of the worst forms of pollution. Plants do not grow as well in muddy water because the silt reduces the depth to which sunlight can penetrate. Without plants to provide food and cover, water soon becomes an unsatisfactory habitat for fish. Farmers and ranchers can help to control silting by maintaining protective vegetative cover on all land that is subject to erosion.

WILDLIFE AS A FARM CROP

There is a limit to the wildlife population that a given unit of land or water will support. This limit is called the *carrying capacity*. All wildlife beyond the carrying capacity is surplus. The surplus population of game animals and furbearers should be harvested and used; otherwise this resource is wasted. Animals that are not valued for their fur or meat have their place in the balance of nature and usually are controlled by limitations of their habitat.

Farmers and ranchers have a very special role in wildlife conservation. They own much of the land on which the nation's wildlife resources must be produced. The cooperation of the families who operate the farms and ranches is necessary if any program of habitat restoration is to be successful.

In most communities, particularly those near large cities, there is not enough wildlife for all the people who want to hunt and fish. Wildlife regulations are necessary to give everyone a fair share of the wildlife crop. Restrictions on hunting and fishing seasons and on the amount of game that may be taken are designed to protect the breeding reserve so that new crops of wildlife will be produced year after year.

Some farmers and ranchers get good supplementary income from hunting leases. Many sportsmen are willing to pay for the privilege of hunting deer, antelope, elk, or wild turkey.

Sportsmen sometimes share with landowners the cost of improving the wildlife habitat. This is a common practice where upland game birds and fish are the principal kinds of game. More often the sportsmen get permission to

This farm pond has a vertical-drop outlet near the center of the pond. This outlet carries away the overflow without the hazard of breaking the dam or without causing sediment to muddy the water. (*Courtesy* Caterpillar Tractor Company)

hunt on farms and ranches operated by their friends. The sharing of game may help to preserve goodwill between sportsmen and landowners.

Federal and state conservation agencies help farmers and ranchers manage the wildlife crop. Technical assistance includes the rearing of fish for stocking ponds and lakes, the transplanting of wildlife from overpopulated areas to suitable habitats that do not have enough breeding stock, and the integration of wildlife conservation with other practices in the farm conservation plan. The local game warden, the county agricultural agent, and the technicians of the Soil Conservation Service are good sources of helpful information.

Left: Baby elk need food, water, and shelter every day in the year. (Washington)
Right: The California mule deer in the Sequoia National Forest are at home at the edge of the woods. (California) (*Courtesy* U. S. Forest Service)

Five groups of animals make up a large part of the wildlife crop. These are: grazers and browsers, fowl-like birds, furbearers, waterfowl, and fish. Reptiles and amphibians are of minor importance.

The white-tailed deer, mule deer, Columbian black-tailed deer, pronghorned antelope, and American elk are grazers and browsers. The white-tailed deer is the most abundant and most widely distributed animal in this group. The mule deer ranks second in numbers. These two species make up about 85 percent of the population of big game animals in the United States.

Some of the grazers and browsers are not abundant enough to be con-

sidered major game animals in parts of their range. The harvesting of surplus elk and antelope, for example, is carefully supervised by state conservation departments.

The fowl-like birds include grouse, quail, wild turkey and pheasant.

The wild turkey, ruffed grouse, blue grouse, Franklin grouse, spruce grouse, woodcock, and mountain quail are generally considered forest birds, although they may occur in mixed habitats. Prairie chickens, ring-necked pheasant, Gambel quail, sage grouse, and sharp-tailed grouse prefer grasslands.

The bobwhite quail is one of the most important game birds in the eastern half of the United States. The bob-

Left: Cottontail rabbits are good game animals. (Colorado) *Center:* The chipmunk is friendly and does no harm. (California) *Right:* All woodchucks around the farm are bad because they eat crops. (Pennsylvania) (*Courtesy* U.S. Forest Service)

A muskrat would build a grass house, with the entrance under water, beyond the next clump of marsh grass. (*Courtesy* U. S. Fish and Wildlife Service)

white is well adapted to farming communities where pastures, meadows, woodlands, and croplands occur in a mixed pattern of land use.

The ring-necked pheasant, a native of Asia, was introduced into the Northern Great Plains during the first decade of this century and soon became a very important game bird in that region. Pheasants are also abundant in most of the eastern United States.

The muskrat is the most important furbearing mammal in the United States. The swamplands of the South produce most of the muskrat pelts for the commercial fur market, although muskrats are found in several geographical regions.

Other furbearers include the mink, weasel, raccoon, skunk, opossum, fox, and badger. In recent years there has not been a strong market demand for pelts from any of these animals but mink.

Waterfowl probably can be considered a farm wildlife crop only in the Lake States, in parts of the upper Northern Great Plains where lakes are numerous, and in certain areas near the Gulf,

Above: Bears can look sweet, but they also steal lambs. (Montana) (*Courtesy* U.S. Forest Service) *Below:* The pilot blacksnake is not poisonous. It is a friend of the farmer because it eats rodents and insects. (*Courtesy* Ralph J. Donahue)

Atlantic, and Pacific coasts. During the autumn migration there is some water-fowl hunting on inland lakes.

On many farms and ranches fish are a by-product of soil and water conservation practices. A well-managed farm pond may produce several hundred pounds of fish per year, depending upon the area of the water. The banks of a newly constructed pond should be stabilized with vegetation before the water is stocked with fish.

PREDATORS AND RODENTS

Predators and rodents are considered together because they have a unique role in the balance of nature. The predators serve to control the rodents.

A *predatory* animal is one that preys on other animals, although this general definition is not entirely satisfactory. Farmers and ranchers usually think of predators as those animals that destroy livestock and poultry. For example, coyotes and bobcats occasionally kill young lambs and calves. Foxes, weasels, skunks, and raccoon often invade poultry yards. Some kinds of hawks catch poultry, particularly when rodents are scarce.

Farmers and ranchers are justified in destroying individual predatory mammals, birds, and snakes that persist in preying on livestock, poultry, and game birds. However, it is a mistake to condemn all species that are potential predators just because a few individual animals are rascals.

Hawks perform a valuable service by controlling rabbits, field mice, rats, and other small rodents on farms and ranches. The coyote is the natural enemy of the prairie dog, cottontail rabbit,

jackrabbit, and ground squirrel. The pocket gopher, one of the most unpopular rodents on grasslands, occasionally makes a meal for a badger, coyote, or weasel, although the gopher spends most of its time underground and must be dug out of its burrows.

The populations of predatory animals usually fluctuate with their natural food supply. Hawks and coyotes are most numerous where rabbits and field mice are abundant.

The predatory species usually become a problem only when the balance of nature is upset. The food habits of the raccoon, skunk, and most snakes are generally beneficial or harmless, but these and other animals with similar food habits sometimes destroy poultry and birds. Skunks, foxes, and snakes are, however, natural enemies of ground-nesting birds such as the bobwhite quail and the meadowlark.

Above left: This Florida fishpond provides pleasure in fishing and food for the table. (*Courtesy* Soil Conservation Service)

Above right: Wild geese feeding in pens on Gambill Wildlife Refuge near Paris, Texas. (*Courtesy* Texas Highway Department)

Below: Porcupines do more harm than good by chewing holes in cabins and by girdling some kinds of trees. (Wisconsin) (*Courtesy* U.S. Forest Service)

QUESTIONS

1 What are the essentials of a good wildlife habitat?

2 How can different kinds of wildlife share the same habitat and not compete seriously for food?

3 What is the relationship between erosion control and wildlife habitat?

4 What undesirable forms of wildlife live in your community?

5 How has agriculture changed the kind and amount of wildlife?

6 How do the mule deer and white-tailed deer differ in their water requirements?

7 In the amount of food eaten, how many deer are equal to one cow?

8 What is meant by "carrying capacity" of land or water?

9 How does soil fertility affect the quantity and quality of wildlife food and cover?

10 Why is wildlife considered a farm crop?

ACTIVITIES

1 Observe the woods, grasslands, streams, fields, and fence rows in your community. Write a brief summary of your observations, indicating whether you consider conditions satisfactory for wildlife.

2 Outline plans for improving the wildlife habitat in your community. Consider each of the three essentials of a good wildlife habitat.

3 Select the species of birds, mammals, and fish in which you are most interested, and do additional reading to learn more about these animals.

4 Look for the tracks of wild animals along the banks of a stream. Try to identify the tracks. Look for dens and den trees, and observe signs that may indicate the kinds of animals that use them.

5 Ask your State Wildlife Conservation Commission (or the corresponding agency in your state), a local game warden or wildlife conservation agent, or the county agricultural agent for a copy of the state hunting and fishing regulations. Learn the hunting and fishing seasons and limits on the amount of different species of wildlife that can be harvested legally.

23 GRASSES AND LEGUMES

Grasses of one kind or another provide much of the hay and pasture for livestock. These versatile plants also make our lawns, parks, and golf courses beautiful and protect the soil against erosive raindrops and blasting winds.

Even before the time of Christ, legumes were recognized as valuable for soil improvement, livestock feed, and human food. Alfalfa, which originated in Persia, was mentioned in very early recorded history as valuable feed for chariot horses in the Persian Army. The value of alfalfa for livestock forage was known then and has never been surpassed by any other plant. Soybeans were used for human food in China and elsewhere in ancient times and are still an important food today.

GRASSES

Plants in the grass family are distinguished by their jointed stems, sheathing leaves, flowers borne in spikelets or bracts, and fruit of a seedlike grain.

Another way to characterize a grass is to say that it has stems solid at the joints, one leaf at each joint, and leaves arranged on the stem like steps on a telephone pole.

Some grasses are best for northern lawns, others for western lawns, and still others are adapted for southern lawns. One grass will grow best only where the winter temperatures are mild; other grasses seem to defy the cold. Wet places are best for one grass, while another grass will grow only where the soil is quite dry. Some grasses creep and crawl; others grow in upright bunches.

What is the best grass? There is no "best" grass, except as you specify "best

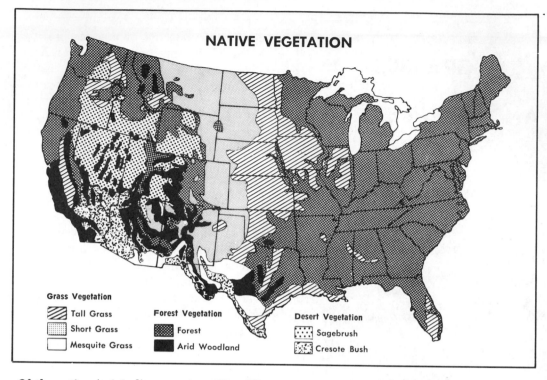

NATIVE VEGETATION

Grass Vegetation
- Tall Grass
- Short Grass
- Mesquite Grass

Forest Vegetation
- Forest
- Arid Woodland

Desert Vegetation
- Sagebrush
- Cresote Bush

Of the native (original) vegetation, 800 million acres were in forest, 700 million acres were in grass, and 400 million acres were in desert. (*Source:* Carnegie Institution of Washington, D.C.)

for what?" An answer to this question will be attempted as we describe 17 of the most important grasses of the United States.

Alta fescue (*Festuca elatior var. arundinacea*). Alta (reed) fescue, a variety of meadow fescue, is a deeply rooted perennial bunchgrass. It flowers in June and July on stems 3 to 4 feet high.

Alta fescue grows in damp pastures and other wet places throughout the far West and the Northeast. It is now planted in the South as a winter grass. It can be distinguished from meadow fescue because it is taller and has broader leaves that are darker green on the upper surfaces.

Alta fescue and meadow fescue were introduced from Europe probably at about the same time. A strain of Alta fescue, Kentucky 31, is now receiving the most attention. Kentucky 31 is a selection from tall fescue that was found in 1931 on a Kentucky farm, where apparently it had been growing for approximately 50 years.

Alta fescue is adapted to a wide variety of soils and in general to the same region as meadow fescue. It does best on fine-textured soils that contain a large amount of organic matter, but it grows fairly well even on excessively drained soils.

The exact forage and pasture value of Alta fescue has not been determined, but it usually is considered fairly low. In the Pacific Northwest, Alta fescue is widely used in pastures for cattle and sheep. It is highly desirable grass for winter pastures in the South. Farther north it serves well for pastures on soils of low fertility. It is quite drought-resistant and aggressive; it competes strongly with companion legumes.

Bermudagrass (*Cynodon dactylon*). Bermudagrass is now common in all

Kentucky 31 fescue, a selection from alta fescue, grows in soils of fairly low fertility. (Florida) (*Courtesy* Soil Conservation Service)

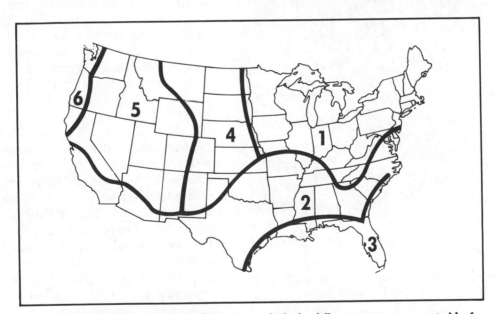

Climatic regions of the United States in which the following grasses are suitable for lawns: REGION 1. Common Kentucky bluegrass, Merion Kentucky bluegrass, red fescue, and Colonial bentgrass. Tall fescue, bermudagrass, and zoysiagrass in southern portion of the region. REGION 2. Bermudagrass and zoysia-grass. Centipedegrass, carpetgrass, and St. Augustinegrass in southern portion of the region with tall fescue and Kentucky bluegrass in some northern areas. REGION 3. St. Augustinegrass, bermudagrass, zoysia grass, carpetgrass, and bahiagrass. REGION 4. Nonirrigated areas: Crested wheatgrass, buffalograss, and blue gramagrass. Irrigated areas: Kentucky bluegrass, and red fescue. REGION 5. Nonirrigated areas: Crested wheatgrass. Irrigated areas: Kentucky bluegrass and red fescue. REGION 6. Colonial bentgrass and Kentucky bluegrass. (*Courtesy* U.S.D.A.)

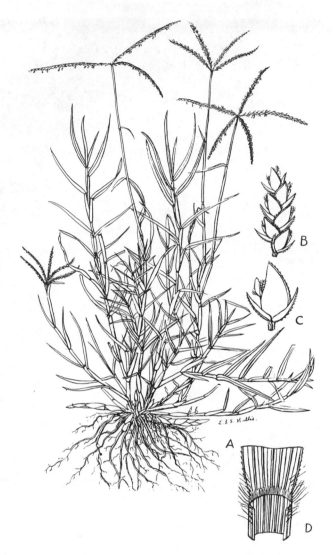

Bermudagrass is one of the best grasses for the humid South. (*Courtesy* U.S.D.A.)

moist and fertile soils in India.) The date of its introduction from India is not known, but official reports mentioned the importance of Bermudagrass as early as 1807. Now this species occurs in all parts of the southern half of the United States and to some extent in the Northeast and Northwest. Bermudagrass is variously called wiregrass, dog's tooth grass, devil grass, dub, and hariali.

This grass is a long-lived perennial with a rapidly spreading habit of growth. Bermudagrass is propagated by runners, by underground rootstocks, and by seed. The runners vary on the average from a few inches to several feet in length and under favorable conditions may grow 20 feet in one season. The underground rootstocks are thick and white. The erect, flowering branches are 6 to 12 inches high, depending upon the productivity of the soil. The leaves are short, flat, bluish-green, and 1 to 4 inches long.

Bermudagrass will grow well on almost any soil that is fertile and not too wet. It seems to do better on fine-textured soils such as loams and clays than on coarse-textured soils like the sands. It thrives in warm or hot weather and does not survive heavy freezing—although it has lived through temperatures of 10° F. in the vicinity of the District of Columbia. Since it often is killed by freezing, the grass is usually considered a pest in lawns in Virginia and Maryland.

Bermudagrass may be propagated by seed or by pieces of sod. Because the seeds are small and light, a well prepared, firm seed-bed is necessary. Spring seedings, 5 to 7 pounds to the acre, are usually most successful. The seed should be covered lightly with a cultipacker or a light harrow.

tropical and subtropical regions. (One of the authors, Dr. Donahue, found Bermudagrass in Brazil in 1943 growing wild on the banks of the Madeira River, 1,500 miles upstream from the mouth of the Amazon River. In 1955 he found it in a field in southern Greece, and in 1956–1966 encountered it on nearly all

For best pasture growth, acid soils should be limed. Apply lime and fertilizer on the basis of soil tests. (*Courtesy* Texas Agricultural Limestone Association)

Many methods are used in planting sod pieces. The common practice in pasture establishment is to plow furrows 5 feet apart, drop pieces of sod 3 feet apart in the furrow, and cover them by plowing. Deep planting is important if the sod is not watered when set, because the sod may dry out. Rolling or cultipacking the soil after plantings is also desirable. The same method is used in sodding a lawn, except that the sod pieces are set one foot apart.

A complete fertilizer containing nitrogen, phosphorus, and potash, such as a 15-15-15 or 20-10-5, should be applied at the rate of 500 pounds per acre just ahead of sodding or seeding. For rapid establishment, nitrogen fertilizer should be applied in midsummer at the rate of 100 pounds of ammonium nitrate per acre.

The principal use of Bermudagrass is for pasture and lawns, but it is often used for hay. It is palatable and nutritious even after it has frosted in the fall. However, many farmers hesitate to plant Bermudagrass because it is difficult to control in cultivated crops.

Bermudagrass responds to tillage. Old undisturbed Bermudagrass pastures that have become weedy and unproductive may be renovated by tillage. The best method of maintaining a productive pasture is by shallow plowing every 3 to 5 years. Fertilizing and maintaining legumes in the Bermudagrass also aids in keeping Bermudagrass land productive.

Improved varieties are more vigorous in growth than common Bermudagrass. They are more disease-resistant and superior in other characteristics. The new varieties produce seed sparsely and must be planted vegetatively with sod pieces.

Of the two varieties most widely used now, Coastal Bermudagrass is more productive over the region than is Suwannee Bermudagrass. The Suwannee strain shows promise in Florida. Fine-leaved strains such as Tiflawn, Ormond, and Everglades No. 1 are being developed for home lawns and golf courses.

Big bluestem (*Andropogon gerardi*). Big bluestem is a native bunchgrass with a vigorous and extensive root system and short underground stems. It occurs with other prairie species mostly in the east-ern part of the Great Plains, where moisture conditions are most favorable. To some extent it grows also in all the eastern and southern states.

Growth starts in the late spring and continues all summer, even during periods when rainfall does not appear to be adequate. The grass is a giant among grasses, often attaining heights of 6 feet or more at maturity. The leafy forage is highly palatable to all classes of livestock and makes a good quality of hay if mowed before the seed heads form.

Because of its deep and strong root system, big bluestem gives the soil excellent protection against water and wind erosion. Usually it is seeded in mixture with species with which it grows naturally.

One important use of big bluestem is to retire cropland to meadows and pastures. It builds up the organic-matter content of the soil very rapidly through the decay of the roots and tops.

Big bluestem does not produce seed every year because the necessary combination of plentiful moisture and moderate temperatures seldom occurs. But when grown in rows, cultivated, and fertilized with nitrogen, this grass regularly produces as much as 200 pounds of seed per acre. The seed matures in late September or October.

Stands of big bluestem should be examined with great care before seed harvest to determine the set of seed. Under the best of conditions, a seed fill seldom exceeds 60 percent, and a fill of 20 percent usually justifies harvesting the seed.

The seed may be harvested satisfactorily with ordinary binders or with small-grain combines. When processed with a hammermill and cleaned with a

Major and minor distribution of big bluestem (*above*) and little bluestem (*below*). (*Courtesy* U. S. D. A.)

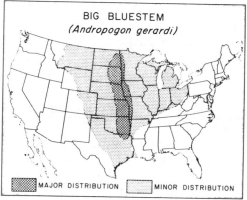

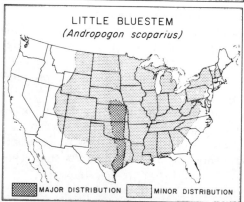

fanning mill, the harvested seed should have a purity of at least 40 percent and a germination of about 60 percent. A pound of pure seed contains 140,000 to 170,000 seeds.

Blue gramagrass (*Bouteloua gracilis*). Blue gramagrass is a native perennial bunchgrass adapted to a wide range of soils and climates. In the southern part of its range it grows with buffalograss; in the northern part of its range it grows with the wheatgrasses.

Blue gramagrass seldom grows taller than 20 inches. It produces excellent yields of highly nutritious forage during the summer, and any forage remaining in the fall makes a good winter pasture if allowed to cure while standing. All kinds of livestock readily eat the forage. This is one of the most dependable native grasses for range grazing and for stabilizing the soil against erosion by water or wind. In cool dry places where very little water is available for irrigation, blue gramagrass can be used as a lawn.

Blue gramagrass is readily established from seed. The amount of seed produced depends upon the right combination of moisture and cool temperatures at the time of seed formation. Because blue gramagrass grows naturally over large areas, almost all the seed used for planting has been collected from natural stands; but seed collected in one area should be used only in that area.

The seed ripens rapidly as it nears maturity in early August, and after maturity the seed heads shatter readily. Under field conditions seed heads may contain from almost none to 75 percent fertile seed. To be satisfactory for harvesting, at least 25 percent of the seed should be plump. Seed yields of as much as 200 pounds to the acre have been obtained.

Buffalograss (*Buchloe dactyloides*). Buffalograss is the most abundant native grass in the central part of the short-grass region. Generally it grows 4 to 6 inches high and produces narrow leaves less than ⅛-inch wide and 3 to 6 inches long. It spreads by surface runners and rapidly forms a dense, matted, perennial sod, even when growing conditions are not those considered ideal. During the growing season the foliage is grayish green, but it turns the color of fresh straw when dry weather or frost retards its growth.

Growth begins in the late spring and continues through the summer. All livestock like its forage. Its palatability, abundance, and adaptation to a wide range of soil and climatic conditions make buffalograss one of the most important forage species of the Great Plains. On sandy soils, however, it will not do well.

Buffalograss withstands more heavy grazing than any native grass in its region. On ranges consistently subjected to severe use, it often survives as a nearly pure stand. Because of its aggressive spread under use and its relative ease of establishment, buffalograss is ideally suited for erosion control as well as for grazing.

Buffalograss plants are unisexual; that is, about half are female and produce seed burs while the other half are male and produce pollen only. When plants are grown under cultivation primarily for seed production, it is customary to grow ten female plants to one male plant. This proportion of female to male plants greatly increases the amount of seed produced per acre.

Left: Canada wild-rye grows in all states outside the Southeast. *Right:* Dallisgrass is a good pasture grass for the Southeast, the irrigated Southwest, and the far West. (*Courtesy* U.S.D.A.)

Pasture or range establishment by the use of sod pieces is effective. Sod pieces about 4 inches in diameter are planted at 3-foot intervals on a well-prepared seedbed. In semiarid regions, good lawns may be established by planting the sod pieces about one foot apart.

Canada wild-rye (*Elymus canadensis*). Canada wild-rye is widely distributed in the Great Plains, the Pacific Northwest, and the Rocky Mountain states, but grows in all states outside the Southeast. It is a very vigorous perennial bunchgrass with blue-green seed heads that grow 3 to 5 feet tall. The leaves are broad, flat, and rough, 6 to 12 inches long and ½ inch or more in width. The mature seed heads are a beautiful dark purple in color. They average 6 inches in length and have sharp awns or barbs.

Wild-rye begins growth about a week later in the spring than do most grasses but continues to grow throughout the summer if moisture conditions are favorable. It may start growing again in the fall after a summer drought if enough moisture is available.

The palatability of young forage is fair, but at maturity the plants become harsh and woody. The young seedlings are exceptionally vigorous and quickly form a good protective cover, which is useful for grazing and for soil and water conservation.

Canada wild-rye is useful in seeding mixtures, especially with grasses that do not rapidly produce a good ground cover. Hay of good quality can be made if the wild-rye is harvested just as the seed heads are emerging.

Centipedegrass (*Eremochloa ephiruroides*). This grass spreads rapidly by short creeping stems that form new plants at each node. It forms a dense, vigorous turf that is highly resistant to weed invasion. Although some seed is available, centipedegrass is usually established vegetatively.

Centipedegrass is generally considered the best low-maintenance lawn grass in the southern part of the United States. It requires less watering, less fertilizing, and less mowing than other southern lawn grasses. It is seldom damaged by disease or insects, but may be severely damaged by salt water spray. This grass is sensitive to a lack of iron. An annual application of a complete fertilizer improves the quality of centipedegrass lawns. Although it is drought resistant, centipedegrass should be watered during dry periods.

This grass should not be planted in farm lawns; it may escape into pastures and destroy their grazing value.

Dallisgrass (*Paspalum dilatatum*). Dallisgrass is a southern grass, but it can be grown as far north as Washington, D. C. It thrives in the irrigated sections of the milder parts of the Southwestern and Western states. Within its range, this vigorous bunchgrass is widely adapted. It was introduced into the

United States about 1875, probably from South America.

A moist and fertile soil is required for dallisgrass, and it grows best in soils high in organic matter. Because it is a bunchgrass and seldom forms a dense sod, it is an excellent grass to grow in mixture with legumes and other grasses. Seeded alone, dallisgrass often fails to make a perfect stand.

Dallisgrass produces abundant seed, but the germination is often poor because of the fungus disease "ergot," which attacks and destroys many of the seeds. This disease is poison to livestock, but mowing controls the fungus.

Dallisgrass is especially well adapted to the lower Gulf Coast where it makes excellent pasture for year-long grazing. When seeded with Ladino clover, locally called white Dutch clover, the combination produces 12 months of nutritious and abundant forage.

This grass is a weed in southern lawns, but cattlemen wish that more such weeds would grow in their pasture.

Kentucky bluegrass (*Poa pratensis*). When Daniel Boone explored Kentucky, Kentucky bluegrass was not there to meet him. A native of the Old World, Kentucky bluegrass occurs naturally over the more humid parts of Asia and Europe. It was brought from Europe by the early American colonists as one of the species contained in mixed grass seeds. Kentucky bluegrass found the soil and climatic conditions favorable and soon "grew wild."

Kentucky bluegrass has creeping underground stems, each bearing a tuft of leaves at the tip. Unlike some other grasses, Kentucky bluegrass blooms but once each season. When not in bloom it can usually be recognized by the leaves, which are V-shaped in cross-section, and by the peculiar leaf tip, which resembles the front of a boat.

Kentucky bluegrass is well known for the beautiful lawns it makes (especially improved varieties such as Merion) and for the highly nutritious spring pasturage it furnishes. Because of its low summer forage production, it is seldom planted now in the best pastures.

Kentucky bluegrass is widely scattered over the central and northeastern United States. (*Courtesy* U.S.D.A.)

Kentucky bluegrass occurs throughout the northern half of the United States, except where the climate is too dry. In the mountains and in the Pacific Coast lowlands, it extends farther southward. More recently Kentucky bluegrass has been found in all states, but it is not common in the South.

Japanese lawngrass (*Zoysia japonica*). This low-growing perennial spreads by aboveground runners and shallow rootstocks to form a dense turf resistant to weed and insect damage.

Japanese lawngrass grows best in the region south of a line drawn from Philadelphia, Pennsylvania, westward to San Francisco, California. It will survive in the region north of that line but its use there, except in some localities, is impracticable because of the short summer growing season. The grass turns the color of straw when the first killing frost occurs in the fall, and it remains off-color until warm spring weather.

Common Japanese lawngrass is coarse in texture. It is somewhat undesirable for home lawn use but is excellent for large areas such as airfields and playgrounds. Meyer zoysia, a selection of common Japanese lawngrass, is more desirable than Japanese lawngrass for home lawns. It is more vigorous, retains its color later in the fall, and regains it earlier in the spring. Meyer zoysia sod is available from a number of nurseries. There is no seed.

Although Japanese lawngrass will survive in soils of low fertility, it makes best growth when given liberal applications of complete fertilizers having a high nitrogen content. It is relatively drought tolerant in the humid regions. It is highly resistant to wear and will withstand close clipping.

Japanese lawngrass may be established by sprig planting the stems, by spot sodding, or by seeding. Three to four growing seasons are generally required to get complete ground coverage.

Emerald zoysia is a hybrid between Japanese lawngrass and mascarene grass that has proven superior to Meyer zoysia in the southern part of the United States. The grass is fine leafed, dense growing, and dark green in color.

Little bluestem (*Andropogon scoparius*). Little bluestem is a vigorous, long-lived, native bunchgrass of wide distribution over the United States. It is the most prevalent native grass in the Great Plains, flourishing particularly in the Flint Hills section of Kansas and Oklahoma, where it supplies dependable grazing and hay. Little bluestem also is scattered throughout the South and East. In many areas it is not considered a major forage because of its low palatability and because of the poor quality of its mature forage.

Little bluestem is, of course, smaller than big bluestem. The two usually are found in close association, but little bluestem is more drought-resistant and is therefore better adapted to sites that receive limited moisture.

Growth begins late in the spring and the plants at maturity are 1 to 3 feet tall. The leaf blades are less than ¼ inch wide and grow from 4 to 8 inches long. The leaves, flattened at the base, are light green until the plants reach maturity, when they develop a beautiful reddish-brown color.

Because it is a long-lived, vigorous bunchgrass and because of the wide range of soils on which it thrives, little bluestem has great value for range graz-

Major and minor distribution of buffalograss (*above*) and blue grama (*below*). (*Courtesy* U. S. D. A.)

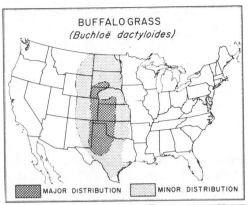

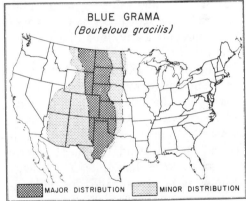

ing and erosion control. It is suitable for use in crop rotations and in mixtures for regrassing abandoned cultivated land.

Mutton bluegrass (*Poa fendleriana*). The range of mutton bluegrass extends from North Dakota to Idaho and California, and east to western Texas. Mutton bluegrass is a perennial bunchgrass. The erect stems vary in height from 6 to 24 inches and are roughened below the flower cluster. The tufts or bunches range up to a foot in diameter. The firm, stiff blades are folded or rolled in and are rarely flat.

Mutton bluegrass occurs in open grasslands, in sparse stands of aspen and coniferous timber, and on rocky slopes. Its range in the far West varies from 4,000 feet to timber line at 12,000 feet. It occurs on dry, southern exposures, but mainly on rich, well-drained clay loams of limestone origin. In the northern parts of its range, this grass often occurs on lower slopes, but in the Southwest it grows at higher elevations.

Mutton bluegrass is resistant to drought and fire. All stockmen know it because of its exceptionally early spring growth and because it starts to bloom immediately after the melting of the snow. Local stockmen often call it "wintergrass," "winter bluegrass," or "muttongrass." The flowering period varies from March to early June; seed ripens from July to November.

Mutton bluegrass is one of the more important native range grasses because it is very palatable, highly nutritious, and has a wide distribution. It rates as an excellent early spring forage for cattle and is very good for sheep; hence the name *mutton* bluegrass. The foliage becomes rather harsh and dry with increasing maturity. For this reason the palatability decreases somewhat as the season advances, although it is grazed well throughout the summer. In the fall, when more tender and succulent forage is scarce, cattle readily eat the standing, air-cured herbage.

Orchardgrass (*Dactylis glomerata*). Orchardgrass is a long-lived perennial bunchgrass. It grows in large masses, but the tussock-forming habit may be decreased by regular mowing, careful grazing management, and by seeding with a legume, such as Ladino clover, lespedeza, or alfalfa. Orchardgrass does not produce stolons or underground "runners" and for that reason never

forms a dense sod. The unique cluster formation of the flower head cannot be mistaken for that of any other cultivated grass. Its leaf blades are distinctly folded inward, making them V-shaped in cross section.

Orchardgrass was first cultivated in Virginia, having been introduced from Europe about 1760. Now it is widely distributed over the United States. Its persistence, leafiness, and productiveness under a wide range of soil and climatic conditions make it a desirable pasture grass. The most extensive acreage is in the region extending from southern New York State to southern Virginia and westward from the Atlantic Coast to eastern Kansas and southeastern Nebraska. It does not appear to be well adapted in the states bordering either the Gulf of Mexico or Canada.

Orchardgrass flourishes on rich soil, but it also succeeds on sandy soils of medium fertility and on moist, clay soils. It is one of the best cultivated grasses for planting in shady places; where adapted in orchards and woodland pastures, it becomes very abundant. It is quite cold-resistant; it starts growth in the early spring and continues growth until the first severe frosts. Smooth bromegrass and timothy are more resistant to winterkilling, but neither can equal orchardgrass in summer production of leafy pasturage.

Good results have been obtained with a combination of orchardgrass and Korean lespedeza in regions where lespedeza is adapted. Both the grass and the legume thrive on soils of medium fertility, although both respond to improved liming and fertilizing practices. The bunch type of growth, once considered objectionable, makes orchardgrass an ideal companion crop for Korean lespedeza or other legumes. Ladino clover is another excellent companion crop for orchardgrass.

The newer practice of making grass silage has further demonstrated the usefulness of orchardgrass. When it is grown in combination with legumes such as red clover or Ladino clover, orchardgrass is able to produce the maximum tonnage of high-quality silage early in the season. If this early growth is removed, the orchardgrass will not crowd the legume, and its rapid recovery will produce abundant, high-quality summer pasturage at a time when permanent pastures are not very productive.

Smooth bromegrass (*Bromus inermis*). Smooth bromegrass is now grown in all states outside the South. It came to us from Hungary in 1884. It is a winter-hardy, aggressive, sod-forming, and moderately drought-resistant perennial. In the localities to which it is adapted, few grasses are better for holding the soil in place.

Smooth bromegrass stands are maintained more satisfactorily if a legume, particularly alfalfa or Ladino clover, is grown in mixture with the grass. A legume is desirable because it makes nitrogen available in the soil to the bromegrass, and also lessens the tendency of the stand to become sod-bound.

Smooth bromegrass grows best on deep, fertile soils, under moderate rainfall, and in mild summer temperatures. Its range has been increased by the selection and development of locally adapted strains; consequently the source of the seed is very important.

Growth begins in the early spring and continues until checked by a shortage of moisture, frost, or available plant

Above: Orchardgrass does best from Virginia northward to Maine and westward to Kansas.

Below: Major and minor distribution of western wheatgrass. (*Courtesy* U. S. D. A.)

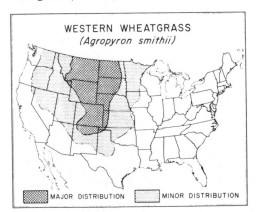

WESTERN WHEATGRASS
(*Agropyron smithii*)

MAJOR DISTRIBUTION   MINOR DISTRIBUTION

Timothy grass grows in the North on acid soils of low fertility where many other grasses seem to fail. (*Courtesy* U. S. D. A.)

food, especially nitrogen. Plants grow to heights of 3 to 4 feet. The leafy vegetation is relished by all classes of livestock at all stages of plant growth. When cured properly, smooth bromegrass makes very good hay.

A new variety, Polar bromegrass, has proved to be the most winter-hardy as well as the most productive bromegrass in Alaska.

St. Augustine grass (*Stenotaphrum secundatum*). This is the best shade grass of the southernmost states. It is a creeping perennial which spreads by long runners that produce short, leafy branches. It can be grown south of Augusta, Georgia, and Birmingham, Alabama, and westward to the coastal regions of Texas. It is established vegetatively. Seed is not available.

St. Augustine grass will withstand salt water spray. It grows best in moist, fertile soils. It produces good turf in the muck soils of Florida. Liberal applications of high-nitrogen fertilizers are necessary, especially in sandy soils.

St. Augustine grass can be seriously damaged by chinch bugs, and it is susceptible to armyworm damage and several turf diseases.

Timothy (*Phleum pratense*). Timothy came to us from Europe sometime around 1700. It is a biennial bunchgrass that grows 20 to 40 inches tall, with spike-like heads.

In 1800, timothy was the most important hay grass in the United States, and for 150 years its supremacy in this country was never threatened. Now there are many grasses more productive and more palatable than timothy.

Timothy grows better on clay loams than on sandy soils. It grows in every state but it is best adapted to the cool humid climate of the Northeastern states, the North Central states, the valleys of the Rocky Mountains, and the Coast Region of the Pacific Northwest.

Timothy is commonly sown with such clovers as medium red, mammoth, Ladino, or alsike, or with alfalfa. Such a mixture is higher in protein and also helps to maintain a better soil productivity than timothy alone.

A decrease in the quality of hay occurs as the season advances. With advancing maturity of the plant, the percentage of protein decreases while the less valuable crude fiber increases. For these reasons, timothy should be cut when it is first beginning to bloom in order to get the greatest value per acre in high-quality hay.

Western wheatgrass (*Agropyron smithii*). Western wheatgrass is distributed throughout the nation except in the humid South and is most abundant in the northern and central parts of the Great Plains. It is a vigorous, perennial, sod-forming, native grass with seed stalks reaching a height of 2 to 3 feet. Leaves are long and narrow, growing 8 to 12 inches long and less than a quarter of an inch wide. The leaf blades do not droop as on most grasses but are stiff and erect. The entire plant is usually covered with a grayish bloom, which makes western wheatgrass easy to recognize at a distance.

Although western wheatgrass is adapted to a wide range of soils, it seems to prefer fine-textured soils in shallow lake beds or along streams. Under these conditions, western wheatgrass may be found in almost pure stands. It also occurs in nearly pure stands on abandoned cultivated fields where the original wheatgrass was not entirely eliminated by cultivation. These "go-back-to-grass" fields are dependable for the production of grazing, hay, or seed.

Western wheatgrass has characteristics that make it exceedingly valuable for use in revegetating overgrazed ranges and abandoned fields for forage or for erosion control. Its hardiness, drought-resistance, and capacity to spread rapidly by means of underground "stems" give it an outstanding usefulness for soil and water conservation,

especially in semiarid regions. It is excellent for sodding waterways, terrace outlets, and gullies, and for the planting of strips in strip cropping.

Abundant forage is produced and is relished by all classes of livestock until it becomes somewhat harsh and woody during the late summer. In the fall, standing mature plants of western wheatgrass cure well into a palatable and nutritious forage that provides excellent winter grazing. Leafy, high-quality hay may be cut when the grass is still young and succulent.

The seed of western wheatgrass, like that of many other native grasses, has a low germination percentage immedi-

This crested wheatgrass is similar to western wheatgrass. *Left:* proper grazing. *Right:* excessive grazing has stunted both the tops and the roots. (*Courtesy* J. I. Case Co.)

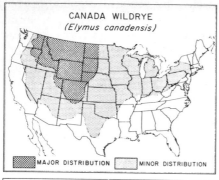

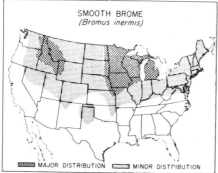

Major and minor distribution of Canada wild-rye (*above*) and smooth brome (*below*). (*Courtesy* U. S. D. A.)

The best soybean seeding rate is one which results in approximately 8 to 10 seeds per foot of row. This means planting about one seed per inch of row, or a seeding rate of 50 to 70 pounds of seed per acre, depending on row width and size of bean. Soybeans should not be planted more than 1½ inches deep; shallower if possible. (*Courtesy* American Soybean Association)

ately after harvest. The seeds will sprout better if they are first put in dry storage for 6 months to a year.

The sod-forming habit of western wheatgrass provides another means of propagation. The usual procedure is to plant sod pieces 4 feet apart in the row, with the rows spaced 4 feet apart. Sod pieces should be 4 inches square. This is an effective method of establishing a dense sod cover for a diversion channel, terrace, water outlet, or contour strip.

Its growth characteristics, drought-resistance, winter hardiness, and wide adaptation to soil and climatic conditions make western wheatgrass one of the best grasses for revegetation and general use on the farm and ranch.

LEGUMES

Legumes are productive, protective, and edible. From legumes comes a large part of the best hay, pasture, and silage for livestock. Legumes help to cover the land scars that are made when beating raindrops splash the soil into flowing muddy water. Food from legumes includes green beans, dry beans, peas, and many soybean products.

Many legume flowers resemble butterflies. The seeds are borne in pods like those you have seen on bean plants. Members of the large legume family vary in size from small clover plants to locust trees. The most unique characteristic of the legumes is that bacteria on their roots have the ability to fix nitrogen from air in the soil.

FIXATION OF NITROGEN

Legumes are the principal plants capable of cooperating with certain bacteria in obtaining nitrogen from air and making it available for plant growth. This is one reason for the increase in corn yields when corn follows clover or alfalfa. The availability of nitrogen also is a factor in the rapid growth of grasses in mixture with legumes.

The air around us is approximately 80 percent nitrogen and 20 percent oxygen. The oxygen in the air is readily available to plants and animals. Take a deep breath now and see how much better it makes you feel. When you breathe deeply, the oxygen in the air is exchanged for the carbon dioxide in the lungs. The new supply of oxygen makes you feel more alert and energetic. Plants also use oxygen when they breathe. This oxygen is used by plants and animals in the "burning" of their food.

The use of nitrogen is a different matter. When you inhale air you are taking in nitrogen, oxygen, and a tiny amount of carbon dioxide. When you exhale you are releasing the same amount of nitrogen, less oxygen, and more carbon dioxide. This process is much the same for all animals. Neither animals nor plants can use nitrogen from the air when they breathe.

You may think that because plants and animals cannot use nitrogen from the air they do not need nitrogen. The truth is that all living things need nitrogen, although few of them can use nitro-

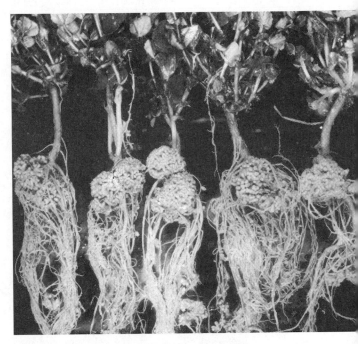

The nodules (small lumps) on these Austrian winterpeas are the homes for billions of bacteria that get their nitrogen for body proteins from the air. (*Courtesy* The Nitragin Co., Inc.)

gen from the atmosphere. Legume bacteria have the unique characteristic of being able to use atmospheric nitrogen to build body protein. However, the legume bacteria can utilize nitrogen from the air only when a legume plant acts as host.

Legume bacteria and legume plants need each other, as we need our friends. The bacteria attach themselves to the small roots of legumes, and the cells in the roots grow around the bacteria. The tiny projections thus formed on the legume roots are called *nodules*.

In their little nodule homes the legume bacteria get essential nutrients from the legume host while they use nitrogen from the air in the soil to build body proteins.

As the bacteria die, their body proteins decay and release fixed nitrogen for direct use by the legume host. Grasses and other plants growing in association with the legumes also are able to use some of this nitrogen. And some of the fixed nitrogen is available for the crop that follows the legume. For example, when vetch, a legume, is plowed under and followed by corn, some of the atmospheric nitrogen that was fixed in body proteins of the bacteria is used by the growing corn plants.

The amount of nitrogen fixed by legume bacteria can only be estimated. According to some calculations, approximately 2 million tons of atmospheric nitrogen is fixed by legume bacteria in one year. This is a tremendous amount of nitrogen, but there is no danger of exhausting all of the nitrogen in the atmosphere.

Above every square inch of land or sea there are 12 pounds of nitrogen. Using some simple arithmetic, we can multiply 12 pounds by 144, the number of square inches in a square foot, and find that there are 1,728 pounds of nitrogen above each square foot of the earth's surface. To compute atmospheric nitrogen on an acre basis, multiply 1,728 pounds by 43,560, the number of square feet per acre. This computation will reveal that there are more than 75 million pounds of nitrogen above every acre.

Thus, although legumes fix large amounts of useful nitrogen, the amount is seen to be very small in relation to the total volume of nitrogen in the atmosphere. The 2 million tons of nitrogen that legume bacteria fix each year

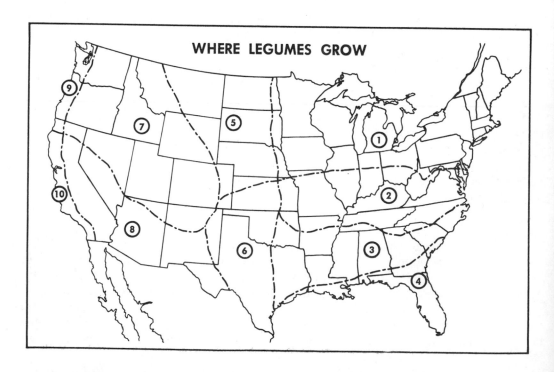

WHERE LEGUMES GROW

are actually equal to the atmospheric nitrogen above approximately 53 acres.

THE LEGUMES AND WHERE THEY GROW

Certain nitrogen-fixing legumes are adapted to the soils, climate, and forage needs in your community. Some of these plants have been developed to resist drought, insects, diseases, heat, and winter-killing.

Some of the important small-seeded legumes are described in this chapter. Use the map as a guide in learning the geographical distribution of the alfalfas, clovers, lespedezas, trefoils, soybeans, and vetch.

Alfalfas. The largest acreages of alfalfa are in the central and northern parts of the United States, although some alfalfa is grown in every state. The wide distribution and high nutritive values of the different varieties of alfalfa make this one of the most important groups of forage plants.

Alfalfa grows best on fertile, well drained, deep loam soils. Acid soils are not satisfactory, as alfalfa requires a good supply of calcium. All varieties of alfalfa also require an abundance of phosphorus, potassium, and other nutrients. Some important varieties are Williamsburg, Vernal, Lahontan, Caliverde, Atlantic, Buffalo, Narragansett, Ranger, and Talent. (See the map on p. 328.)

Williamsburg (Region 1, 3). This variety is popular in Virginia and adjacent areas because it persists through the summer. One weakness of Williamsburg alfalfa is its susceptibility to bacterial wilt.

Vernal (Regions 1, 5, 7). Because of its winter-hardiness and resistance to bacterial wilt, Vernal alfalfa is well adapted to the northern half of the United States. It grows well in bromegrass mixtures and makes better yields than Ranger alfalfa.

Lahontan (Regions 8-10). Resistance to bacterial wilt and stem nematodes makes this legume well adapted to Nevada and California. Lahontan alfalfa is not adapted to the humid East where leaf and stem diseases cause serious damage. In the areas of adaptation this variety yields well and is resistant to winter weather.

Caliverde (Region 10). This high-yielding variety is well adapted to the Central and Lower valleys of California. An outstanding characteristic is resistance to common leaf spot, bacterial wilt, and downy mildew.

Atlantic (Regions 1, 2). Atlantic alfalfa yields better than most other varieties in the East. It has some resistance to bacterial wilt, is vigorous, and recovers quickly after cutting.

Buffalo (Regions 1-3, 6-10). In adapted areas this variety maintains good stands longer than do varieties that are more susceptible to bacterial wilt. Fast recovery after spring cutting is one good characteristic of Buffalo alfalfa.

Narragansett (Region 1). This hardy hybrid is adapted to Rhode Island and adjacent areas where bacterial wilt is not a serious threat to the alfalfa crop. In this region, Narragansett alfalfa is a long-lived legume that makes good yields of hay.

Ranger (Regions 5, 7, 9). Resistance to bacterial wilt is the main characteristic of Ranger alfalfa, a variety adapted to

Alfalfa makes up about 42 percent of all hay crops. This one-man hay baling and loading operation was in Maine in 1971. (*Courtesy* State Department of Agriculture, Augusta, Maine)

the northern United States. This variety is winter-hardy but may be killed by severe winter weather. Ranger alfalfa outyields other varieties that are more susceptible to wilt.

Talent (Regions 7, 9.). This variety is vigorous enough to crowd out grass and weeds. An important characteristic of Talent alfalfa is its ability to start early in the spring and grow late in the

fall. This variety is moderately winter-hardy.

About 42 percent of the total acreage of hay crops is alfalfa hay, which includes alfalfa and alfalfa mixtures used for hay and for dehydrating. Alfalfa is common in all areas of the United States except the Southeast, where the humid climate and sandy soils are not favorable for production. Soils with adequate lime

Potash helps build a sturdy root system in a crop. This young alfalfa shows what happened with and without potash when used with band seeding. Potash helps build alfalfa quality and insure more frequent cutting. (*Courtesy* Potash Institute of North America, Inc., Atlanta, Georgia)

are the most favorable for growing alfalfa. In most areas, the crop is harvested two or three times a year, and yields range from 2 to 3 tons per acre. Annual yields are particularly high in the Southwest because of the long growing season and the common use of irrigation water. In 1962, the California yield was 5.2 tons and the Arizona yield was 4.8 tons per acre. Four states cut more than 2 million acres in 1962: Wisconsin, Minnesota, Iowa, and South Dakota.

Clovers. Approximately 250 true clovers are known to grow in the United States. The annual clovers grow best where the climate is cool and moist. An abundance of available calcium, phosphorus, and potassium is required for these legumes. Unfavorable climate, diseases, and insects are the worst enemies of the clovers. Sweetclover is not a true clover, but it is included with the clover group because of its value for soil building and temporary pasture.

Ladino clover is one of the best clovers for pasture in humid or irrigated regions. (*Courtesy* B. A. Brown, Connecticut Agricultural Experiment Station)

Red clover (Regions 1-4). Red clover is one of the best forage and soil-building legumes in the eastern half of the United States. This legume makes one to three cuttings a year for two years, depending upon soil and precipitation.

Red clover may be sown alone or in grass-legume mixtures. It is often seeded with winter or spring grain crops. Kenland, Pennscott, and Midland are the three main varieties of red clover. Farmers should select the variety that is recommended for local conditions.

White clover (Regions 1-4, 7-10). This is an excellent legume for pasture and for soil building. For use as a winter annual, white clover is seeded in the fall on established warm-season grasses. In some areas white clover is seeded with other legumes and grasses. Ladino clover is one of the best varieties for all of the white clover regions except 3 and 4. Pilgrim is a good variety for some of the Northeastern and Middle Atlantic states. Fall seedings of white clover are recommended in regions 3 and 4. Local recommendations should be followed in selecting a variety for any community.

Crimson clover (Regions 1, 3, 4, 9, 10). This upright winter annual is best adapted to the Southern states but also is grown in the Pacific Northwest. Crimson clover is used for hay, pasture, and soil building. It is a good green manure crop and makes a protective winter cover for the soil. The best time to cut crimson clover for hay is when the flowers are beginning to open. Crimson clover is adapted to a wide range of soil conditions, but does not grow well on wet, clay soils.

Alsike clover (Regions 1-4, 7-9). This perennial legume is suitable for wet soils. Alsike clover is particularly well adapted to the Northern states. It usually is grown with grasses or with other legumes. The upright plants attain a height of 1 to 3 feet. In addition to being a good soil-builder, alsike clover is used for hay, pasture, and silage.

Persian clover (Regions 3, 4). This winter annual is adapted to lowland clay soils in the Southern states. In addition to being valuable for pasture and hay, Persian clover is a good cover and green manure crop.

Subclover (Regions 9, 10). This winter annual is adapted to the Pacific Northwest. Subclover is planted alone or with grasses. In mixture with grasses, subclover makes good pasture, hay, and silage. Leaf growth varies from 2 to 8 inches. It is very important to select the right variety from among the many varieties of subclover.

Strawberry clover (Regions 7-10). Strawberry clover is a perennial that is adapted to the wet soils in the Western states. This legume is tolerant of salty and alkaline soils. It survives both hot summers and below-zero winter temperatures. Strawberry clover is used mainly for grazing.

Hop clover (Regions 2, 3). Small hop clover is a winter annual that is widely distributed in the South and in the Pacific Northwest. Its tolerance of unfavorable soil and climatic conditions is an important characteristic. Large hop clover, also a winter annual, is more productive than small hop clover. The establishment and growth of large hop clover requires a greater supply of mineral elements. The hop clovers are valuable for spring pasture. Grasses grown with hop clovers benefit from the increase in soil fertility.

Above: Sweetclover is excellent for grazing and soil building. (Kentucky) (*Courtesy* Kentucky Agricultural Experiment Station) *Below:* The sericea lespedeza on the left was not grazed because the cattle preferred the sand lovegrass on the right. (*Courtesy* Oklahoma State University Extension Service)

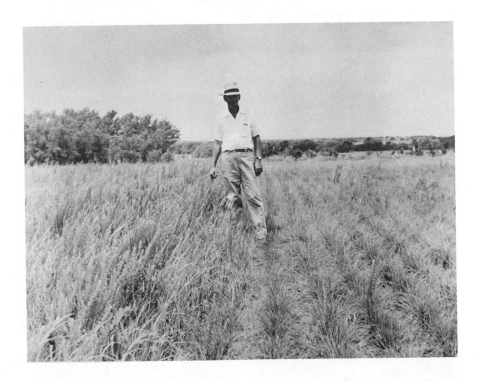

Sweetclover (Regions 1, 5, 6). There are approximately 20 species of sweetclover, none of which is native to the United States. These legumes apparently originated in southwestern Asia Minor. In the Corn Belt, sweetclover is grown in rotation with corn and small grains. When grown in this cropping system, sweetclover provides temporary grazing in spring and early summer before being plowed under as a green manure crop.

Sweetclover requires at least 17 inches of effective precipitation to make good growth. This means that the total annual precipitation must be considerably more than 17 inches. Sweetclover is drought-resistant after it becomes established.

Madrid is a yellow-blossom, biennial sweetclover highly recommended because of its large yields and the ease with which it can be established. Important varieties of biennial white sweetclover include Spanish, Willamette, and Evergreen. The biennial white sweetclovers are particularly useful in the Corn Belt, the Great Plains, and the Intermountain Region. Two popular varieties of annual white sweetclover are Hubam and Floranna.

Lespedezas. The lespedezas are used mainly for pasture. These soil-building legumes also make good yields of nutritious hay. The most important lespedezas are common, Korean, and sericea. Common and Korean are annuals; sericea is a perennial.

Common lespedeza (Regions 3, 4). This legume is a good pasture plant for the lower South. After the grasses become scarce the lespedeza continues to provide grazing. Kobe is the most important selection from common lespedeza.

Korean lespedeza (Regions 1, 3). Climax, Rowan, and Iowa 6 are leading varieties of Korean lespedeza. Climax, a late-maturing variety adapted to the deep South, yields more hay than ordinary Korean lespedeza. High resistance to powdery mildew and moderate resistance to root-knot nematode are characteristics of Rowan. In the upper half of the lespedeza belt, Iowa 6 is a well-adapted legume. Its early maturity permits it to be grown farther north in the Corn Belt.

Sericea lespedeza (Regions 2-4). This tall perennial is adapted to the east-central region of the United States. It is grown from the Atlantic Ocean westward to northeastern Oklahoma and eastern Nebraska. The northern limit of the range of this lespedeza is near the Missouri-Iowa line. Arlington sericea is a new variety noted for its vigor and its high yield of forage.

When given a "cafeteria" style of grazing, cattle will eat first the forage that they like best and are sometimes reluctant to eat this legume. However, they can acquire a liking for sericea lespedeza, and when they do they may prefer it to some other forage.

Trefoils. The trefoils are used mainly for pasture, although they are grown as hay crops in some areas. Big trefoil and birdsfoot trefoil are the two species of agricultural importance in the United States. Both are deep-rooted perennials. The fine-stemmed, leafy plants grow upright in thick stands, but single plants lie close to the ground. The colorful yellow flowers are borne on long flower stalks. The trefoils can be grown in some areas that are not suitable for clovers and alfalfa. Trefoils are grown alone or in mixtures of grasses and legumes.

Big trefoil (Regions 3, 4, 9, 10). This species is better adapted to acid soil but does not survive severe cold or drought as well as birdsfoot trefoil. Big trefoil is particularly useful for developing acid cutover timberlands into highly improved pastures. This is one of the main uses of big trefoil in certain areas in the Pacific Northwest and on the Coastal Plains in the Southeast.

Birdsfoot trefoil (Regions 1, 2, 7-10). This legume gets its common name from its spreading seed pods, which resemble a bird's foot. Birdsfoot trefoil has a wider geographical distribution and a larger acreage than big trefoil. Mixtures of birdsfoot trefoil and certain grasses make good permanent pastures on the less fertile soils in the Northeast, the Corn Belt, and the Pacific Coast states. Birdsfoot trefoil and Kentucky bluegrass grow well together in the regions where bluegrass is grown. One important characteristic of birdsfoot trefoil is that it can be grown on soils that are not fertile enough for some of the other legumes, although it responds to applications of lime and fertilizer.

The broad-leaf type of birdsfoot trefoil is adapted to drier, less fertile soils. This type is grown on some of the western irrigated lands. The narrow-leaf type of birdsfoot trefoil has a shallower root system that makes it better adapted to moist, fertile soils. This type is best for some of the wet, clay soils in the Northeast. The newer varieties of birdsfoot trefoil include Empire, Viking, Cascade, and Granger. With the exception of Empire, the seed supply of the new varieties is very limited.

Soybeans as a special legume (Regions 1-3). The soybean is a special kind of legume because:

1. It is raised in such large quantities.
2. Its importance has increased tremendously in recent years.
3. It is grown extensively as a hay crop.
4. The seed has special usefulness as human food and as an edible oil.

The food products made from soybean seed include meal, flour, vegetable oil for shortening, margarine, and salad oils. Many of the oils are treated in a different way and used for making paints and varnishes. Soybean-oil meal is a high protein concentrate for livestock feed.

Soybean production has been increasing at a phenomenal rate in the United States, more than doubling during the past decade. Soybeans are grown primarily for beans which are processed for oil and meal, although a small acreage is grown as a forage or green ma-

Cultivating soybeans, a very important cash crop, in Central U.S.A. (*Courtesy* National Committee on Boys' and Girls' Club Work, Inc.)

Potash-starved (above) and potash-fed (below).

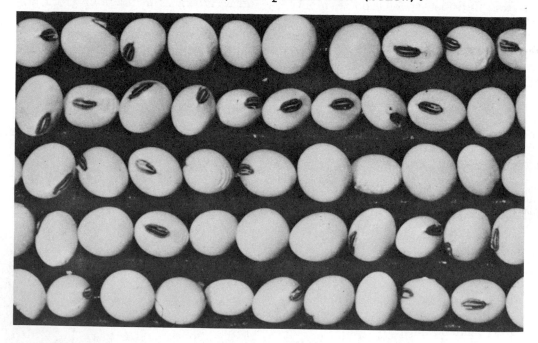

Potash helps build soybean quality. Potash also helps reduce the number of shrunken, moldy, discolored seed, boosts seed weight and germination rate, and reduces moisture buildup in seed. (*Courtesy Potash Institute* of North America)

nure crop in some areas. The main soybean-producing area is in the North Central States, although the South Central and South Atlantic States are becoming increasingly important producers. Illinois is the leading state followed by Iowa. These two states accounted for nearly 30 percent of the nation's total production in 1962. Other top-ranking states include Arkansas, Minnesota, Missouri, Indiana and Ohio.

Vetches (Regions 2, 3, 7). Hairy vetch is the most common of the vetches. It is grown mainly as a soil-building legume in the South and in the Northwest. It grows fairly well in cool weather and makes at least fair growth on poor soils. In the Cotton Belt, vetch can be planted at about the time of the last picking of cotton in the fall and will usually make good growth in time to plow under before cotton planting time in the spring.

QUESTIONS

1 What is the best grass for lawns in your community?

2 Contrast little bluestem and big bluestem.

3 Where does smooth bromegrass grow?

4 Explain how western wheatgrass can supply good winter grazing even though the tops are killed by frost.

5 Which legume makes the best hay?

6 How much nitrogen is there above every square inch of the earth's surface? How much above every acre?

7 Name the common legumes that grow in your community.

8 Where is the center of production of alfalfa?

9 Which variety of alfalfa is best for your community?

10 Why are soybeans considered a special kind of legume?

ACTIVITIES

1 Arrange a field trip or class demonstration for practice in identifying grasses. If a field trip is possible, start individual and class collections of grasses. Attach a tag or label to each plant as the species is identified. After the plants are dried, mount them on cardboard and exhibit them.

2 Find a place where grasses are growing in association with legumes. Note the color and condition of the grass and legume plants. Try to find the same kind of grass plants growing alone. Explain how grasses and legumes are beneficial to each other.

3 Dig up several kinds of legume plants. Look for the tiny nodules on the roots. Explain the special work of the bacteria that causes nodules to

form. If some of the plants do not have nodules, explain some of the possible reasons why nodules did not form.

4 Through field trips, class demonstrations, and individual study, learn to identify the important legumes in your community.

5 By observation and by talking with farmers, find out which species of legumes are popular in the community, the purposes for which the different legumes are used, and the methods by which the legumes are seeded.

24 CORN AND SORGHUMS

Corn is the leading crop in American agriculture both in value and acreage grown each year.

The sorghums are to the Great Plains what corn is to the Corn Belt. Sorghums seem to grow best in semiarid regions were there is not enough timely moisture for corn.

CORN

In 1969, 54.6 million acres of corn were harvested for grain, producing 4,578 million bushels. The acreage for grain comprised 85 percent of the corn grown for all purposes. The other 15 percent of the acreage of corn in 1969 included 0.9 million acres harvested for green forage (cut and fed green) and 7.8 million acres for silage. Nearly three-fourths of the corn-for-grain acreage lies in the Corn Belt, with Iowa the leading state and Illinois second. Corn-for-grain estimates began in 1919. The largest acreage of record, 97 million, was harvested in 1932. Acreage gradually declined, except for larger wartime plantings, and was down to 72 million acres by 1950. The downtrend continued to about 63 million acres in 1957 and 1958 since allotments were in force in commercial counties for producers desiring price support. With the discontinuance of allotments in 1959, acreage returned to the 1950 level of 72 million, but declined with the advent of the Feed Grain Program in 1961. The average yield per acre has trended upward since 1940 with more rapid increases since the mid-1950's, resulting in a new record being established nearly each year since 1955. The 1969 average yield was 83.9 bushels per acre.

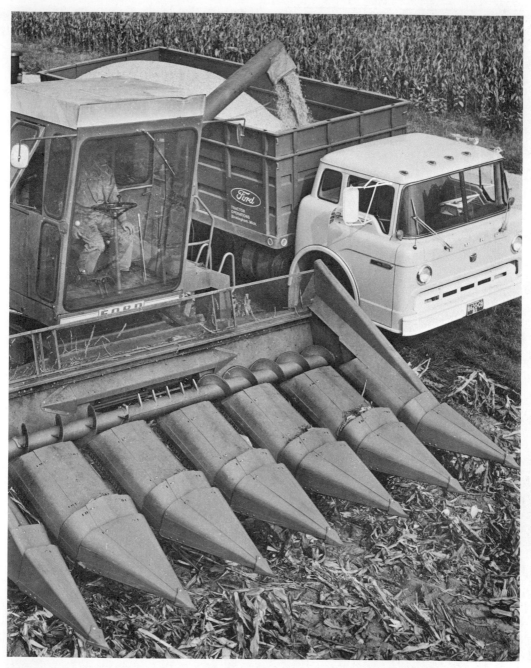

Harvesting corn for grain has become highly mechanized and very efficient. This six-row combine cuts the stalks, shells the grain from the cob, and blows the grain into a truck, as both the combine and truck operate at about the speed of walking. (*Courtesy* Ford Tractor Operations)

USES OF CORN

Sweet corn, popcorn, and field corn are the three main types of corn. Sweet corn provides tempting roasting ears to brighten our summer meals. (Sweet corn can even be grown in Alaska, under a clear plastic mulch.) Popcorn is popular at ball games, movies, and carnivals.

Field corn is used mainly for human food and livestock feed. In dairy regions most of the corn is cut, chopped, and stored in silos. The harvesting of corn for silage is an important method of preserving corn for winter feeding.

In addition to being used directly as human food and livestock feed, corn is the raw material for many industrial products. The main reason for this extensive use is that a kernel of corn is about 80 percent starch. Corn starch is high in quality, abundant, and relatively cheap in price.

Other corn products include hominy, grits, corn meal, corn flakes, sugar, syrup, oil, alcohol, and acetone. It is easy to understand why so many valuable products can be made from corn. The composition of a corn kernel is 84 percent dry matter and 16 percent water.

CORN HYBRIDS

The production of hybrid corn has increased rapidly during the past 20 years. In 1933, hybrid seed corn was planted on only one-tenth of 1 percent of the total corn acreage. Now almost all of the corn that is planted is hybrid.

Corn hybrids were developed through the practical application of the science of genetics. This science is based upon principles of heredity. The hybrid corn plant inherits certain characteristics from each of the two parent strains of corn that are crossed to produce an improved strain. After hybrid corn has been grown one year it should not be used to plant the next year's crop. A corn hybrid must be produced anew each generation; its value is in the high yield rather than in reproduction.

Corn hybrids are produced by controlling the process of pollination of the parent strains of corn. A corn "family" is kept pure by transferring pollen (male parts) to the silks (female parts) of the same stalk of corn. This is repeated for several years. Another corn "family" is kept pure by making it pollinate itself, a process known as self-pollination. Then the pollen from one corn family is placed on the silks of the other corn family. The kernels that develop are hybrid corn. This process is called a single cross.

CHOOSING AN ADAPTED VARIETY

Ordinary varieties of corn, sometimes referred to as old-fashioned varieties, are said to be *open-pollinated.* That is, they were allowed to cross freely with each other.

Farmers once selected seed corn in the field before the crop was harvested. The best ears were picked from sturdy stalks. The development of corn hybrids has made this method of seed selection obsolete.

It is well to remember that not all corn hybrids are good in all places. Some of the hybrids that are well adapted to the Corn Belt yield poorly in the South. Likewise, warm-weather hybrids are not satisfactory in cool climates. Certain hybrids seem to do better in acid soils,

Corn is chopped and blown into a trailer. When full, the trailer is hauled to the silo into which the chopped corn is blown. After fermentation in the silo, the corn silage is excellent feed for cattle. (*Courtesy* Ford Tractor Operations)

while other hybrids are adapted to alkaline soils. There is now a hybrid that is bred to produce well in your community, and that is the one to use. Since there are hundreds of corn hybrids, the practical way to find out which ones are adapted to your locality is to ask your county agricultural agent or your vocational agriculture teacher.

CLIMATE AND CORN CULTURE

Corn requires approximately 20 inches of water during the growing season to produce satisfactorily. This means 20 inches of water available in the soil for growing plants to use, not 20 inches of precipitation. Over most of the Corn Belt the average annual precipitation is 40 inches, although small areas in the region have 30 to 35 inches.

For maximum yields the 20 inches of water must be available in the soil during the growing season as follows:

MONTH	AVAILABLE WATER
May	3 inches
June	4 inches
July	7 inches
August	4 inches
September	2 inches

During years with normal precipitation the Corn Belt soils have a reserve supply of moisture at the beginning of the growing season. Without reserve moisture in the soil, the growth of corn is likely to be retarded in July unless irrigation water is applied.

Although a corn farmer has no control over weather and climate, he can do certain things to make conditions

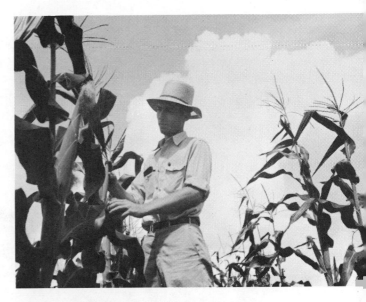

An adapted corn hybrid for this Florida farm would not be suited to farms in Iowa. (*Courtesy Soil Conservation Service*)

favorable for the best growth of corn. He can help corn to grow better in critical periods by these practices:

1. Use minimum tillage to encourage more water to enter the soil.
2. Fertilize and lime the soil in accordance with the needs of the land as indicated by a soil test.
3. Select the variety of corn that has proved itself best adapted to local soil and weather conditions.
4. Drain land where water stands for long periods in the spring. This will permit the soil to warm up faster and give the corn a better start. Drainage also encourages a deeper root system.
5. Practice a good system of crop rotation that includes sod crops. A newer practice is to plant corn continually, closely spaced (more than 25,000 stalks per acre), but to use heavy

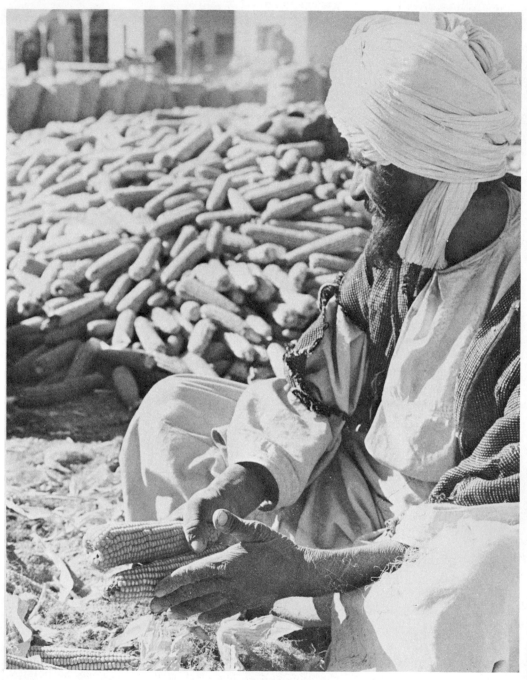

A farmer in Afghanistan (western Asia) is shucking an improved hybrid corn which was made possible by a project financed by the International Development Association, a unit of the United Nations. (*Courtesy* Kay Muldoon)

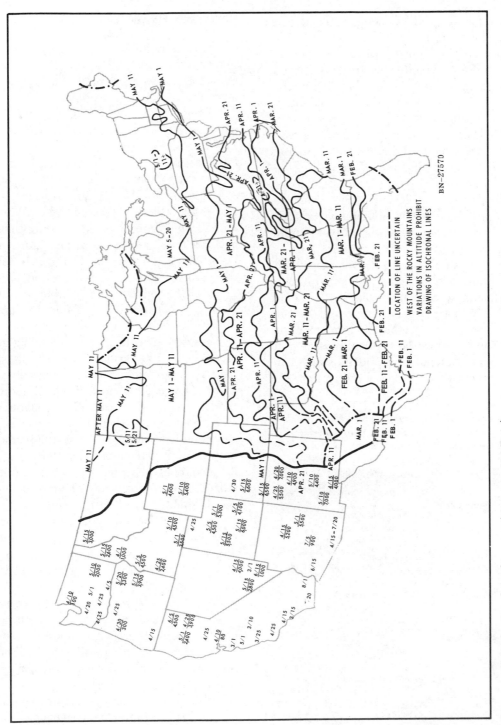

Recommended planting dates for corn. (*Courtesy U.S.D.A.*)

fertilization and to incorporate all corn residues into the soil.

6. Apply as much barnyard or poultry manure as possible. In places where manure is available at reasonable cost, it should be used in amounts of 20 tons or more per acre per year on fields that are to be planted to corn. Less chemical fertilizer can be used in decreasing amounts as recommended by your Extension Service.

7. Terrace the field if erosion is a hazard, practice stubble-mulch farming, and on sloping fields cultivate on the contour to conserve water.

8. Irrigate if there is not adequate rainfall. A rule-of-thumb technique is to irrigate corn if it wilts by 10 A.M.

9. Cultivate shallow to avoid cutting corn roots. A more modern technique is to apply herbicides according to recommendations and thus eliminate the necessity of any cultivation.

American Indians taught the Colonists how to fertilize corn. A fish was placed beside each hill of corn at planting time.

LIME AND FERTILIZERS

The most suitable soil for corn is only slightly acid. If a test shows the corn soil to be moderately to strongly acid, the soil should be limed.

Corn often needs large quantities of commercial fertilizers. Although fertilization rates vary according to differences in soil and climatic conditions, on the average an acre of corn may need 1,000 pounds of a 10-10-10 fertilizer (10 percent nitrogen, 10 percent phosphate, and 10 percent potash) at planting time and 200 pounds of ammonium nitrate or other suitable nitrogen fertilizer after the corn is up. Such a fertilization program may produce over 100 bushels of corn per acre if other conditions are favorable.

An acre of corn plants producing at the rate of 100 bushels per acre must have approximately 5 million pounds of water (19 to 24 inches of rain), more than 5,000 pounds of carbon (the amount of carbon contained in 4 tons of coal), 130 pounds of nitrogen (four 100-pound bags of a 32 percent nitrogen fertilizer), 22 pounds of phosphorus (two and one half 100-pound bags of 20 percent superphosphate), and 110 pounds of potassium (two 100-pound bags of 60 percent muriate of potash). The corn plants also require 22 pounds of yellow sulfur, 33 pounds of magnesium (equivalent to 330 pounds of Epsom salt), and 37 pounds of calcium (equivalent to 93 pounds of limestone). Other requirements are small amounts of iron, manganese, and boron, in addition to the trace elements of chlorine, iodine, zinc, and copper.

For more specific recommendations on the use of lime and fertilizers, ask your vocational agriculture teacher or your county agricultural agent about having your soil tested. All state agricultural colleges have a soil-testing service.

PLANTING AND CULTIVATING

Over a period of several years the best corn yields usually are obtained when the corn is planted about 10 days after the average date of the last killing frost in the spring. For example, if May 10 is the average date of the last killing frost in the spring, May 20 is the recommended date for planting corn.

The corn rows should be about 30 inches apart, and the corn plants in the row should be about 7 inches apart or less. Plants should be closer in very productive soils and farther apart in less productive soils. Plant populations of 25,000 plants per acre in good soils are usually satisfactory. The highest yield obtained in 1968 was from 28-inch rows and 29,000 stalks per acre.

Corn should be cultivated as shallowly as possible and only often enough to control weeds. Deep cultivation may injure the roots and permit excessive evaporation. Minimum tillage should be practiced and weedicides used.

HARVESTING

When corn is harvested for silage the whole stalk, including the ears, is chopped and put into a silo for feeding at a later period. Making corn silage is the most efficient way to use all of the feeding value of the crop. As much as one-third more feeding value is obtained from corn when it is made into silage.

Above: Women in Malawi, southern Africa, are proud of the rich harvest of an adapted corn hybrid grown on a new development project financed by the International Development Association. (*Courtesy* Lilongwe Land Development Project, Malawi) *Below:* Corn yields of more than 200 bushels per acre are possible with an adapted variety, adequate plant population, irrigation when rainfall is deficient, and fertilizers applied according to a soil test. (*Courtesy* Robert Lucas, Michigan State University Extension Service)

Above: Manure for next year's crop can be spread any time. Here the manure is being applied to the current year's corn stubble. The field will remain over winter with manure and stubble until time to plant corn in the spring; then in one operation over the field, corn will be planted and fertilized, and weedicides, insecticides, and fungicides will be applied. (*Courtesy* New Idea Machinery Company) *Below:* Eighty percent of all corn grown is harvested for grain, as shown here. (*Courtesy* John Deere and Company)

Hawaii has intensified its grain sorghum production to help the beef cattle industry compete with grain-fat beef from other states. (Hawaii) (*Courtesy* University of Hawaii)

In a few places over the nation, corn is "hogged off." This means turning hogs into a field where the stalks are still standing and the ears of corn have not been harvested. This practice is a good one because it reduces the cost of harvesting the corn.

By far the larger part of the corn crop in the United States is harvested with corn pickers. The picker pulls the ears of corn from the stalk, takes off most of the husks, and elevates the ear corn to a trailer that is pulled behind the picker.

So many human hands have been lost in accidents during the operation of corn pickers that a few words of safety are worth remembering. *Always stop the picker before cleaning the husking or snapping rolls or making any other adjustments.*

SORGHUMS

Several drought years which occurred three decades ago gave added importance to the sorghums, particularly as grain crops. An average annual precipitation of 15 to 30 inches apparently is best for sorghums.

The four general classes of sorghums are (1) grain sorghum, used for grain; (2) sweet sorghum (sorgo), used for syrup; (3) forage sorghum, used as forage for livestock; and (4) broomcorn, used to make brooms.

Sorghums harvested for grain totaled 13.5 million acres in 1969. The trends in acreage since 1949 have been erratic and have varied from 5.3 million acres in 1952 to 19.7 million acres in 1957. The acreage of sorghum for grain increases in dry years and follow-

Above: Improved varieties of grain sorghum are being tested at a research station on a government farm in Pirang, Gambia, western Africa. (*Courtesy* Roy L. Donahue) *Below:* Americans working abroad must involve the people in that country for continuity of the practices demonstrated. Frank Shuman (left) in Central India is shown here with J. S. Patel, formerly Agricultural Commissioner with the Indian central government, and in 1964–1968 vice chancellor of a new university, the Jawaharlal Nehru Agricultural University, Jabalpur, Madhya Pradesh, India, established in cooperation with the University of Illinois under a United States Agency for International Development contract (the crop is grain sorghum). (*Courtesy* Frank Shuman)

ing dry years because it is more drought resistant than corn. Prices received by farmers have varied from $1.58 a bushel in 1952, when acreage was least, to the lowest price of 84 cents a bushel in 1960. In 1957 when the greatest acreage was harvested the average price was approximately 97 cents a bushel.

Sorghums harvested for forage totaled 2.8 million acres in 1969, with a total production of 5.5 million tons and a farm price of $18 per ton.

There were 766,000 acres of sorghums for silage in 1969, with a total production of 9 million tons.

GRAIN SORGHUM

The grain sorghums yielded more than 743 million bushels of grain in 1969. In order of production, the yielding sorghum-producing states ranked in the following order: Texas, Kansas, Nebraska, California, Oklahoma, and New Mexico.

Grain sorghums, unlike corn, will produce a satisfactory crop in regions of low and uncertain rainfall. Grain sorghum is somewhat unique in this respect. On first thought, you may guess that sorghum requires less water per acre than corn, but this is not true. Sorghum and corn have the same water requirements per acre. The grain sorghums are able to produce a satisfactory crop even when drought periods occur because the plants are able to become dormant (inactive in growth) when there is not enough moisture. When rains occur the sorghums can resume growth and produce a crop of grain.

It is interesting to compare the trend in acreages of corn and grain sorghum in a typical semiarid region. The Southern Great Plains Field Station in Woodward County, Oklahoma, started its research with sorghums and other field crops in 1914. The reports show that Woodward County receives an average of nearly 24 inches of precipitation. Since 1915 the extremes in annual precipitation have been as low as 15 inches and as high as 41 inches.

In 1900 there were nearly 14,000 acres of corn and less than 1,000 acres of grain sorghum in Woodward County. Fifty years later there were only 3,000 acres of corn and 17,000 acres of grain sorghum.

The trend in acreages is much the same in all areas of the United States where the growing season is warm

Grain sorghum in the Southern Great Plains is now almost entirely the double dwarf type, about 3 feet high. It can be harvested easily with a combine, as shown here. (*Courtesy* John Deere and Company)

Above: A sorghum-sudangrass cross which grows well in western India was imported from Texas. (*Courtesy* George Moore) *Below:* Half of the world's acreage of grain sorghum is in India. Here an Indian uses a sling-shot to keep birds away from the grain until it is harvested. (*Courtesy* Roy L. Donahue)

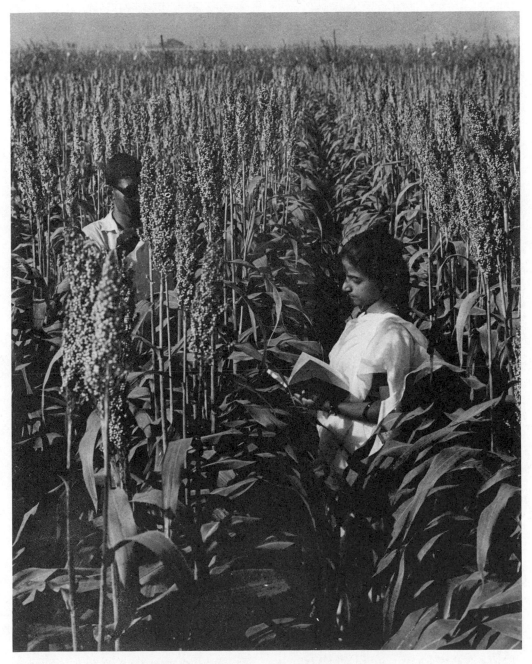

The Rockefeller Foundation is helping India to develop and test hybrid grain sorghums that are higher yielding as seen here in Central India. Grain sorghum is the number one feed grain in acreage in India and is also a food for people. (*Courtesy* The Rockefeller Foundation)

enough for grain sorghums. One year of dry weather in the humid regions causes many farmers to turn to the sorghums the next year as a more dependable source of livestock grain.

The grain sorghums probably will become more important on Corn Belt farms. In recent tests in Iowa, grain sorghums yielded 80 bushels per acre when corn yielded 60. Sorghums have made silage yields of 20 to 30 tons per acre on Missouri and Iowa farms.

Feeding experiments in South Dakota indicate that grain sorghum is worth 98 percent as much as corn for feeding hogs. As feed for dairy or beef cattle, sorghum silage is 90 percent as effective as corn on a weight basis. In other words, 1,000 pounds of sorghum silage is equivalent to 900 pounds of corn silage.

Grain sorghums are easy to produce. They can be planted with a corn planter or grain drill. The grain of short-stalk types usually is harvested with a combine. The best yields are obtained when the sorghums are grown in rows like corn rather than broadcast. Among the early-maturing varieties are Early Kalo, Colby, Norghum, and Reliance. Medium-early varieties include Plainsman, Martin, Westland, and Midland. A late variety is Dwarf Kafir 44-14. Grain sorghum hybrids are designated by numbers rather than by names, and the lower the number of the grain sorghum hybrid the earlier the maturity.

The soil should be at least 70° F. to a depth of 2 inches before grain sorghums are planted. In most localities the planting dates for sorghums are about two weeks later than for corn. Early tillage reduces soil moisture, helping to get the soil warm by planting time.

A well-pulverized mellow seedbed is best for the small sorghum seed. If all weeds are destroyed before the sorghum is planted, less cultivation will be needed later.

A planting rate of 1½ to 2 pounds of good treated seed per acre is satisfactory where the rainfall is light. About four times that amount of seed is recommended for irrigated land or for fertile soil in areas where rainfall is abundant. The seed should be covered to a depth of 1 to 1½ inches. Treating seed with a recommended fungicide helps to prevent kernel smut. Plant populations on the best soils should be about 100,000 plants per acre.

Sorghums usually do not need as much nitrogen fertilizer as corn, although the sorghums respond better to phosphorus fertilizer. The phosphorus should be placed below the seed at planting time and the nitrogen applied after the crop is up. Soil should be analyzed to determine the kind and amount of fertilizer needed. In general, sorghums and wheat are fertilized in much the same way.

Grain sorghum is harvested when the seed is hard and fully colored. At harvesting time the moisture content of the grain should be about 13 percent or less.

SWEET SORGHUM (SORGO)

The sweet sorghums (sorgo) are grown over a much wider area than are the sorghums used for grain, forage, and brooms. Some varieties of sorgo are grown from the Gulf of Mexico to the Lake states. In recent years 40 states have reported the production of

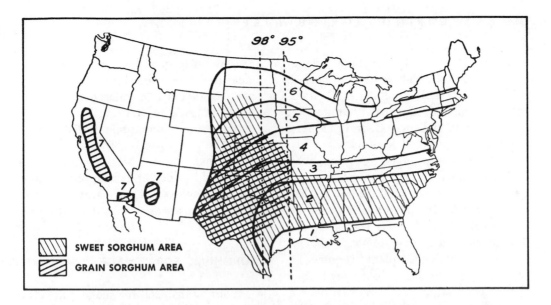

REGION 1—Gooseneck and Honey sorgos: Japanese sugarcane and Napier grass are more productive than sorghums for forage. Shallu is the leading grain sorghum. REGION 2—Sumac, Orange, Honey, Gooseneck, White African, Sugar Drip, Rex, Colman, and Sapling sorgos. The grain sorghums hegari, Schrock (Sagrain), Ajax, darso, Blackhull kafir, and Grohoma are suitable dual-purpose types. REGION 3—Sumac, Atlas, Kansas Orange, and Sourless (African millet) sorgos. Hegari, Blackhull kafir, darso, several strains of Dwarf Blackhull kafir, Hydro kafir, Red kafir, Pink kafir, Sunrise kafir, and occasionally Spur feterita are grown for forage and grain. REGION 4—West of the ninety-eighth meridian: Sorgos: Early Sumac, Leoti, and Sourless; grain sorghums for forage: Pink kafir, Dwarf (Western) Blackhull kafir, Dawn kafir, Sunrise kafir, and Freed. East of the ninety-eighth meridian: Sorgos: Atlas, Kansas Orange, Waconia Orange; grain sorghums for forage: Blackhull kafir. Pink kafir. REGION 5—Sorgos: Black (Early and Minnesota) Amber, Early Sumac, Red Amber, Dakota Amber, Waconia Amber, Fremont, Atlas, Kansas Orange. Grain sorghums for forage: Greeley, Cheyenne, Freed, Dawn kafir, Pink kafir, Highland Improved Coes, and Early Kalo. REGION 6—Sorgos: Black (Early and Minnesota) Amber, Dakota Amber, Red Amber, Fremont. Grain sorghums for forage: Cheyenne, Freed, Greeley, Highland Improved Coes. REGION 7—Sorgos: Honey, Atlas, Gooseneck. Grain sorghums for forage: hegari.

Map showing sorghum regions in the United States. The varieties named for the various regions are recommended on the basis of the normal planting date. When it becomes necessary to plant after the most favorable date it is then often desirable to use an earlier-maturing variety. (*Source:* Farmers' Bul. 1844. U.S.D.A.)

some sorgo, but the greatest production is in the South.

Most fields of sorgo are grown to supply syrup for home use. For that reason the sorgo crop averages less than an acre on each farm. The syrup usually is called sorghum syrup, although the name sorgo syrup is less likely to be confused with the other three classes of sorghums. The simple term *molasses* is most commonly used.

In areas where sorgo syrup is used extensively the production of sorgo increases rapidly during wars and other periods when sugar is scarce. The record for the annual production of sorgo syrup was established in 1920, when 50 million gallons of syrup valued at $1.00 a gallon were produced. The average annual production in recent years has been 5 million to 6 million gallons. Approximately half of the total production comes from Alabama, Georgia, Mississippi, Tennessee, and Kentucky.

If sorgo is so good for syrup, why is it not a good source of sugar, as sugar beets are? This is a good question. Also, you may ask why sugarcane cannot be used to produce syrup as well as sugar. The answers are very simple.

The sugarcane plant requires a warmer climate than does sorgo. For that reason sorgo can be grown over a much wider area and farther north. The syrup made from sorgo is darker and more flavorful than sugarcane syrup.

The possibility of making sugar from sorgo is now being investigated, and some successes in the experiments may be announced soon. There are several problems to be solved before sorgo can be used economically to make sugar. One problem is that the sweet juices of sorgo contain enough starch to interfere with the crystallization of the sugars. This characteristic is desirable in making syrup, but it is objectionable in the manufacture of sugar.

A second problem that must be overcome in the commercial production of sugar from sorgo is the low yields per acre. This problem is being solved with hybrid varieties, better use of fertilizer, and other improved cultural practices.

Sorgo generally reaches a height of 10 feet when it is mature. The leaves are then removed and fed to livestock, the stalks are cut, and the seed heads are also fed to livestock. The stalks are then squeezed through a rotary press that extracts the juices. The juices are collected in a vat and boiled until enough water evaporates to give the proper density, as measured by a special hydrometer. (You may recall that a similar hydrometer is used to test the freezing point of the antifreeze in automobile radiators.)

The sorgo stalks from which the juice has been pressed are often shredded, baled, and sold on the market as "bagasse." This material is used in greenhouses as a substitute for peat.

SORGHUMS FOR FORAGE

Sorghums used for silage, hay, grazing, or fodder include tall grain sorghums, Sudangrass, and sorgo. Sorgo is a desirable forage sorghum to chop for silage because it yields abundantly and is very palatable due to its sweet sap.

In a 5-year test at the Kansas Agricultural Experiment Station, sorgo and corn were compared as to yield and nutritive value per acre in the production of milk or beef. Sorgo yielded 18 tons of silage per acre per year, and corn yielded 11.8 tons. When fed to dairy cows, an acre's yield of sorgo produced 3,750 gallons of milk, while an acre's yield of corn produced 2,750 gallons. When fed to beef cattle, an acre of sorgo produced 1,850 pounds of beef, and an acre of corn produced 1,315 pounds.

SUDANGRASS

Sudangrass was introduced into the United States from Africa in 1909. This member of the sorghum family now grows in the South, the Midwest, and the West, and to some extent in the Northeast.

Sudangrass tolerates droughty conditions. Its rapid growth from late plantings makes it an ideal emergency forage for hay, pasture, or silage. For these reasons it is one of the most popular annual, warm-season grasses in the United States.

Sudangrass grows 4 to 7 feet tall, depending mainly upon available water and soil fertility. The stems, which arise from a clump, are relatively fine and erect. Seldom do they become larger than a pencil. The numerous leaves are long and narrow. The seed heads are borne on loose-bending branches 6 to 18 inches long.

In areas of limited soil mosture, Sudangrass usually is grown alone. Where rainfall is adequate, Sudangrass is sometimes grown with soybeans.

Any conditions that slow plant growth, such as dry weather, may cause the formation of prussic acid, which is poisonous to livestock. Short, stunted growth usually contains more prussic acid than does normal growth.

There are several new varieties of Sudangrass. One variety is resistant to certain leaf diseases. Another variety known as sweet Sudan combines the sweet-stem characteristic of sweet sorghum with the fine stems and leaves of common Sudangrass. Still another variety is nearly free of poisonous prussic acid.

Sudangrass is an excellent "catch crop" to be used for late summer grazing or for hay. The term "catch crop" means that Sudangrass is ideal as an emergency forage crop because it can be planted late and grows quickly for use as pasture or hay.

Sudangrass grows best on fertile soil but can withstand periods of dry weather better than most emergency crops. It is planted extensively over the nation, especially in the Southern Great Plains where the supply of soil moisture usually is not sufficient for other crops.

BROOMCORN

Broomcorn is a kind of sorghum that bears the seed heads in terminal straw-like bunches. The grain is found along these straws.

In harvesting broomcorn the seed head, which is borne on top like all sorghums, is cut and the grain removed. The empty seed heads are put in bales and sold as "straw" for the manufacture of brooms.

QUESTIONS

1 What are the three main types of corn?

2 What hybrid corn is best adapted to your community?

3 What precipitation during the growing-season is necessary to have a satisfactory corn crop?

4 How would you fertilize corn?

5 What is the best date for planting corn in your community?

6 Explain why sorghums are to the Great Plains what corn is to the Corn Belt.

7 Grain sorghums and corn have similar water requirements per acre. Why will grain sorghum grow in a drier climate?

8 Why is sorgo not used for making sugar?

9 Of what special value is Sudangrass?

10 How is broomcorn used?

ACTIVITIES

1 Conduct a practice in seed identification and grain judging as a class activity.

2 Collect and identify weeds from local corn or sorghum fields.

3 Make a class collection of corn and sorghums grown in your community.

4 Plan a crop rotation that includes corn, sorghum, or both crops.

5 Compare the prices per bushel of corn and grain sorghum in your community.

25 SMALL GRAINS

Small grains have been a major part of agriculture since the beginning of civilization. Rice was the main food crop in China for many centuries before Christ, and is today the main cereal for 60 percent of the people in the world. The Egyptians were raising grains before the first great pyramids were built. The ancient Phoenicians and Syrians also produced grain crops on their hilly fields. The Bible contains many statements about the importance of grain, the sowing and harvesting of grain crops, and the great famines that resulted from crop failures.

For more than a century wheat has been the main cash crop in most of the Northern Great Plains. Almost every crop rotation in the Corn Belt includes wheat and oats in combination with grasses and legumes. Barley and rye are grown in the Northern Great Plains, the Corn Belt, and in most of the regions in which mixed farming is the principal type of agriculture. In the South, particularly in Arkansas and in the coastal areas of Texas and Louisiana, rice is one of the major crops. Rice also is grown in California and Mississippi.

WHEAT

Wheat is perhaps the world's most important grain because it is the crop from which most of our bread is made. In 1969 approximately 350 million tons of wheat were produced throughout the entire world.

On a world basis, the United States is the leading wheat-producing nation. Other leading wheat-producing nations are Russia, China, Canada, France, and Australia.

In humid areas and under irrigation, wheat usually responds to a fertilizer, especially to nitrogen. *Left:* Not fertilized. *Right:* Fertilized according to results of a soil test. (*Courtesy* Allied Chemical Corporation)

In our nation, 40 states report the production of wheat. Six of the leading wheat states are Kansas, North Dakota, Texas, Oklahoma, Montana, and South Dakota.

Wheat is one of the oldest crops known. Lake Dwellers in the Stone Age depended upon wheat as a cereal for bread, while China reported wheat as early as 2700 B.C. Wheat is mentioned in the Bible, in Genesis.

KINDS OF WHEAT

The two general kinds of wheat are hard (winter) and soft (spring).

Hard wheat is dark in color and high in gluten, which makes a "strong" flour suitable for yeast breads. Hard wheat is higher in protein than is soft wheat, averaging approximately 13 percent protein. Soft wheat averages 10 percent protein.

Soft wheats are white or pale and low in gluten. These wheats are used primarily for making biscuits, crackers, pastries, and breakfast cereals.

Winter wheat. Winter wheat is widely grown throughout the United States, with the heaviest concentration in the central and southern parts of the Great Plains. Five Great Plains states—Kansas, Oklahoma, Nebraska, Texas and Montana—harvested 60 percent of the winter wheat acreage in 1969. Among the states, Kansas led in wheat production, with 9.8 million acres harvested, 27 percent of the United States total. Oklahoma, with 4.1 million acres harvested, and Texas, with 2.9 million acres, were the second and third ranking winter wheat states. Winter wheat is also known as hard wheat.

Winter wheat is planted in the fall of the year. When weather conditions are favorable for early fall growth, much of the winter wheat in the Great Plains area is grazed in the fall prior to going into dormancy and again in the late winter and early spring when new growth starts. Winter wheat harvest begins in the southernmost producing areas in the late spring and quickly spreads northward with the start of combining, usually extending well into the summer months in the northern tier of States.

Spring wheat. Spring wheat acreage harvested in 1969 amounted to 7.5 million acres and accounted for 21 percent of the total United States wheat acreage. North Dakota, the leading spring wheat state, had slightly over one-half of the 7.5 million acres of spring wheat harvested in 1969. Montana was the second leading state in spring wheat acreage with 1.1 million acres harvested; South Dakota was third with only a few thousand acres less. Durum wheat, used in making macaroni and spaghetti, was harvested from 3.3 million acres, representing nearly one-third of the total spring wheat acreage. Of the total durum acreage harvested, North Dakota had 82 percent. Durum wheat is also known as soft or spring wheat.

Spring wheat is planted in the late spring and harvested late in the summer. In the West North Central and Northwestern States, where spring wheat is primarily grown, a high proportion of the total rainfall comes during the summer months. The favorable seasonal distribution and greater effectiveness of the precipitation make it possible to produce spring wheat with a relatively small total annual precipitation.

In 1966, 17 states reported that fertilizer was used on wheat, to the extent of 1 percent of the wheat fields in Colorado to 100 percent of the fields in Michigan. Eleven states reported that 50 percent or more of the wheat fields were fertilized, whereas 6 states reported that less than 50 percent were fertilized. The rates of fertilizer per acre on wheat varied from 65 pounds of nitrogen (N) and 16 pounds of phosphorus (P_2O_5) in

The Rockefeller Foundation has helped the wheat-producing countries of the world to increase yields per acre by developing a light-insensitive dwarf-size plant in Mexico. This wheat has a giant ability to respond to present high inputs such as nitrogen fertilizer without lodging (falling over) (India). (*Courtesy* The Rockefeller Foundation)

Idaho to 8 pounds of nitrogen and 10 pounds of phosphorus per acre in Idaho. In Ohio, 42 pounds of potassium (K_2O) per acre was used in addition to nitrogen and phosphorus. (Source: Statistical Reporting Service, United States Department of Agriculture, Washington, D. C., 1966.)

CULTURE OF WHEAT

Winter wheat should be seeded at the average date of the first killing frost in the fall. This seeding date avoids the damage inflicted by the Hessian fly. Spring wheats should be sown in the spring as soon as the ground can be prepared. Wheat grown in the more humid areas usually responds to nitrogen fertilizer.

Wheat is normally harvested with a combine just as the seed matures. In the southern Great Plains most of the wheat is harvested in June. The period from July 15 to August 15 is the main harvest season in the northern Great Plains. Corn Belt farmers harvest their wheat early in July.

Most wheat is sold and then stored in huge elevators until it can be made into flour and other useful products.

OATS

Oats are grown throughout the world in cooler regions. Each year approximately 60 million tons are produced. The United States produces 38 percent of the world's supply of oats. The countries ranking next are Russia, 20 percent;

All of the small grains respond in almost all regions to liberal applications of manure. Because of the smallness of the seed the manure should be finely divided when spread, as shown here. (*Courtesy* New Idea Machinery Company)

Canada, 11 percent; Poland, 5 percent; France and West Germany, 4 percent.

Oats, the second major small grain produced in the United States, is most important as a feed for livestock. In 1969, 18.5 million acres were harvested, producing over 909 million bushels of grain. Minnesota, Iowa, Wisconsin, South Dakota, and North Dakota are the leading oat-producing states.

Wheat is grown in all countries that have cool winters. Here in central India, wheat responds by a 250 percent yield increase to a complete fertilizer containing nitrogen, phosphorus, and potassium. (*Courtesy* Frank Shuman)

Acreage has shown a variable trend in recent years, from 40 million acres harvested in 1954 to less than 18.5 million acres in 1969. Yields, however, have been increasing and now average approximately 53 bushels per acre. Relatively few oats are produced in the southern and western parts of the United States.

In the northern half of the United States, oats are planted in the spring as early as the ground can be worked. Much of the oat acreage in the Corn Belt is planted early in March. Fall-seeded oats should be planted about a month before the average date of the first killing frost in the fall.

Oats usually respond to a complete fertilizer and to an additional applica-

Grain elevators serve as storage bins for wheat until it is needed for making flour and other products. (*Courtesy* Atchison, Topeka, and Santa Fe Railway)

tion of a nitrogen fertilizer after the plants become well established.

Harvesting methods vary, depending upon the use for which the crop is intended. When used for hay, oats should be mowed when the seed is in the milk stage. Oats harvested for grain usually are harvested when the seed is in the firm dough stage.

BARLEY

Barley is grown throughout the world as a grain crop and as a source of alcohol. China leads in the production of barley, followed by Russia, the United States, Canada, and the United Kingdom.

More than 9.6 million acres of barley were harvested in 1970. While year-to-year fluctuations have occurred, acreage is still at about the same level as in the late 1920's. Yields, however, have been steadily increasing and in 1970 the average was 43 bushels per acre—65 percent more than the 1949 yield.

The major barley-producing state is North Dakota, where almost one-fourth of the United States acreage is sown. This state, and California and Montana, the second and third ranking producing states, account for about 53 percent of the nation's annual acreage and production. Only minor acreages are grown in the eastern and southern parts of the United States. Most barley is fed to livestock, although about 20 percent of the crop is used for malting, essential in brewing and distilling alcoholic products such as beer.

Barley should be planted approximately two weeks before the average date of the first killing frost in the fall.

This field cultivator is preparing soil for a small grain. Note the good tilth and the desirable crop residues from last season's crop. (*Courtesy* Allis-Chalmers)

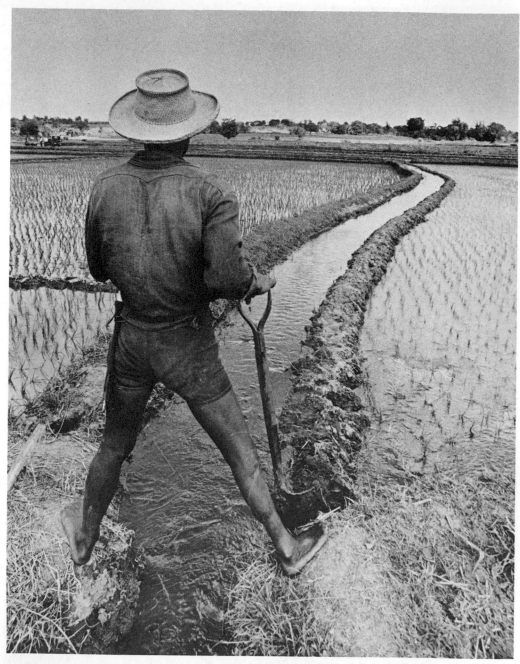

Wetland rice is grown in water at a constant depth of 2 to 4 inches until the rice is almost ready for harvest. This Philippine rice farmer has been helped by the International Bank for Reconstruction and Development (World Bank). (*Courtesy* Yutaka Nagata, United Nations)

Spring-planted barley should be planted as early as weather and soil conditions permit.

Barley is harvested with a combine when the grain is firm and ripe, in a manner similar to that of wheat.

RICE

Rice is the principal food crop of more than half of the world's population. All tropical countries produce rice, but the center of production is in Asia. China raises 30 percent of the world production, India 21 percent, Pakistan 8 percent, Vietnam 4 percent, and the United States 2 percent.

Successful rice culture is dependent upon high temperatures during the growing season, a dependable fresh water supply for the irrigation period, soils that are comparatively level and underlain with impervious subsoil, and good drainage. Areas which meet these requirements are the Coastal Prairie region of southwestern Louisiana and southeastern Texas, eastern Arkansas and northwest Mississippi, and the central valleys of California (particularly the Sacramento Valley). Production in the United States is confined mainly to these three regions.

The acreage of rice harvested in 1970 totaled 1.8 million acres. The peak year was 1954, when 2.6 million acres were harvested.

This well-prepared seedbed for small grain has many medium-size clods (peds). This permits rainwater to soak into the soil without causing a surface crust. (*Courtesy* John Deere)

When ready for harvest, rice in India is cut with a hand sickle and carried on the head to a well-drained place where the rice is dried and threshed. (*Courtesy* Mary M. Hill, International Development Association)

The average yield per acre has increased sharply in recent years, averaging 4,566 pounds per acre in 1970. During the preceding four years the yields of rough rice in the United States had exceeded 4,000 pounds per acre.

ADAPTATION OF RICE

Rice requires abundant moisture, high summer temperatures, and soils that can be irrigated. Like other crops, rice grows best on fertile soils. This crop can be grown only on areas with the following characteristics:

1. The summer temperature should average about 82° F., sometimes reaching 100° F. in midsummer. Spring and fall temperatures should average 70° F. The relative humidity must be high.
2. A dependable supply of fresh water must be available for irrigation, as all U.S. rice is grown under shallow water throughout most of its culture.
3. Soils must be level and must have a tight subsoil that will not permit water to escape downward as deep seepage.

CULTURE OF RICE

Rice culture is hard on the physical condition of the land. For this reason, crop rotations are necessary to keep the soil in good physical condition. After rice is grown one or two years the land should be used for pasture crops two or three years.

The land must be plowed and harrowed for rice, just as for any other crop. Then the land is leveled and numerous small levees are built so that water will cover the rice at a uniform depth when the field is flooded.

Rice is commonly seeded on dry land, as wheat and other small grains are sown. In recent years airplanes have been used successfully in seeding operations.

A complete fertilizer is often used on rice. One interesting fact concerning the fertilization of rice is that the ammonium form of nitrogen is more effective than the nitrate form. Rice is the only crop that responds better to the ammonium form than to the nitrate form of nitrogen.

As soon as the rice is about 6 inches high it is flooded with 2 inches of water. As the plants grow taller, the water level is gradually raised to about 5 inches. This water level is maintained until two weeks before the rice matures. Then the land is drained and the rice is combined the same as wheat and other small grains.

RYE

Rye has a special place among the small grains. It will grow on soils that are low in nutrients and that are droughty. Of all small grains, rye ripens earliest in the spring. Rye also can withstand winter cold better than most small grains.

Czechoslovakia, the United States, France, Spain, and Canada are the leading rye-producing countries.

Rye was harvested from 1.49 million acres in 1970, with a total production of 39 million bushels. However, this was well below the peak in 1919, when 7.2 million acres were harvested. It should be noted that a large acreage

Rye is a very hardy cool-season small grain that protects the soil against erosion in winter and supplies supplemental grazing, as shown here. (*Courtesy* Holstein-Friesian Association of America)

of rye, in addition to the acreage harvested for grain, is used as a winter cover crop. This often serves as pasture and then is turned under as a green manure crop. In 1966, only 17 percent of the acres planted with rye were harvested for grain.

Rye is usually planted in the fall and harvested during the late spring and early summer months. Most of the rye is produced in the Northern and the Central Plains. Four states—South Dakota, North Dakota, Nebraska, and Minnesota —had 57 percent of the 1969 production.

Only secondary acreages of rye are grown in the eastern and southern parts of the United States.

Rye seed is used for food and for the manufacture of alcohol, while the young plants provide fall and winter grazing and serve as winter cover to protect the soil.

Rye should be seeded early in the fall or very early in the spring. The crop responds to applications of fertilizer but will produce a fair yield without it. When harvested for grain, the crop is combined the same as wheat.

QUESTIONS

1 Why is wheat an important crop?

2 Describe two kinds of wheat and their uses.

3 When should winter or spring wheat be seeded in your area?

4 What is the average date of the first killing frost in your community?

5 Name the five states ranking highest in wheat production.

6 In how many states are oats grown?

7 When should fall barley be planted?

8 Describe a good rotation system for rice.

9 Why does rye have a special place among the small grains?

10 What other farm enterprises are combined with the production of rice?

ACTIVITIES

1 Make a list of foods you ate today that came from wheat.

2 Look at the maps in the chapter on climate to find out how the length of the growing season affects grain farming.

3 Refer to the climate maps to determine when to plant rye, oats, and barley.

4 Describe an ideal soil for rice.

5 Find out whether rye is grown in your community. Describe the soil requirements of this crop.

26 COTTON

Cotton is grown in southern areas of the United States, mostly south of the 36th parallel. Cotton belongs to the mallow family and requires a long frost-free season of at least 200 days. Summer temperatures must average 77° F. or more, and the average annual rainfall must be at least 20 inches unless irrigation is practiced. Autumn rainfall should not be more than 10 inches, as wet weather seriously interferes with the harvesting of cotton. Under tropical conditions, plants continue to grow each year and develop into trees. In the United States, cotton is grown as an annual from seed planted after soils become sufficiently warm. However, some cotton in Arizona is grown from surviving stubble of the previous year's crop and is called "stub cotton."

Planting gets underway in the Lower Valley of Texas the latter part of February and moves north across the Cotton Belt as the season advances. The bulk of the United States crop is planted during April, but in some late years planting is not completed until around mid-June, especially in the Plains areas of Texas. Although the first bale is ginned in June from the early-planted cotton in the Lower Valley of Texas, the bulk of the United States crop is harvested in October and November except in the Plains areas of Texas. In that area, strippers are used to harvest the crop after the first freeze with peak ginning during December in most years.

Most cotton grown in the United States is Upland cotton with a staple length of 1 inch or longer. Some extra-long-staple cotton (American-Egyptian), which has a staple of 1½ inches or longer, is grown in Texas, New Mexico, and Arizona under irrigation.

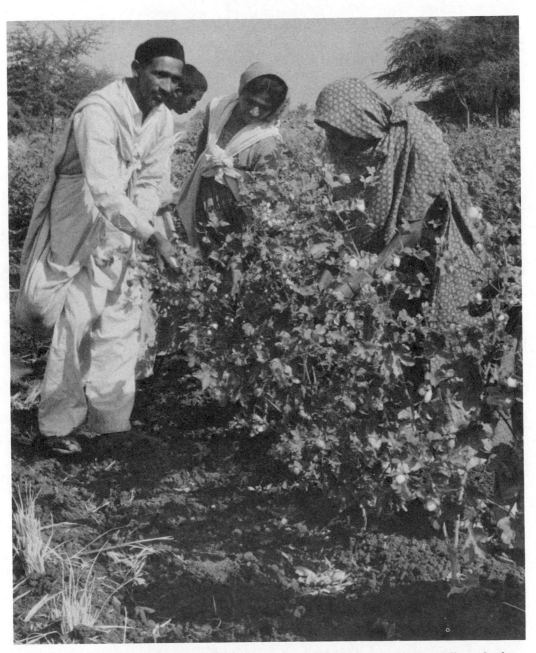

This is tropical India where American-type cotton is grown successfully and where a large cotton cloth industry has developed. Cotton is being picked by hand and put in bags hung over the shoulder (*Courtesy* Roy L. Donahue)

Cotton is an important cash crop in all tropical and subtropical regions of the world. Here a farmer in Afghanistan, eastern Asia, is picking the lint and seed from a boll of cotton. The project was assisted by the International Development Association. (*Courtesy* Kay Muldoon, International Development Association)

TRENDS IN ACREAGE AND PRODUCTION

Cotton acreage harvested in 1970 was 11.2 million, half of the acreage 15 years ago. Yield per acre of lint cotton in 1970 averaged 441 pounds. The total farm value of cotton in 1970 was approximately 1.4 billion dollars.

States ranking highest in cotton production in 1970 were, in order from high to low, Texas, Mississippi, Arkansas, and California. Other states with warm temperatures and a long growing season also produce some cotton.

CULTURE OF COTTON

Cotton is planted approximately two weeks after the average date of the last killing frost in the spring. A more scientific method of determining when to plant cotton is to measure the soil temperature at a soil depth of 2 to 4 inches each day between 7:00 A.M. and 9:00 A.M. until the soil temperature is 65° F. or more for three successive days.

Cotton does not require as much from the soil as does corn, but cotton responds to commercial fertilizers, especially in areas of high rainfall. Five hundred pounds per acre is a common application of a complete fertilizer. The recommended amount and grade of fertilizer for each field should be determined on the basis of a soil test.

Usually cotton is thinned after the plants are well established. Weeds are controlled by frequent and shallow cultivations (1 to 2 inches).

The more progressive cotton farmers use a minimum amount of mechanical cultivation and a maximum amount of flaming (burning with an open flame), herbicidal petroleum oil, or chemical herbicides such as treflan, cotoran, caparol, dacthal, parmex, herban, DSMA, MSMA, and/or lorox. These chemicals are "tricky" and must be applied exactly as recommended.

INSECT AND DISEASE CONTROL

Cotton appears to have more than its share of insects and diseases. The more serious insects are the boll weevil, bollworm, and leaf worm. Thrips, nematodes, aphids, and spider mites are usually minor pests.

There are always insects in cotton, and the time to dust or spray for their control depends upon the kinds of insects and their relative numbers. A systematic way to determine when to start applying insecticides to control boll weevils and/or bollworms is recommended by the North Carolina State University in *Leaflet CPG* Number 7, April, 1971, as follows:

At least once each week after cotton begins squaring, examine 100 squares in each field (or 100 squares for each 10 to 20 acres in extremely large fields). Select squares at random from lower, middle and upper parts of plants as you walk diagonally across a field. Do not check more than one square per plant. The number of damaged squares per 100 examined gives the percent infestation. Start controls when 10% boll

This mechanical cotton picker picks the lint with its seed from the cotton boll, in contrast to the stripper that picks the entire boll with the lint and seed attached. (*Courtesy* John Deere)

weevil or 5% bollworm-damaged squares occur. The number of eggs and small bollworms may be checked by examining plant terminals (upperside of new terminal leaves). The number of eggs and young larvae present per 100 terminals is a good estimate of bollworm damage potential in the field. These counts should definitely be made from late July throughout the growing season. When eggs and five or more small bollworms are found per 100 terminals, damaging bollworm infestations are likely to occur. During late season examine small bolls for bollworm damage.

The control of diseases of cotton consists primarily of treating the soil by mixing the recommended fungicide in the furrow where the seed is to be planted. An area about 2 inches deep and 2 inches wide should be treated, with the seed being planted at the bottom and in the center of the treated area.

The most common diseases of cotton are the ones which are soil-borne: fusarium, rhizoctonia, pythium, and thi-elaviopsis. The fungicides must be used exactly as recommended, including any small type and footnotes in the printed instructions.

IRRIGATION

Because cotton ranks very high as a cash crop, farmers are usually encouraged to irrigate cotton where this practice is needed. It is often difficult to determine when water is needed. One method is to bury several soil moisture tensiometers in the soil in the cottonfield and irrigate when the dials indicate a critically low level of available soil moisture.

The not-so-scientific but keenly observant farmer will start to irrigate when the upper one-third of the leaves on the cotton plant wilt by 10:00 A.M.

When the need for irrigation is indicated, farmers should apply about 2

Plastic syphon tubes in use in cotton irrigation in Texas. (*Courtesy* Soil and Water Conservation Research Division, U. S. D. A., Weslaco, Texas)

inches of water at each application at a rate just fast enough not to cause any runoff.

HARVESTING

In recent years plant breeders have developed cotton varieties that mature at one time, have bolls that are uniformly and conveniently spaced, and that are otherwise adapted to harvesting by a mechanical cotton picker or cotton stripper. Very little cotton is now picked by hand.

The United States Department of Agriculture offers the following suggestions for harvesting and handling cotton:

1. Excess moisture, trash, and dirt in the seed cotton reduce the value of the ginned lint.
2. Ginning cotton in a green, damp, or wet condition lowers the quality as

Above: The worst enemy of cotton, a boll weevil, getting ready to puncture and ruin this cotton boll. (*Courtesy* Shell Chemical Division) *Below left:* The larvae (upper), pupae (lower left), and moth (right) of the cotton bollworm. (*Courtesy* Hercules Powder Company) *Below right:* The cotton leaf worm. *Upper:* pupae (cocoon). *Center:* larvae (caterpillar). *Lower:* adult moth. (*Courtesy* Hercules Powder Company)

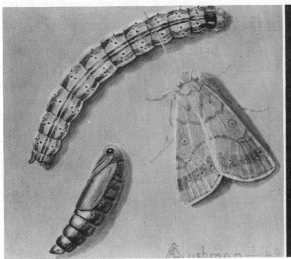
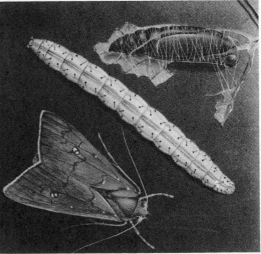

This cotton stripper harvests the entire boll (below) containing cotton seed and cotton lint. At the gin the burrs, seed, and lint are separated and the burrs discarded. (*Courtesy* Texas Agricultural Extension Service)

much as one or two grades. This lowered quality may reduce the market value of the cotton from 1 dollar to 10 dollars per bale.

3. Failure of cotton to mature fully as a result of unfavorable weather, insect infestation, or disease damage is responsible for lowering the grade.

4. Undue exposure in the field makes the cotton discolored, dull, and trashy, and results in heavy losses to farmers.

5. Dry-picked seed cotton gives the best results in ginning. If picked while damp, cotton can be dried artificially at the gin, or it can be dried naturally through proper handling and storage, provided it is not so wet that the seed is soft.

6. Clean picking eliminates much trash and dirt that otherwise would lower the grade of the ginned lint.

7. Proper ginning preserves all the value of the seed cotton that the farmer delivers to the gin. It pays producers, therefore, to insist on good ginning for their cotton.

8. Care in packing and covering bales at the gins, care in handling the cotton all the way from the gin through the process of recompression, and care in sampling the bales as they pass through marketing channels, benefit all branches of the cotton industry from the grower to the person who handles the finished product.

MARKETING

As cotton is picked it is hauled by truck or trailer to the gin. Here the fibers are separated from the seed, and the lint is packed into bales. The seed may be stored at the gin, hauled back to farms, or sent to cottonseed oil mills.

Sometimes cotton is hauled from the gin to farms or warehouses for storage until the farmers are ready to sell their crop. Some of the growers in the Southeast sell cotton directly to spinners.

Many farmers market their cotton through their own cooperative associations. The Department of Agriculture offers a free cotton-classing and market-news service to groups of cotton producers who organize for this purpose.

A cotton buyer may be an agent for large dealers who supply cotton to domestic mills and to exporters, or he may buy for himself and then sell to large firms. The marketing of cotton is a rather complex process that requires the services of cotton buyers, cotton merchants, exporters, and brokers to get raw cotton from producers to spinners.

COTTON MANUFACTURING

The cotton-growing states have most of the cotton mills that manufacture the nation's raw cotton into yarn and cloth. Among the states that do not produce cotton, Massachusetts leads in cotton manufacturing. The principal products of the cotton-manufacturing industry are woven cloth, cotton yarn and thread, and the waste that is sold to other industries for batting, wadding, and mattress felts.

Cotton yarn and cloth are used in the manufacture of shirts, dresses, work clothes, and many other kinds of cloth-

Weaving a fancy design with cotton thread on an automatic loom. (*Courtesy* National Cotton Council of America)

A cotton flannel shirt is durable, washable, and attractive. (*Courtesy* National Cotton Council of America)

ing. Other familiar cotton products include towels, tablecloths, sheets, and pillow cases. Large quantities of cotton also are used in the manufacture of automobile tires, cotton bags, and cotton belts and belting.

The Industrial Revolution in the latter part of the eighteenth century brought about the change from hand looms to power looms and power machinery. Since that time, science has greatly improved manufacturing processes and increased the quantity of goods that each worker and each machine can produce.

Some of the inventions that have revolutionized the manufacture of raw cotton are: Hargreave's spinning jenny, 1764; Arkwright's spinning frame, 1769; Crompton's spinning mule, 1779; Cart-

wright's power loom, 1789; the adaptation of Watt's engine to spinning and weaving machinery, 1785; the cotton gin, 1793; ring spinning, 1831; the automatic loom, 1894; and long-draft spinning, 1913.

COTTONSEED

Cottonseed has become an important product in the United States. Most of the cottonseed is sold to processing mills that produce cottonseed oil, cottonseed cake and meal, cotton linters, and hulls.

Farmers usually sell the cottonseed to ginners or to local oil mills. A ginner accumulates seed until he has a large lot that can be sold to the mill offering the best price. From storage houses at the mills the cottonseed is conveyed mechanically to machines that do the cleaning, delinting, hulling, separating, rolling, cooking, and pressing.

The crude cottonseed oil may be refined at local plants or at large refineries. Refined cottonseed oil is used mainly in the manufacture of margarine, although some is sold as salad and cooking oils.

Cottonseed cake is used mainly in feeding livestock. The cakes are cracked into various sizes or ground into meal for cattle, sheep, and hogs. The meal also is used as a fertilizer. Some cottonseed cake made from high-quality seed is ground into flour to make bread and cakes for persons who should not eat sugars or starches.

Cottonseed hulls are used as a roughage feed for livestock. The hulls also are a source of many chemical compounds.

QUESTIONS

1 What are the climatic restrictions for growing cotton?

2 What is the ideal planting date for cotton?

3 What are the major cotton-destroying insects?

4 Name five rules for harvesting and handling cotton.

5 How much does the United States Department of Agriculture charge for its cotton-market news service?

6 Where are most of the manufacturing plants that make cotton goods?

7 Name three inventions that have revolutionized cotton manufacturing.

8 What is cottonseed oil used for?

9 What is the main use for cottonseed cake and cottonseed meal?

10 What use is there for cottonseed hulls?

ACTIVITIES

1 Make a list of everything you see in the room that is made from cotton.

2 Ask ten store clerks who sell cotton clothing what the advantages and disadvantages of cotton material are.

3 Have a class discussion on the uses and meanings of these words: washable, sanforized, fast colors, drip-dry, and spin-dry.

4 Look on a grocery shelf and make a list of things made from cottonseed.

5 Write a 200-word theme on cotton-growing.

27 TOBACCO

Tobacco is grown on every continent and in 54 countries. The nations ranking highest in tobacco production are the United States, China, India, Brazil, and Turkey. Of the total world production, the United States supplies more than 28 percent and China nearly 20 percent. It is an important crop in Puerto Rico.

Domestic production is concentrated in North Carolina, where 43 percent of the nation's tobacco is grown, and in Kentucky, which produces 21 percent of the tobacco crop. Other states that rank high in the production of tobacco are Virginia, Tennessee, South Carolina, and Georgia. These six states produce nine-tenths of the total United States poundage. Specialty tobaccos are grown in Connecticut, Maryland, Wisconsin, Ohio, Florida, Georgia, Pennsylvania, and Missouri. A total of 21 states produce some tobacco for the market.

Although grown at various latitudes, the crop has little tolerance to frost and freezing temperatures. Tobacco seeds are usually sown during the winter and early spring in protected beds. The seedlings are pulled and transplanted in the spring. Harvesting operations are conducted mostly during summer and early fall. Harvest completion dates are governed largely by latitude, elevation, seasonal conditions, and the nature of the particular type produced in a given belt. The flue-cured and cigar wrapper types are harvested by priming (picking the leaves as they ripen) whereas other types are usually stalk cut.

Varieties of tobacco can be grown successfully from southern Canada to the tropics. But each type of tobacco will grow only in a very restricted soil and climatic region, and each type has a restricted usefulness.

Of all nations, the United States ranks first in tobacco production and technical knowledge of this crop. Here a Peace Corps volunteer is helping a farmer in Dahomey, western Africa, to improve his crop of tobacco. Sorghum is in the background. (*Courtesy* Action-Peace Corps)

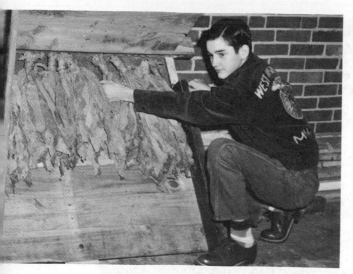

Above: Flue-cured tobacco leaves are tied in bundles and hung across a stick to dry. (*Courtesy The American Farm Youth Magazine*) *Below:* The tobacco cold-frame bed is made at the edge of the woods where the virgin soil is relatively free from diseases that may attack the tobacco seedlings. For protection the seedbed is surrounded by boards 6 to 10 inches high and covered with cheesecloth. (*Courtesy Burley Tobacco Protective Fund*)

Acres of tobacco harvested in 1970 were 899,000, with a yield of 2,120 pounds per acre. The average price per pound received by farmers in 1970 was 72.8 cents and the total farm value of tobacco was 1.4 billion dollars. In cash income to United States farmers, tobacco ranks fifth among all field crops; and among United States agricultural sales to other countries, tobacco ranks fifth.

Tobacco is typically an American crop, having been cultivated by the Indians before the first white men came to America. Indians used tobacco as a stimulant and as part of their tribal ceremonies. Smoking a "peace" pipe was a symbol of friendship which the early settlers and explorers learned from the native American Indians. The word *tobacco* comes from a Spanish word, *tabaco*, which is thought to be derived from an Indian word, *taino*, which meant a roll of tobacco leaves.

The production of tobacco for the world market was first reported in Virginia in 1612. The early American colonists raised much tobacco because it was one commodity they could exchange for manufactured goods, tea, coffee, and sugar from other countries.

Tobacco was in such demand among the colonies that it was often used instead of money. In 1732 the state legislature in Maryland legally recognized tobacco as "money" and permitted it to be used to pay debts, buy groceries, pay state salaries, and to pay the preachers. At that time the legal value of tobacco was set at one cent per pound.

CLASSES AND TYPES OF TOBACCO

The main classes of tobacco are based upon their uses. These classes are: (1) flue-cured; (2) fire-cured; (3) air-cured, light or dark; (4) cigar-filler; (5) cigar-binder; and (6) cigar-wrapper.

Flue-cured tobacco. This type of tobacco, also known as bright tobacco, is grown extensively in southern Virginia, northern and eastern North Carolina, eastern South Carolina, southeastern Georgia, and northern Florida. The bright yellow color of the leaf is due mainly to the kind of soil upon which the crop is grown and to the method of curing. Flue-cured tobacco is adapted to sandy soils with yellow or red sandy loam subsoils containing relatively small proportions of clay. Such tobacco also is grown on the Piedmont soils, which have sandy clay subsoils.

The preferred seedbed for flue-cured tobacco is a well-drained friable soil. When practicable, a woods site is selected for the seedbed because the soil is free from disease. The forest growth or other vegetation should be removed and the seedbed plowed or spaded to a depth of 4 or 5 inches. The seedbed should be prepared at a time when the soil is not too wet; this avoids puddling the soil.

A commercial fertilizer analyzing about 20 percent phosphate and 10 percent each of nitrogen and potash should be raked into the seedbed. The fertilizer should be applied at the rate of one pound per square yard before seeding.

Because the seed is so small, it should be mixed with a large volume of corn meal or sifted sand to obtain even distribution. About 1 bushel of the mixture should be used for each ounce of seed. A heaping teaspoonful of seed is enough to sow 25 square yards of seedbed. This area should produce enough plants to set an acre in the field, although in common practice a seedbed of 50 to 100 square yards is planted for every acre to be used for the tobacco crop. The seed should be covered very lightly.

The seedbed is surrounded by a coldframe made of boards set on edge to a height of 6 to 10 inches. Cheesecloth is placed over the frame before the plants come up. The coldframe helps to protect the plants from frost and low temperatures.

The tobacco field should be plowed and harrowed before transplanting begins. Low ridges are then thrown up for the rows. The plants are set 20 to 28 inches apart in rows 4 feet apart. Fertilizers must be used because the soils that produce the best tobacco leaves are naturally infertile.

It is not possible to recommend a generalized fertilizer to use for all tobacco because of the contrast in types of tobacco grown, the nutrient level of the soil, and the specific growing season. A soil sample should be sent to the state soil testing laboratory to obtain this information. Certain principles of fertilizing tobacco, however, can be stated.

It is better to use potassium sulfate instead of potassium chloride. The sulfate form results in a tobacco ash that does not fall readily from a cigarette or cigar (long ash). It is recommended that at least one half of the nitrogen be used in the nitrate form. (For example, ammonium nitrate has half of the nitrogen in the nitrate form. Potassium nitrate and sodium nitrate have all of their nitrogen in the nitrate form. Ammonium sulfate has no nitrate; all of its nitrogen is in the ammonium form.)

Deficiency symptoms of tobacco have been recognized at various times

Tobacco leaves are picked selectively as they become "ripe" and are carried to sheds for air-curing. (Puerto Rico) (*Courtesy* Puerto Rico Economic Development Administration)

for each of the following: boron, calcium, copper, iron, manganese, magnesium, sulfur, nitrogen, phosphorus, potassium, or zinc.

Planting begins about the first of April in South Carolina and Georgia. In the western sections of North Carolina the planting season may extend into May or even into June.

Cultivation begins as soon as the plants start to grow and continues as long as the size of the plants permits. The first cultivation should be fairly deep, while the succeeding cultivations should be shallow and frequent.

The tobacco plants are topped when the flower heads begin to appear. Farmers in the Piedmont section remove all but the lower 10 to 14 leaves on the plant, depending on the vigor of the plant and the fertility of the soil. In the Coastal Plain section the plants are topped higher, with 14 to 18 leaves remaining. Suckers must be removed from the plants at intervals of 7 to 10 days.

Flue-cured tobaccos are harvested when they are thoroughly ripe. At this time there are numerous light-yellow patches on the leaf surfaces, and even the green portions should be of a light tint. Bundles of leaves are tied together, hung across sticks, and stored in barns to dry. Each stick is about 4½ feet long and holds 25 to 30 bundles of leaves. The harvesting, curing, and grading of tobacco for market are processes that require much experience and skill.

Special problems of nitrogen. There are very special problems associated with the use of nitrogen on flue-cured tobacco. The following information on the use of nitrogen is from North Carolina State University *Extension Folder* No. 279,

June 1969, "Nitrogen Fertilization of Flue-Cured Tobacco."

An increase in the supply of nitrogen, from deficient to excessive, increases leaf size but decreases body and thickness.

Under field conditions, within commonly used rates, an increase in nitrogen, with medium to high soil moisture during the growing season will produce thinner bodied leaf.

As the nitrogen rate is increased maturity is delayed. Excessive delay in maturity increases the likelihood of leaf diseases such as brown spot.

As the nitrogen rate is increased from deficient to adequate to excessive, the cured leaf color tends to range from yellow to orange to brown.

For proper maturity, it is essential that nitrogen absorption decrease rapidly soon after topping. When the plant leaf has reached maximum size, the readily available nitrogen in the soil should be nearly exhausted.

An inadequate amount of nitrogen throughout the growing season has occasionally given a lower quality leaf than when the same amount of nitrogen was absorbed early. Thus, tobacco may be expected to perform better if a high percent of the total available nitrogen is present in early stages of plant growth and to diminish rapidly during later phases.

The accumulation of nicotine (an alkaloid), which is synthesized in tobacco roots, is regulated more by the nitrogen supply than by any other plant nutrient, since nitrogen is a part of the nicotine molecule. As the supply of available nitrogen increases from deficiency to excess, the nicotine of the leaf increases. This is due in part to delayed maturity.

Increased rates of nitrogen reduces sugar content.

Recent investigations suggest a definite correlation between nitrogen deficiency and the production of gray or toady tobacco in some varieties and pale and slick tobacco in others.

As nitrogen rates go from deficient to adequate, tobacco yields increase. Small quantities of nitrogen above that needed for maximum yields will have little consistent influence on yield, but very high rates will reduce yields.

Fire-cured tobacco. Almost all of the fire-cured tobaccos are grown in central Virginia and in western Kentucky and Tennessee. These tobaccos have a thick leaf and a dark color. Their distinctive flavor is the result of the open fires used in curing. The greater portion of the fire-cured tobacco is exported, although some is used in the making of snuff and in wrapping chewing tobacco.

Fire-cured tobaccos are grown on loam soils having a high percentage of silt and clay. These soils are not suitable for other types of tobacco. The culture

A good field of tobacco in Connecticut planted on the contour for better soil conservation and higher tobacco yields. A sod crop (S) and a cotton crop (C) are included in the strip cropping. (*Courtesy* Soil Conservation Service)

of fire-cured tobaccos is similar to that of other types.

Air-cured tobacco. The two classes of air-cured tobacco are *light* and *dark*. White Burley and Maryland are the main types of light tobacco. Dark tobacco consists mainly of two types, One-Sucker and Virginia Sun-Cured.

White Burley tobacco grows best in the limestone soils in the bluegrass section of Kentucky, in eastern Tennessee, and in southern Ohio. This type is very important for cigarettes and is used in the manufacture of chewing tobacco.

The Maryland type of leaf is produced extensively in several counties in southern Maryland. This type of tobacco has a thin leaf, is light in color, and burns well. Its principal use is for the manufacture of blended cigarettes. A large part of the crop is exported.

The dark air-cured types of tobacco are grown in that part of Kentucky and Tennessee lying between the Burley district and the dark-fire-cured sections. The dark air-cured tobaccos are grown for the domestic manufacture of chewing and smoking tobacco and for export.

Cigar-filler. Most of the domestic cigar-filler leaf is grown in the Lancaster district in Pennsylvania and in the Miami Valley in Ohio. This type of tobacco is grown mainly on loam soils that are well adapted to general farming. The plants are spaced wider apart and are topped lower than the plants in the binder and wrapper districts. The best results are obtained when the tobacco is grown in rotation with other crops.

Pennsylvania Broadleaf is grown in the Lancaster district. Here the silt-loam soils are of limestone origin. The Zimmer Spanish variety is grown in the Miami Valley in Ohio.

Cigar-binder. Cigar-binder leaf tobacco is grown in Connecticut and in southern and southwestern Wisconsin. The soils on which this tobacco is grown vary from sandy loams to clay loams, and all must be well drained.

The main types of cigar-binder tobacco are Connecticut Havana Seed, Connecticut Broadleaf, and the Comstock Spanish and Wisconsin Havana Seed.

Cigar-wrapper. This type of tobacco is produced in the Connecticut Valley; Gadsden County, Florida; and Decatur County, Georgia. The high-grade cigar-wrapper tobacco is grown under artificial shade.

The shade is provided by a wire frame about 8 feet high supported by posts and covered on the top and sides with cloth or slats. In the Connecticut Valley only cloth is used as shading material.

On farms in the Connecticut Valley the rows are spaced 40 inches apart. In the southern area the row spacing is 48 to 54 inches. In both sections the plants are set 10 to 15 inches apart in the row.

In the Connecticut Valley the tobacco fields are treated with commercial fertilizers and manure. In the Florida-Georgia district the available manure is applied at the rate of 10 to 15 tons per acre. Commercial fertilizers also are used.

The plants are topped very high, and the flower head is removed before the blossoms open. The first picking of about three leaves is made at the time of topping. Subsequent pickings of three to four leaves each are made at increasing intervals. The interval between the fourth and fifth pickings is seven to ten days or longer.

TOBACCO DISEASES AND INSECTS

Tobacco is subject to many diseases and to a few insect pests.

Some of the most troublesome diseases are root rot, tobacco wilt, black shank, root knot nematode, mosaic, leaf spot diseases, and blue mold.

The chief insect enemies are aphids, flea beetles, wireworms, cutworms, hornworms, and budworms. Of all insect enemies, the hornworms probably are the most destructive.

QUESTIONS

1 Name the two states that lead in the production of tobacco.

2 Why is tobacco called a typical American crop?

3 Name the classes of tobacco.

4 Why is a site in the woods selected for a tobacco seedbed?

5 What is a coldframe?

6 Where is fire-cured tobacco grown?

7 Where does White Burley tobacco grow best?

8 Name five tobacco diseases.

9 Name five insect enemies of tobacco.

10 How far from where you live is the nearest tobacco field?

ACTIVITIES

1 Make an outline map of the United States and show by dots the location of tobacco production.

2 Write an imaginary story about how the "pipe of peace" prevented an Indian war.

3 Consult some standard references on the harmful effects of tobacco. Have a debate on this subject.

4 Ask your local doctor to explain just what tobacco does to your body.

5 The tobacco plant is so sensitive to photochemical air pollutants that it is used to monitor levels dangerous to man and animals. In June 1969 the United States Department of Health, Education, and Welfare published "Tobacco, a Sensitive Monitor for Photochemical Air Pollution," National Air Pollution Control Administration Publication No. AP-5, which explains how to grow tobacco plants and how to use them to detect levels of toxic ozone in the air. As a special individual or class exercise, grow some tobacco plants and use them to detect air pollution in places suspected of having polluted air.

28 PEANUTS, SUGAR BEETS, AND FLAX

PEANUTS

Production of peanuts is confined mainly to the Southern States because a long growing season is required. There are three main producing areas: the Virginia-Carolina area, where large-seed bunch and runner types predominate; the Southeast, where the crop is mainly small-seed runner and Spanish bunch types; and the Southwest, where the acreage is almost exclusively planted to Spanish bunch types. Some acreage of Valencias, a roasting type, is planted in New Mexico. Georgia, Texas, Alabama, North Carolina, Oklahoma, and Virginia lead in peanut production.

The acreage harvested for nuts has been fairly stable in recent years. The average acreage is approximately 1.5 million acres, with an average yield of 2,000 pounds per acre in 1970.

The peanut (*Arachis hypogaea*) is an annual herbaceous legume. Peanuts also are known by the common names of "goober" and "groundnuts." Peanuts apparently were growing in Brazil, Mexico, and Central America before the arrival of the Europeans. This crop was brought to North America during colonization.

Peanuts require a growing season of 120 to 140 days, moderate rainfall, abundant sunshine, and relatively high temperatures. Light-colored sandy soils are best for commercial peanuts. Dark-colored soils stain the hulls but are satisfactory for growing peanuts for hog feed. Clay soils can be used if crop is to be hogged off.

DESCRIPTION OF PEANUT PLANT

Peanuts grow 1 to 2 feet tall, with angular hairy stems and spreading branches. Some varieties have short up-right branches; other varieties have prostrate branches which spread along the surface of the ground. Small yellow flowers develop at joints where leaves are attached to the stem. After the flowers are pollinated, the fertilized flowers are extended on a "peg" which grows downward and into the surface of the soil where the peanut develops. Peanuts must have loose soil to become established and to grow well. The small taproot has many strong branches, with only nodules on the roots. The nodules contain nitrogen-fixing bacteria which enable the peanut plant to manufacture some of its nitrogen from the air, beginning 30 days after planting. The inoculation of the seed with special bacteria helps to increase the number of nodules on the roots.

SOIL AND SITE SELECTION

Peanuts produce best on deep, well-drained loamy sands, sandy loams, or sandy clay loams of the soil series of Tifton, Norfolk, Greenville, Carnegie, Magnolia, and Faceville. The soil should have a pH of 5.8 to 6.2 (slightly to moderately acid). If the soil is more acid than this, lime should be applied according to the results of a soil test recommendation.

To reduce diseases, peanuts should follow a grass or grass-like crop such as rye or corn. One year in three is as often as peanuts should be planted on the same field. Peanuts should not follow cotton or soybeans.

PLANTING, CULTIVATION, AND HARVESTING

Peanuts should be planted soon after the average date of the last killing frost.

Peanuts are a legume and can manufacture some of their own nitrogen with the aid of special bacteria that live in the pea-size "factories" (nodules) shown here on peanut roots. Before planting, peanuts should be treated with special nitrogen-fixing bacteria. (*Courtesy* Nitragin Company)

The planting season varies from about April 10 along the Gulf Coast to about May 10 in the Virginia-North Carolina district.

For large-pod varieties, 110 to 120 pounds of shelled seed per acre are planted. Typical planting rates for small-pod varieties are 75 to 90 pounds of shelled seed.

Rows should be 2 to 3 feet apart. Plants of the runner varieties should be 4 to 5 inches apart in the rows, bunch varieties 3 to 4 inches, and Spanish varieties 2 to 3 inches. Planting is usually done by machines, although small areas may be planted by hand. Peanuts should be planted 1½ to 2 inches deep on sandy soil, and 1 to 1½ inches deep on clay soil.

Fertilizer should be applied in amounts indicated by soil tests. On some thin sandy soils, 300 to 500 pounds of a complete fertilizer may be a satisfactory rate for each acre of peanuts. Fertilizer should be broadcast and plowed in prior to planting. The fertilizer should not be in contact with the seed, as seedlings are readily injured by fertilizer.

The peanut is the main high-protein food of most people in the tropics. This smiling French-speaking boy-with-a-hoe is in a peanut field in Dahomey, western Africa, where more than 90 percent of all tillage is done by hoes made by village blacksmiths. Note the handle cut from the nearby woods. Africa is a major exporter of peanuts. (*Courtesy* United Nations)

The cultivation of peanuts is similar to the cultivation of corn. The peanut crop should be harvested before frost. Tractor-mounted peanut shakers are used in harvesting. Potato diggers equipped with elevators may be used in harvesting. Combines may be used after the plants are removed from the soil.

Peanuts are used for food (raw or roasted peanuts), candies and brittles, peanut butter, and oil. Peanut seed is an excellent source of vitamin B but contains only small amounts of vitamins A, C, and D. The seeds yield 40 to 50 percent or more of oil. The meal left as a by-product in the extraction of oil is used as feed for livestock, especially for cattle. Peanut hulls are sometimes used as bedding for livestock, and are finally returned to the soil with manure.

World trade in peanuts depends largely on the European demand for oil. India, China, Burma, Sumatra, Java, and Africa are leading exporters of peanuts. The United States imports peanuts to supplement its own production.

SUGAR BEETS

Sugar beets are grown in the North Central and western areas of the United States. Most of the acreage is grown under irrigation in the Mountain and Pacific States, although considerable acreage is grown in the more humid areas of the North Central States. Sugar beets are grown to some extent in 22 states, although most of the production is in California, Colorado, Michigan, North Dakota, Utah, Idaho, Wyoming, Montana, Minnesota, Nebraska, and Washington. This crop has been grown successfully under irrigation on the High

Sugar beets are relatives of table beets and have been developed for a high percentage of sugar in the white fleshy root. The white fleshy root grows mostly under ground. (*Courtesy Farmers and Manufacturers Beet Sugar Association, Saginaw, Michigan*)

Plains of Texas and may become much more important in that area.

The sugar beet crop requires plentiful and well-distributed moisture and extended moderately cool weather during the growing season. Each sugar beet plant requires about 15 gallons of water in the growing season. At harvest, the root may be 75 to 80 percent water, and the top may be 90 percent water.

Sugar beet acreage has increased sharply in recent years. In 1970, 26 million tons of beets were harvested from 1.42 million acres. Most beets are seeded from March to May and harvested from September to December. October is the peak of the harvesting season. California is the only state in which beets are

A one-man sugar beet harvester. (*Courtesy* Farmers and Manufacturers Beet Sugar Association, Saginaw, Michigan)

planted in the fall for harvest the following spring.

There are some differences of opinion regarding the exact origin of the sugar beet, but apparently this plant grew wild in prehistoric times. When found as a wild plant in Europe and North Africa, the sugar beet was an annual. After being introduced into temperate areas it developed into a biennial plant. As a biennial, the sugar beet stores sugar in the root the first year and produces seed the second year.

When Spanish explorers came to the Santa Clara Valley in California, the Indians were making a kind of candy from the juice of wild sugar beets. In the middle of the eighteenth century it was found that beet sucrose could be crystallized. By the end of the century the commercial production of beet sugar was considered feasible. As early as 1830 there was some interest in the production of beet sugar in the United States. The Beet Sugar Society of Philadelphia

sent a representative to Europe to learn production methods. For forty years between 1838 and 1879, most efforts to manufacture sugar from beets failed in the United States. By this time the beet sugar industry was successfully established in Germany.

A sugar beet plant (*Beta vulgaris*) consists of a long silvery-white root from which sugar is made and a leafy top. The latter is a good livestock fodder when properly harvested and cured. The fine rootlets of the deep-feeding sugar beet may extend downward 3 to 4 feet. The root may be 4 to 6 inches in diameter at the widest part. Sugar beet roots average about 2 pounds in weight and contain 14 to 17 percent sucrose.

Sugar beets grow best in deep friable soils that are high in organic matter. This kind of soil is easily cultivated and permits rapid growth of the beet root. Sugar beets tolerate a relatively high salt content in the soil, but are very sensitive to high acidity and low boron content. A deficiency of boron slows plant growth and often causes black lesions in the root.

PLANTING, CULTIVATING, AND HARVESTING

Sugar beets require a variable growing season of 20 to 30 weeks depending on the region. Planting begins soon after the seeding of spring grain. A deep seedbed of well-pulverized soil is necessary for sturdy plants. Fall plowing is desirable if wind erosion is not a hazard.

Special planting machines place the seed ¾ to 1½ inches deep in rows 18 to 22 inches apart. The seeding rate is 10 to 25 pounds of seed per acre. Processed and treated seed may be planted at the rate of 6 to 8 pounds per acre. Anyone planning to grow beets for the first time should contact the local county agricultural agent or the State Agricultural University for information on rate of planting and sources of good seed.

When the plants have 4 to 8 leaves, the stand is thinned to one plant every 10 to 12 inches in the row. The sugar beet plants are very small and weak when they first appear. Weeds should be controlled by the proper use of a recommended herbicide or by shallow cultivation. The kind and extent of the weeds to be controlled may affect the choice of control method.

Crop rotations help to maintain soil fertility and high crop yields. The rotation should include a legume. One popular four-year rotation includes two years of alfalfa. Sweetclover is a good legume for crop rotations in the northern Great Plains. A typical rotation is small grain —sweetclover—potatoes—sugar beets.

Fertilizer should be placed 1½ to 2 inches below the level of the seed, and directly under or ½ to 1 inch from the row. Sugar beets respond well to the application of manure to plowed fields. An even better response is obtained when the manure is supplemented with superphosphate. Some soils need small applications of borax to supply boron.

The sugar beet root is extremely perishable, especially if it is alternately frozen and thawed. Unless special storage facilities are available, the sugar beet crop must be processed between harvest time and early freezing weather.

A large field of sugar beets ready for harvest. (*Courtesy* Farmers and Manufacturers Beet Sugar Association, Saginaw, Michigan)

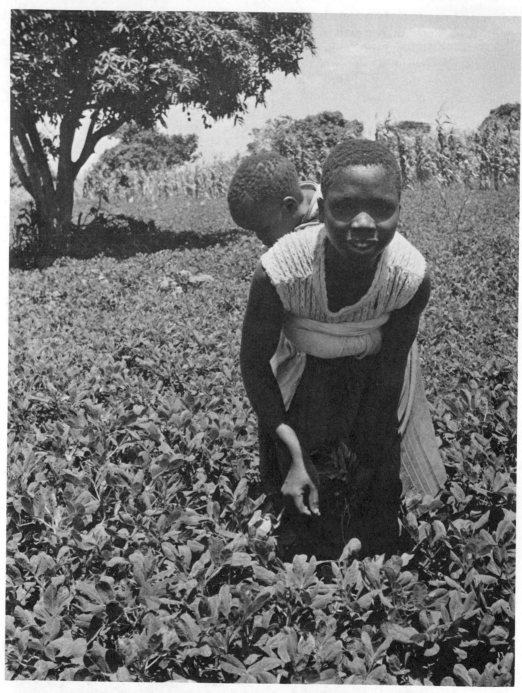

A young African mother pulls weeds from a field of healthy peanuts in Malawi, southern Africa. (*Courtesy* International Development Association)

A sugar beet factory may operate only about 100 days per year, although refineries operate throughout the year.

Special harvesters do the complete job of topping and digging sugar beets. The harvester usually conveys the beets into a truck or trailer, and the harvested beets are then stockpiled near a railroad loading point. Average yields are 14 to 16 tons per acre, although much larger yields are produced under intensive farming conditions.

Sugar beets now provide about 25 percent of the sugar produced in the United States. Beet sugar is sold mainly in the western half of the nation. High transportation costs seem to limit the beet sugar market. Seventy-five percent of our sugar comes from sugar cane. This sugar is chemically the same as beet sugar.

Most European sugar comes from sugar beets, except for the small amounts manufactured from Spanish sugar cane. France, Germany, Great Britain, Italy, and Russia are major beet sugar producers in Europe.

FLAX

North Dakota, South Dakota, and Minnesota produce most of the flaxseed in the United States. Some flax is grown in Texas. North Dakota had more than half of the 2.7 million acres of flax harvested in 1969. In this country and in adjacent parts of Canada, flax is grown primarily for seed. The seed yields linseed oil used in the manufacture of paint and other industrial products.

Yields vary considerably from year to year in the main producing areas. This variation is due mainly to fluctuations in weather, particularly rainfall. Yields of flaxseed in recent years have varied from 5.2 bushels per acre in 1957 to 13.5 bushels per acre in 1969.

Uses of flax. Flax has been cultivated since prehistoric times. It is believed to have originated in western Asia but is now unknown in its wild state. Flax may have grown in Chaldea as early as 3000 B.C. The ancient Egyptians and Hebrews used linen cloth, and the Romans imported linen from Spain before the Christian Era. Remains of flax plants have been found in the ruins of stone-age dwellings in Switzerland.

Linen was one of the first textiles to be used in the New World. In the early self-sufficiency period in the United States, the making of cloth from flax fiber was a household industry. Flax is now grown throughout the world and is next to cotton in importance as a fiber plant. Some of the leading nations in the production of flax fiber are the Soviet Union, Hungary, Belgium, Holland, France, and Ireland. The regions of western Europe near the sea produce the highest quality of flax fiber, although the Soviet Union produces the greatest volume. In the United States, some flax is grown for fiber in the Willamette Valley in Oregon.

More than a century ago flax growing followed the westward development of agriculture. From New York and Pennsylvania, flax was introduced into Ohio, Indiana, Illinois, Iowa, Minnesota, the Dakotas, and Montana. Pioneer farmers had to turn to new land as the fatal wilt disease developed in soils which frequently produced crops of flax. Flax became a permanent crop in the North Central States after wilt resistant varieties were developed.

Flaxseed production in the United States increased rapidly during World War II. This was mainly the result of the growing demand for domestic fats and oils, and especially for flaxseed and linseed oil.

In addition to being used in paint and varnish, linseed oil is used in the manufacture of enamels, oilcloth, patent leather, and waterproofing for raincoats and tarpaulins. In some countries flaxseed is the source of edible oil and some of the oil used in the manufacture of soap.

After the oil is pressed out of the flaxseed, the linseed cake or meal is used as livestock feed. This material is high in protein and is also considered beneficial to the digestive system.

Low quality flax fiber is used in making toweling, matting, and rugs. Straw from seed flax is used in making paper pulp, tow (a fiber), binder twine, bagging, insulating wallboards, and some upholstery products. Flax straw is also used in making cigarette paper.

DESCRIPTION OF FLAX PLANT

The flax plant (*Linum usitatissimum*) is an annual which has a distinct main stem and short tap root. At maturity, flax plants may vary from 12 to 40 inches in height, depending upon such factors as soil and weather. A light soil permits slender root branches to extend to a depth of 3 to 4 feet. The flax flower has 5 petals and a 5-cell boll or capsule which contains 10 seeds when filled. In different varieties of flax the colors of petals may be blue, pale blue, white, or pale pink. Flowers usually open at sunrise and fall in a few hours.

Because flax is normally self-pollinated, there is little natural crossing of varieties.

PLANTING AND HARVESTING

Most of the flax varieties grown in the United States are wilt resistant and classified as the short-fiber type. A variety known to be wilt resistant and adapted to local conditions should always be selected. Clean, plump, disease-free seed is essential to the successful production of flax. Dirt, chaff, and shrunken seed may carry the wilt disease.

The seedbed should be disked, harrowed, and rolled to provide a firm, level surface. A grain drill may be used to plant flax, or the seed may be broadcast and harrowed in. The planting rate varies from approximately 28 pounds of seed per acre in the drier states to 42 pounds per acre where there is high rainfall or irrigation. The county agricultural agent or a reputable seed dealer can provide information on the best rate of planting for a specific community. If flax is to be grown for fiber, the amount of seed planted on each acre is about twice the amount required if a seed crop is to be grown. For either crop, local recommendations should be followed.

In addition to increasing fiber yields, thick planting helps to control weeds. Flax should follow a legume or a clean-cultivated crop, as the flax plants are not able to compete with weeds. The young flax plants are resistant to frost, and planting may be done two or three weeks *before* the average date for the last killing frost in the spring. In hot, dry weather the early seedings have an advantage over late seedings.

Land preparation for peanuts, sugarbeets, flax, and other high-value-per acre crops is especially critical to success of raising profitable yields. Here a 6-bottom plow is plowing about 10 inches deep to create a favorable seed and plantbed. (*Courtesy* John Deere)

When flax is grown under irrigation, the shallow roots respond best to light applications of water. Irrigation should cease when the crop begins to ripen. After that time, irrigation tends to keep the plants blooming and delays maturity.

Most of the flaxseed crop in the United States is harvested with combines, just as wheat is harvested. Special pulling machines are used to harvest the fiber crop. The plants dry in the field before being threshed. Threshing is done in a way to prevent breaking the straw. The resins in flax fiber may be removed by exposure to weather or by chemical methods. After the flax straw has been properly weathered or chemically treated, it is subject to a mechanical process which separates the fibers from bark and stem. The processed straw is then baled for manufacture.

QUESTIONS

1   What are the suitable climatic conditions for growing peanuts?
2   What are typical planting rates for large and small pod varieties of peanuts?
3   List some uses of peanuts and peanut seeds.
4   What conditions of climate and elevations are required for growing sugar beets?
5   What is the best kind of soil for sugar beets?
6   How are sugar beets planted and cultivated?
7   How are sugar beets harvested?
8   For what purpose is flax grown in this country and adjacent parts of Canada?
9   What procedure is followed in planting flax?
10   How is flax harvested?

ACTIVITIES

1   Dig up a peanut plant or other legume and look for nodules on the roots. What special organisms live in the nodules?
2   Use encyclopedias and other references to obtain more information on sugar beets, peanuts, and flax.
3   North Dakota ranks first in flax production and Georgia in peanut production. Compare the average climate in these two states. (See Chapter 10, "Climate and Weather.")

4 See how many kinds of sugar and how many peanut products you can find in a grocery store.

5 Plant some beet seed, preferably sugar beet seed, in flower pots or cans and watch the little plants grow. Do the same with peanuts and flax if seed can be obtained.

29 FRUITS AND VEGETABLES

The production of fruits and vegetables is widely distributed throughout the United States. Much of the commercial production is in areas in which one or the other kind of crop is the principal farm enterprise. Sometimes both fruits and vegetables grow well in an area which has the right combination of soils and climate. In some parts of the nation where markets are favorable, fruit or vegetable farms are interspersed among other types of farms.

Climate determines where many fruits and vegetables grow. The commercial production of winter vegetables and citrus fruits is limited to the most southern parts of the nation. The climate of Hawaii favors the growing of pineapples and various tropical vegetables. The best areas for deciduous fruits are those where early fall and late spring frosts are least likely to occur. These fruits can also be grown in northern areas near large bodies of water which reduce the risk of frost damage.

The profitable production of fresh fruits and vegetables requires good management and continuous and vigorous attention to marketing. This is an intensive type of farming in which land values are high and much capital is required for labor, seed, fertilizers, equipment, and the control of diseases, weeds, and insects. Vegetables must be planted at the right time to be ready for market when prices are best. Good planning is necessary for efficient production.

The major producers of important fruits and vegetables are as follows:

Apples—Washington, New York, Michigan, Virginia, and California
Cherries—Michigan, New York, Oregon, California, and Washington

Citrus fruits—Florida, California, Arizona, and Texas

Cranberries—Massachusetts, New Jersey, Wisconsin, and Washington

Grapes—California, New York, Michigan, Pennsylvania, and Washington

Peaches—California, South Carolina, Georgia, New Jersey, and Pennsylvania.

Pears—California, Washington, Oregon, Michigan, and New York

Strawberries—California, Oregon, Michigan, Washington, Louisiana, and New Jersey

Vegetables for fresh market—California, Florida, Arizona, Texas, and New York

Vegetables for processing—California, Wisconsin, Oregon, Washington, and New Jersey

The production of all vegetables is now about 20.4 million tons. This production has remained almost constant for the past 10 years. The relative tonnage of vegetables for the fresh market and for processing has remained unchanged although production for processing has been variable. During the past 15 years the per capita consumption of all vegetables has changed only slightly. The per capita consumption of fresh vegetables has decreased, whereas the per capita consumption of canned and frozen vegetables has increased.

The production of the 20 major fruits is approximately 22 million tons. This is a slight increase over the production at the beginning of this decade.

Rainfall, summer and winter temperatures, plant diseases, and insects largely determine what varieties of fruits, vegetables, and nuts can be grown in a given locality. It is possible to find vari-

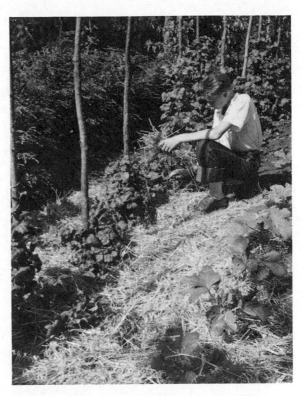

In almost every part of the nation some type of fruit or vegetables can be grown in the home garden. (*Courtesy* U. S. D. A.)

eties that are adapted to almost every home garden in a particular region. Five regions are considered here: Northeastern and North-Central; Southeastern and Central Southern; Central Southwestern; North Great Plains; Pacific Coast, Arizona, Alaska, and Hawaii.

NORTHEASTERN AND NORTH-CENTRAL STATES

There are several kinds of fruits that can be grown in home gardens where spraying is not commonly practiced. In the order of their adaptability, nine of

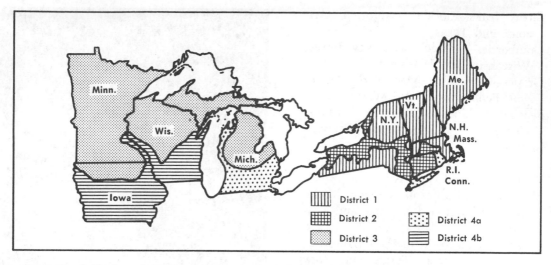

Agricultural districts of the Northeastern and North-Central states. (*Source:* Leaflet 227, U.S.D.A.)

DISTRICT 1—growing season of 90 to 150 days; moderate summer temperatures; low winter temperatures. DISTRICT 2—growing season of 150 to 180 days; fairly high summer temperatures or modified by lake or ocean. DISTRICT 3—growing season of 90 to 150 days; severe winters. DISTRICT 4a—growing season of 150 to 180 days; fairly high summer temperatures, similar to District 2. DISTRICT 4b—more severe winters than District 4a.

Wild blueberries are being harvested with a hand rake in Maine. Tame blueberries and huckleberries have been developed for planting in the "kitchen" gardens and for commercial production in the Northeastern and North-Central states. (*Courtesy* Maine Department of Agriculture)

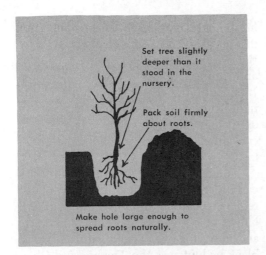

Set tree slightly deeper than it stood in the nursery.

Pack soil firmly about roots.

Make hole large enough to spread roots naturally.

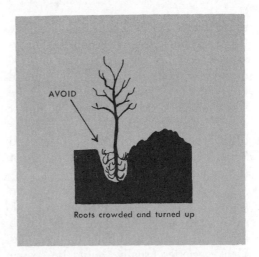

AVOID

Roots crowded and turned up

A fruit tree should be planted in a hole that is 6 inches deeper and 6 inches wider than the spread of the roots. Set the tree approximately 1 inch deeper than it stood in the nursery. Place the topsoil around the roots and tamp with the hands. Then complete the filling of the hole with the subsoil. After planting, put some mulch on the soil around the tree and soak thoroughly. After planting, each tree should be circled with a cut tin can and 36 inches of 2-inch mesh chicken wire to protect against mice and rabbits. (*Source:* Home and Garden Bulletin No. 11, U.S.D.A.)

these fruits are: strawberries, raspberries, sour cherries, grapes, plums, pears, sweet cherries, blackberries, and apples. Peaches, blueberries, and cherry-plum hybrids can be grown under some conditions. Currants and gooseberries can be grown wherever quarantine regulations permit. These fruits are limited mainly to areas where white pine trees are not important because the currant and gooseberry plants are hosts to the white-pine blister rust.

Proper spraying is beneficial to fruit trees and grapes in all parts of the region. Home gardens near commercial orchards and vineyards should be sprayed to prevent the spread of diseases and insects.

Strawberries are the most widely adapted to this region. They ripen before

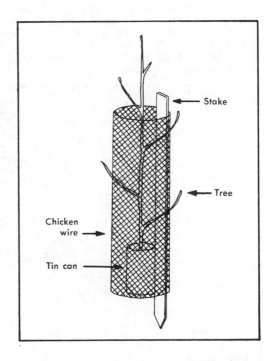

Stake

Tree

Chicken wire

Tin can

other fruits are available, have a fine flavor, and are rich in vitamin C. Strawberries can be kept frozen for several months without destroying the vitamin C content. These advantages make strawberries a desirable part of almost every home fruit garden. Two varieties of everbearing strawberries, Gem and Superfection, can be grown throughout the region.

Red and purple raspberries can be grown in all districts of the region. Black raspberries can be grown everywhere except in northern Minnesota. Red and black varieties usually should be grown in separate gardens because of a virus disease that kills black raspberries.

Strawberries, raspberries, plums, and grapes ripen for home use from June until the first killing frosts in the fall. By growing more than one variety of certain fruits, the family can have a continuous supply throughout the growing season.

Black walnuts, chestnuts, and filberts can be grown in the southern part of the region, including roughly the southern halves of both Michigan and Wisconsin, and most of Iowa. Two good varieties of black walnuts are the Thomas and Ohio.

Agricultural districts of the Southeastern and Central Southern states. (*Source:* Leaflet 219, U.S.D.A.)

DISTRICT 1—relatively high areas; growing seasons ranging from 150 to 180 days; temperate climatic conditions prevailing; suitable for growing standard northern fruit varieties. DISTRICT 2—growing season ranging from 180 to 200 days; many standard southern fruit varieties not grown in District 1 thrive. DISTRICT 3—upper boundary corresponds roughly with the northern limit of the Cotton Belt; pecans, muscadine grapes, and many other desirable fruit varieties may be grown. DISTRICT 4—southern part of the Coastal Plains area; characterized by a hot, humid climate during the growing season; typically southern fruits, including muscadine grapes and figs, thrive best. DISTRICT 5—citrus fruits are grown principally, but other southern fruits may be grown advantageously in the home garden. DISTRICT 6—hot, humid area; only semitropical fruits are adapted.

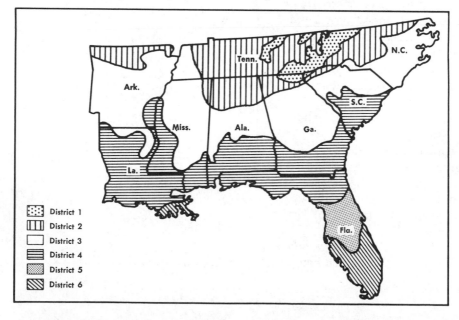

Varieties of chestnuts adapted to the region include the Chinese, Carr, and Hobson. The Bixby and Buchannan are good varieties of filberts. Black walnuts should be planted 40 feet apart, chestnuts 30 feet, and filberts 15 feet.

SOUTHEASTERN AND CENTRAL SOUTHERN STATES

In the order of their adaptability where spraying is not commonly practiced, the best fruits and nuts for home gardens in this region are: grapes (muscadine), pecans, figs, strawberries, blackberries, blueberries, pears, bunch (American) grapes, peaches, plums, apples, and raspberries.

Muscadine grapes are widely adapted, except in the northern districts of the region. The bunch grapes arc better adapted to the northern areas. The muscadine grapes produce heavily without spraying. In addition to providing fresh fruit over a long period, these grapes are used for preserves, beverages, and jelly.

Pecans are widely adapted and are grown extensively for the nut crop and as shade trees. Figs grow in most of the region. This fruit should be planted near a building or in an area that is kept in grass. Under other conditions the fig trees may be killed by root-knot nematodes.

In most of this region strawberries, trailing blackberries, figs, and grapes provide fresh fruit from April or May until frost. A greater variety of fruit is available during much of the year if the home garden is large enough to include blueberries, pecans, pears, peaches, and plums.

To compete on United States and World markets, tomatoes must be packed to avoid damage in shipment and to appeal to the eye upon arrival. (*Courtesy* Appleton Associates, Massachusetts)

The Young and Boysenberries (trailing blackberries) can be grown in much of the region except in the high mountains and in central and southern Florida. These vigorous varieties produce abundantly one year after planting.

CENTRAL SOUTHWESTERN STATES

The best fruits and vegetables for home gardens in the southern part of this region are: grapes (muscadine), pecans, figs, dewberries, strawberries, blackberries, bunch grapes, peaches, and plums. Several citrus fruits, oriental persimmons, guavas, pomegranates, and many other fruits can be grown under the more subtropical conditions.

Fruits that can be grown without much spraying in the northern part of the region are: strawberries, bunch grapes (American), sour cherries, plum and cherry hybrids, plums, peaches, and

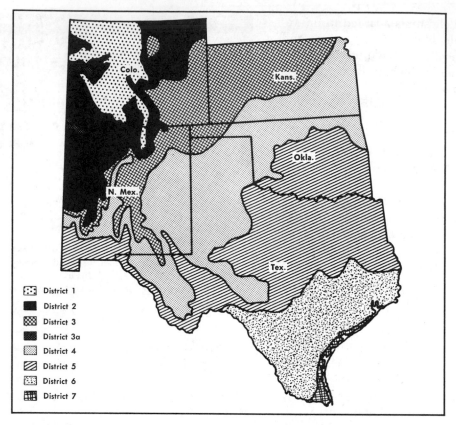

Agricultural districts of the Central Southwestern states. (*Source:* Leaflet 221, U.S.D.A.)

DISTRICT 1—low winter temperatures; growing season less than 90 days; not adapted to the growing of fruits except for selected sheltered valleys and sunny slopes where especially hardy and early-maturing varieties may be grown with winter protection. DISTRICT 2—growing season of 90 to 150 days; restricted rainfall makes irrigation necessary or desirable; fruits that succeed must be fully winter-hardy and early-maturing. DISTRICT 3—growing season 150 to 180 days; restricted rainfall; irrigation in all but the more eastern part necessary or desirable for most fruit varieties; in the more northern parts standard northern varieties are preferred, and in the more southern parts selected northern varieties and representatives of the standard southern groups are the most desirable fruits. DISTRICT 3a—more favorable temperature conditions and water for irrigation make possible the growing of less hardy varieties. DISTRICT 4—growing season 180 to 210 days; in the eastern part where rainfall is sufficient considerable fruit can be grown; much of the area, however, is included in the southern Great Plains where rainfall is low and irrigation is necessary, or at least desirable, for fruit growing. (Soil-moisture conditions determine largely what varieties can be grown successfully.) DISTRICT 5—growing season 210 to 250 days; a considerable variety of fruits adapted to much of this District; in selected locations in the western part vinifera grapes may be grown successfully. DISTRICT 6—growing season 250 days or more; muscadine and southern bunch grapes, figs, dewberries, and other fruits as well as pecans, do well in much of this District; along the southern border even hardy citrus fruits may be grown. DISTRICT 7—frosts rare; subtropical fruits grown.

apples. All of these fruits are widely adapted. As in all other parts of the region, spraying is beneficial.

The usefulness of home gardens can be extended with a simple irrigation system. Small gardens can be watered with a sprinkling can. Large gardens should be adapted for row irrigation.

NORTHERN GREAT PLAINS

The states included in this discussion of fruits and vegetables adapted to the Northern Great Plains are North Dakota and South Dakota, and the following areas in other states: the western half of Nebraska, northeastern Wyoming, and the eastern three-fourths of Montana.

The Northern Great Plains often have the hottest as well as the coldest temperatures in the United States. The growing season varies from 120 to 140 days, and the average annual precipitation from 15 to 20 inches.

The hardiest varieties of the following fruits have been grown successfully in the Northern Great Plains: apples, crabapples, pears, apricots, plums, cherries, currants, gooseberries, grapes, raspberries, and strawberries.

Site selection. Fruits should be planted on gently sloping land that has good air drainage as well as good internal soil drainage. Fruits should not be planted in frost pockets or on stony, wet, excessively dry, or salty or alkali soils. If the site will produce good field crops it should be suitable for growing the recommended fruits. The fruit garden should be protected by buildings or windbreaks from the strong west winds, hot south winds, and cold north winds. Only the east side need not be protected.

Nursery stock. In all regions it is a good practice to buy or order nursery stock from a local nursery rather than from one far away. Locally grown stock is adapted to the heat and the cold, the winds, and the varying length of days in different seasons.

Pruning, soil cultivation, and protection against insects, diseases, and rodents are much the same in all regions.

PACIFIC-COAST STATES, ARIZONA, ALASKA, AND HAWAII

In almost every district in this region some fruits can be grown without much spraying. The following fruits are well adapted to most of the region: grapes, strawberries, Young or Boysenberries (trailing blackberries), red raspberries, filberts, Persian (English) walnuts, almonds, plums and prunes, cherries, pears, peaches, apricots, and apples.

In most of the region the fruit plants are not attacked by many of the diseases and insects that cause serious damage in humid regions. In some districts it is necessary to spray apple and pear trees to control the coddling moth, several species of mites, and pear psylla.

Citrus fruits, figs, olives, avocados, and certain other fruits are grown in the warmer parts of southern California and Hawaii.

Alaska. Fruits and vegetables are grown mainly in southeastern Alaska, the south coast, and the southwestern islands. The growing season in these areas averages about 145 days. Strong winds are common. The average annual precipitation is 100 inches.

Hawaii. The climate of Hawaii is tropical. Mean annual precipitation

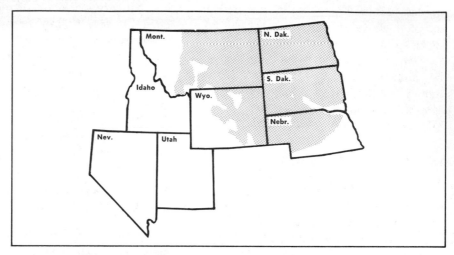

Recommended Fruit Varieties for the Northern Great Plains (*Source: Growing Fruit for Home Use in the Northern Great Plains,* Home and Garden Bulletin No. 11, U.S.D.A.)

Fruit	*Varieties*
Apples:	
Short storage period	Beacon, Erickson, Garrison, Mantet, Red Duchess, Stephens, Yellow Transparent
Long storage period	Haralson, Killand, Patten, Greening, Peace Garden, Thorberg, Wealthy
Apricots [1]	Mantoy, Ninguta, Scout, Sunshine
Cherries:	
Manchu (or Nanking)	Drilea, Orient, seedlings
Sour [1]	Meteor, North Star
Cherry-plums:	
Red flesh	Hiawatha, Sacagawea, Sapa, Sapalta
Green or yellow flesh	Compass, Convoy, Opata
Crabapples	Chestnut, Cranberry, Dolgo, Heart River, Prairie Gold, Red Rover, Rosilda
Currants	Perfection, Red Lake
Gooseberries	Pixwell
Grapes [2]	Alpha, Beta
Pears [1]	Bantam, Golden Spice, Pioneer, Sodak, Seedless, Tait-Dropmore
Plums:	
To be planted with pollinizer varieties	Gracious, Kaga, LaCrescent, Redcoat, Redglow, Tecumseh, Underwood, Waneta
Pollinizer varieties	Chilcott, Chinook, Manet, South Dakota, Toka
Raspberries [2]	Chief, Durham, Indian Summer, Latham
Strawberries: [2]	
June-bearing	Dunlap, Premier, Robinson
Everbearing	Gem, Ogallala, Radiance

[1] For trial
[2] Winter covering recommended

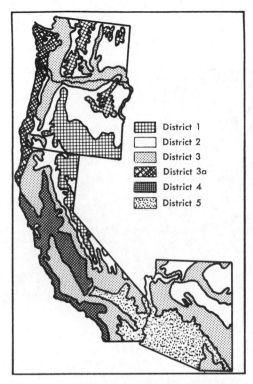

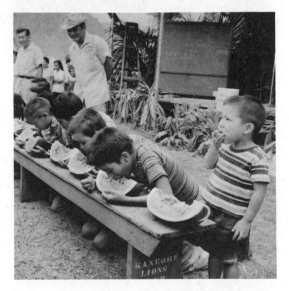

DISTRICT 1—high mountainous parts of Washington, Oregon, and California, which in general are not suited for fruit growing. DISTRICT 2—mountain slopes and elevated areas having a growing season of 90 to 150 days; only especially hardy and, in some parts, drought-resistant varieties can be grown. DISTRICT 3—western and river-valley areas of Washington and Oregon, foothill slopes in California, and parts of Arizona having a growing season of 150 to 240 days, where a large variety of fruits and nuts can be grown; irrigation required east of the Cascade Mountains in Washington and Oregon, in southeastern California, and in Mohave, Yavapai, Gila, and Graham Counties, Arizona. DISTRICT 3a—rather narrow coastal strip of Washington, Oregon, and northern California having the same length of growing season as District 3, but summer temperatures too low for the best growth of many fruits. DISTRICT 4—northern and central river valleys and coastal areas of California having a growing season of more than 240 days; in the interior valleys high summer temperatures prevail, and many fruits and nuts can be grown under irrigation. DISTRICT 5—arid parts of southern California and Arizona having a growing season of more than 240 days and high summer temperatures, where many fruit varieties can be grown under irrigation.

Above left: **Agricultural districts of the Pacific-Coast states and Arizona.** (*Source: Leaflet* 224, U. S. D. A.) *Above right:* **A California citrus orchard that is irrigated according to a tensiometer, a porous clay cup by the trade name of Irrometer.** (*Courtesy* the Irrometer Company, Riverside, California) *Below right:* **Watermelon is a warm-season, tropical, and semitropical crop that is well adapted to gardens and fields in the southeastern and central-southwestern states, and to the warmer parts of the Pacific Coast states, Arizona, and Hawaii.** (East Oahu County, Hawaii) (*Courtesy* University of Hawaii)

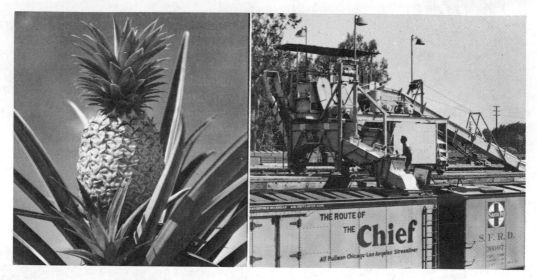

Above left: **The pineapple is well-adapted to the sloping, well-drained, acid soils of high rainfall areas in tropical Hawaii.** (*Courtesy* Dole Pineapple Company) *Above right:* **Commercial fruits and vegetables are being iced with this machine in preparation for shipping to markets across the country.** (*Courtesy* Santa Fe Railroad) *Below left:* **Cutworms eat off plants at the ground, but these pests can be quite easily controlled.** (*Courtesy* Hercules Powder Company)

varies sharply within short distances, due to the effects of mountains, from less than 10 inches to 471 inches, the highest in the world, on Mt. Waialeale, Kauai Island.

SITES FOR VEGETABLE GARDENS

A backyard is a convenient place for the home garden either in town or in the country. Farm gardens usually are located near the house and may contain one-half acre or more. All of the common vegetables, including potatoes, sweet corn, and vine crops, can be grown in the garden. A garden tractor or power tools can be used to work the large garden. The small garden can be kept clean by hoeing.

A small garden close to the house may include lettuce, beets, carrots, radishes, beans, onions, and other vegetables that are consumed while they are fresh. In addition to a small garden plot, a farm family may plant vegetables between rows of young fruit trees.

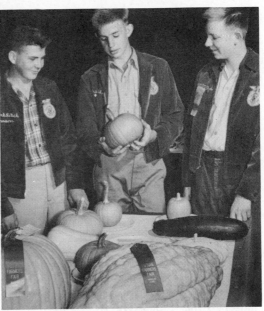

Left: **A small garden usually contains onions be-cause most people like them.** (*Courtesy* J. R. Hepler, New Hampshire Extension Service) *Right:* **Pumpkins and squash can be raised in most gardens.** (*Courtesy The National Future Farmer*)

Among the garden crops that require more space are potatoes, sweet-potatoes, sweet corn, squash and other vine crops, and tomatoes for canning. A large garden may yield a surplus crop of vegetables that can be marketed to supplement the family income. A large garden also permits the planting of individual crops in a different place each year.

It is often necessary to fence the farm garden with close-woven wire to keep out dogs, poultry, and other animals. A well-constructed fence also helps to reduce the damage caused by rabbits.

Rodents are a threat to gardens in every region. Moles and certain species of mice cause considerable damage in the East. These pests can be controlled by trapping, by placing repellents in their runs, or by using poison baits or gases. Poisoning is the most effective way to control ground squirrels and prairie dogs in the West. Cutworms are

always present, but they can be controlled by special poisons.

Garden crops grow best on fertile, deep, friable, well-drained soils. Good drainage and abundant organic matter are important. The organic matter retains moisture and also improves the physical and chemical condition of the soil. The garden site should be free from stones and other materials that interfere with cultivation.

Early crops grow best on a gentle southern slope. Where strong winds are prevalent the garden should be protected by a hedge, board fence, or other windbreak. In the Northern Great Plains the trees that are planted for windbreaks and shelterbelts are good protection for fruit and vegetable crops.

A slight slope improves the drainage of the garden site. There should be no low places where water may lie after rains or where cold air may collect to form frost pockets. The garden should also be protected from drainage from adjacent land. Subsoiling will loosen the soil so that more water is absorbed. Drainage may be accomplished artificially by using agricultural tile or open drainage ditches.

Garden plants need the direct rays of the sun throughout the day. Some plants are more tolerant of shade than others, but no amount of fertilizer, water, or care can replace sunshine.

Tree roots may rob garden crops of moisture and nutrients even when the trees are not close enough to shade the vegetables and fruits. Some of this damage can be prevented by digging a trench 1½ to 2 feet deep between the trees and the garden and cutting all roots that cross the trench. The trench may be lined with waste sheet metal or heavy roofing paper along one side and then refilled. This treatment may be effective for several years. The use of a trench is not practicable where the cutting of large roots would kill valuable shade or fruit trees.

PREPARING AND IMPROVING GARDEN SOILS

Farm gardens are improved by the use of animal manure. A new site usually needs 15 to 20 tons of decomposed cattle or horse manure for each half acre. Poultry manure is higher in plant nutrients, and the rate of application should be one-fourth to one-third that of cattle or horse manure.

Usually it is best to spread the manure several weeks before planting, just before plowing the garden. If thoroughly composted manure is available it should be spread on the surface after plowing and worked into the topsoil.

After the first year the annual application of cattle or horse manure at the rate of 8 to 10 tons per acre or poultry manure at the rate of 2 to 3 tons per acre will help to keep the garden soils fertile.

Artificial manure (compost) can be made from straw, dry or green grass or weeds, or spoiled hay. Chemicals may be added to speed up the process of decomposition. A common mixture for this purpose includes: ammonium sulfate (20 percent nitrogen), 45 parts; superphosphate (20 percent), 15 parts; and ground dolomitic limestone, 40 parts. The mixture is applied at the rate of 150 pounds to each ton of dry vegetable matter, or about 50 pounds to each ton of fresh green material.

Most vegetable gardens can be improved by commercial fertilizers. The kinds and amounts of fertilizers vary according to locality, soils, and the crops to be raised. Good results usually can be obtained by using a fertilizer containing 15 percent nitrogen, 15 percent phosphoric acid, and 15 percent potash. The additional nitrogen required by leafy crops such as lettuce, spinach, kale, and cabbage may be provided by side dressings with ammonium nitrate or urea.

Potatoes, sweetpotatoes, beets, carrots, turnips, parsnips, and other tuber and root crops need more potash than is necessary for other vegetables.

Commercial fertilizers usually are applied a few days before planting, or at planting time. For fairly fertile soils, 300 pounds may fertilize one-half acre; less fertile sites may need 1,000 pounds.

CARE OF THE GARDEN

Weeds must be controlled so that they do not compete with garden plants for moisture, nutrients, and light. The garden should be hoed or cultivated as soon after each rain or irrigation as the soil is dry enough to be worked. Cultivation should be shallow enough to avoid injury to the roots and frequent enough to control weeds.

Where irrigation is not practiced, gardens are sometimes mulched with organic materials or plastic mulches to conserve moisture. A mulch may also be used around unstaked tomato, cucumber, and bush squash plants to keep the fruit from touching the soil. Plastic mulches also have a special value in Alaska, where they increase soil temperatures and thereby lengthen the growing season.

Straw, dried lawn clippings, manure and leaves make a good mulch which should be applied between the rows and around the plants before the soil becomes too dry. It should be about 2 inches deep after it settles—not deep enough to absorb all the moisture from light rains. In humid areas heavy mulching may be harmful by keeping the soil too wet.

A garden needs the equivalent of about an inch of rain a week. This means that an acre of garden needs almost 28,000 gallons of water per week, and a garden 50 by 30 feet needs more than 900 gallons. Watering is necessary when no rains occur for several weeks or when the reserve supply of moisture is depleted. A good soaking once a week is better than several light waterings that do not penetrate the soil.

In tropical and subtropical countries the broad bean (*Vicia faba*) is grown in "kitchen" gardens or for the local market. This bean produces high yields and supplies a high protein food in areas where protein is scarce or very expensive. (Tunisia, northern Africa) (*Courtesy* United Nations)

QUESTIONS

1   What is one of the most popular fruits for the home garden? Why?

2   What is the relationship between white pine trees and the raising of currants and gooseberries?

3   In the region where you live what are the adapted fruits? The adapted vegetables?

4   Where are most of the pecans grown? Figs?

5   Where are English walnuts adapted?

6   Describe the ideal garden soil.

7   Describe a citrus fruit.

8   Which slopes warm up most quickly in the spring?

9   What should you do when tree roots interfere with a garden?

10 How would you make artificial manure (compost)?

ACTIVITIES

1   Arrange a classroom display of the fruits and vegetables grown in your community. Members of the class may bring different products. Discuss the nutritional value of each product, the soil and climatic conditions to which it is adapted, and the methods used in production and marketing.

2   Select one of your favorite fruits or vegetables that is not produced locally. Use bulletins, encyclopedias, and other references to obtain information about this product. Make careful notes so that you can share your information with other members of your class.

3   Plan a home orchard and garden to include several kinds of fruits adapted to local soils and climates. Make a sketch showing the size and location of the area to be used for each kind of fruit. Use catalogs of nursery companies to estimate the cost of trees and other plants.

4   Obtain samples of soil from local orchards and gardens. Examine the soil carefully, observing its color and texture. Look for worms and other living organisms. Try to decide whether each sample is a loam, sandy loam, clay loam, or clay. Send a soil sample to your state soil testing laboratory.

5   Study the photograph on page 415 of the young man in Tunisia, northern Africa, picking beans. Write a short theme on the ideas that flash through your mind as you continue to look at the picture.

APPENDICES

APPENDIX A: BIBLIOGRAPHY, SOURCES OF INFORMATION

BIBLIOGRAPHY

Agriculture and Industrialization, FAO, 1967. Unipub, Inc., Box 443, New York, New York 10016, 129 pages.

Agricultural Conservation Program Accomplishments, 1970, Agricultural Stabilization and Conservation Service (changed to Rural Environmental Assistance Program (REAP) in 1971). USDA, April 1971, Washington, D. C. 20250, 123 pages.

Agricultural Development: A Review of FAO's Field Activities, FAO 1970. Unipub, Inc., Box 443, New York, New York 10016, 194 pages.

Agricultural Statistics, 1970, USDA, United States Government Printing Office, 1970. Paper bound, 627 pages.

Alaska Resource Development Directory, University of Alaska, Publication No. 10, January 1971, 52 pages.

Allen, G. A. Jr., *Sheep Management Schedule,* Virginia Polytechnic Institute, Pub. 365, April 1970, 4 pages.

Allen, W. L., *Sheep Raising in Canada,* Canada Department of Agriculture, Publication No. 1401, 1969, 55 pages.

Almand, J. David, *Wildlife Plantings,* University of Georgia Circular 578, May 1968, 2 pages.

Anderson, D. A. and Smith, W. A., *Forests and Forestry,* The Interstate Printers and Publishers, Danville, Illinois, 1970, 357 pages.

Anderson, G. C., Veatch, C., Newman, J., and Gillespie, W., *Control of Brush in Pastures with Aerial Herbicide Application,* West Virginia University Bulletin 598, February 1971, 14 pages.

Anderson, Jonathan D. and Johnson, Roger G., *Analysis of Optimum Farm Organization in the Red River Valley,* North Dakota State University Bulletin 489, March 1971, 15 pages, Fargo, North Dakota 58103.

Anderson, Wallace L. and Compton, Lawrence V., *More Wildlife Through Soil and Water Conservation,* USDA Bulletin 175, August 1971, 16 pages.

Andrews, Daniel K., *If You Want to Raise Commercial Rabbits,* Washington State University Leaflet, E. M. 3382, March 1970, 2 pages.

Animal Science, University of Maine Bulletin 545, February 1968, 48 pages.

Arbor Day, Kansas State University Leaflet, L-241, April 1970, 12 pages.

Arrington, L. R., *Raising Laboratory Animals and Other Small Stock,* University of Florida Circular 326, June 1968, 10 pages.

Assistance Available from the Soil Conservation Service, USDA Bulletin 345, 1970, 29 pages.

Background on United States Agriculture, Leaflet No. 491, USDA, January 1971, 7 pages.

Barrons, Keith C., *Environmental Benefits of Intensive Crop Production,* Agricultural Science Review, Vol. 9, No. 2, 1971, pages 33-39, USDA.

Beef Cattle Breeds, USDA Farmers Bulletin No. 2228, 1968, 28 pages.

Bergeaux, P. J., *Soils of Georgia,* Bulletin 662, September 1969, 14 pages, University of Georgia, Athens, Georgia 30601.

Bigger Crops and Better Storage: the Role of Storage in World Food Supplies, FAO, 1969, 49 pages, Unipub, Inc., Box 443, New York, New York 10016.

Block, William J., *Rural Zoning—People, Property, and Public Policy,* Montana State University Bulletin 331, April 1970, 32 pages.

Boll Weevil, How to Control It, Farmers Bulletin No. 2147, September 1969, 12 pages.

Bond, Dean, *Peanut Production Guide for Alabama,* Auburn University Circular, P-82, February 1970, 28 pages.

Broodley, James W. and Langhans, Robert W., *Home Greenhouse Gardening Structures,* Cornell University Extension Bulletin 1043, May 1970, 7 pages.

Bosserman, Willard E., *Compost Will Improve the Environment,* Michigan State University Extension Bulletin E-727, October 1971, 2 pages.

Brannon, David H., *English Sparrow Control,* Washington State University, No. E. M. 335, unnumbered, undated.

Breeds of Chickens, Canada Department of Agriculture, Publication No. 1348, 1970, 16 pages.

Brock, Samuel M., *Relations Between Forest Resource Development and Human Resource Development,* Information Series 9, 1970. Office of Research and Development, 17 Grant Ave., West Virginia University, Morgantown, West Virginia 26506.

Brooks, Joe F., *Berries for the Garden,* North Carolina State University Extension Circular No. 510, April 1970, 12 pages.

Brooks, Leroy and Gates, Dell E., *1971 Kansas Field Crop Insect Control Recommendations,* Kansas State University Circular C-431, November 1970, 20 pages.

Brown, Allen J., Gossett, J. W., and Powell, William E., *Livestock Marketing in Alabama,* Auburn University Circular R-24, November 1970, 14 pages.

Burch, Thomas A. and Hubbard, John W., *Minimum Land Requirements for Specified Levels of Income, Major Cotton-Producing Areas, South Carolina,* Clemson University Bulletin 545, March 1969, 76 pages.

Burdett, Robert A., Jr., *What's in the Bag? Read the Tag!* Auburn University Circular, p. 29, December 1970, 6 pages.

Burlison, Vernon H., *Timber: How Much Do I Have?* University of Idaho Bulletin 420, February 1971, 20 pages.

Burton, Clifford, *Judging Dairy Cattle,* Oklahoma State University Circular C-698, January 1971, 16 pages.

Butler, B. J. and Siemens, J. C., *Calibrating and Maintaining Spray Equipment,* University of Illinois Circular 1038, March 1971, 20 pages.

Cabbage Insects, How to Control Them in the Home Garden, USDA Home and Garden Bulletin No. 44, March 1971, 7 pages.

Caley, Homer K., *Health Handbook for Profitable Swine Production,* Kansas State University Circular C-393, December 1970, 16 pages.

Caley, Homer K., *Rabies,* Kansas State University Leaflet L-162, December 1971, 5 pages.

Callahan, Charles H., *Massachusetts Directory of Agricultural Organizations and Home Economic Leaders,* Massachusetts Department of Agriculture, 1971, 56 pages.

Calories and Weight: the USDA Pocket Guide, Home and Garden Bulletin No. 153, 1970, 75 pages, USDA, Washington, D. C. 20250.

Canadian Fur Industry, The, Canada Department of Agriculture Publication No. 1201, 1971, 14 pages.

Career in Montana Cooperative Extension Service, A, Montana State University Circular 287, December 1970, 11 pages.

Carlson, Axel R., *Building in Alaska, Permafrost—A Problem of Building in Alaska,* University of Alaska Publication No. 754, March 1970, 6 pages.

Carlson, Axel R. and Epps, Alan C., *Greenhouses in Alaska,* University of Alaska Publication No. 51, June 1971, 40 pages.

Catalog of State Services and State Programs for Community Development, ORD Information Series 8, 1969, Office of Research and Development, 17 Grant Ave., West Virginia University, Morgantown, West Va. 26506.

Chapman, Louie J., *Cotton Defoliation,* Auburn University Circular P-73, October 1970, 5 pages.

Christ, Ernest G., 1971 *Tree Fruit Production Recommendations for New Jersey,* Rutgers University Leaflet 446-A, March 1971, 50 pages.

Christ, E. G. and Smith, Carter R., *Bush Fruits in the Home Garden,* Rutgers University Extension Bulletin No. 298-C, April 1971, 11 pages.

Clapp, J. G., Jr. and Brim, C. A., *Practices for Efficient Soybean Production,* Folder 286, October 1971, 6 pages.

Clepper, Henry, editor, *Leaders of American Conservation,* The Ronald Press, New York, New York, 1971, 353 pages.

Colyer, Dale and Pohlman, G. Gordon, "Yield and Quality Response of Burley Tobacco to Nitrogen and Potassium," *Agronomy Journal,* Vol. 6, No. 6, November 1971, pages 857-860, American Society of Agronomy.

Commercial Agricultural and Vegetable Seeds, 1970; Pesticides, 1970, Official Inspections, Bulletin 297, January 1971, 75 pages, Experiment Station, University of Maine, Orono, Maine.

Commercial Fertilizers, 1971, Official Inspections, Bulletin 300, October 1971, 35 pages, Experiment Station, University of Maine, Orono, Maine.

Common Forage Legume Insects, Iowa State University, Pm. 463-8, 1970.

Common Fruit Insects, Iowa State University, Pm. 463-10, 1970.

Common Household Insects, Iowa State University, Pm. 463-2, 1970.

Corn Insects Above Ground, Iowa State University, Pm. 463-4, 1970.

Corn Insects Below Ground, Iowa State University, Pm. 463-5, 1970.

Common Small Grain Insects, Iowa State University, Pm. 463-7, 1970.

Common Soybean Insects, Iowa State University, Pm. 463-6, 1970.

Common Tree and Shrub Pests, Iowa State University, Pm. 463-3, 1970.

Common Vegetable Insects, Iowa State University, Pm. 463-9, 1970.

Controlling Erosion on Construction Sites, Agricultural Information Bulletin 347, Soil Conservation Service, USDA, Washington, D. C. 20250, 31 pages.

Coppock, C. E. and Slack, S. T., *The Use of Urea in Dairy Cattle Feeding,* Cornell University Extension Bulletin 1219, January 1970, 16 pages.

Corn Yields as Affected by Soil Slope, Fertilization, Year from Sod, and Rainfall, Clemson University Technical Bulletin 1032, October 1969, 16 pages.

Cotton Production Guides, A series of leaflets on all subjects related to cotton, April 1971, North Carolina State University, Raleigh, North Carolina, 27607.

Cox, F. R., Nicholaides, J. J., Hallock, D. L., Reid, P. H. and Martens, D. C., *Nutrient Concentrations in Virginia-Type Peanuts During the Growing Season,* North Carolina State University and Virginia Polytechnic Institute, Cooperating Technical Bulletin 204, November 1970, 24 pages.

Crawford, Johnny L., *Cotton Disease Control,* University of Georgia, Leaflet No. 93, February 1970, 3 pages.

Cress, Donald C., *For Poultry—Insect and Mite Control,* Michigan State University Bulletin E-405, March 1971, 6 pages.

Cress, Donald C., Potter, Howard S. and Wells, Arthur, 1971 *Control of Insects and Diseases of Commercial Vegetables,* Michigan State University Extension Bulletin 312, March 1971, 32 pages.

Cross, Chester E., et al., *Modern Cultural Practices in Cranberry Growing,* University of Massachusetts Publication No. 39, September 1969, 52 pages.

Cuozzo, Roscoe F., *Rabbit Raising Handbook for Youth,* University of Maine, Bulletin 461, March 1971, 20 pages.

Currin, R. E., Ford, Z. T. and Graham, T. W., *Performance of Tobacco Varieties in South Carolina—1970,* Clemson University Circular 517, February 1971, 13 pages.

Dairy Goats: Breeding, Feeding, Management, American Dairy Goat Association, Box 186, Spindale, North Carolina 28160, June 1969, 77 pages.

Davis, Charles A., Babb, David E. and Ballard, Warren B., *Foods of Mourning Doves in the Mesilla Valley of South-Central New Mexico,* New Mexico State University, Research Report 201, July 1971, 4 pages.

Day, Maurice W. and Rudolph, Victor J., *Development of a White Spruce Plantation,* Michigan State University Research Report No. 111, March 1970, 4 pages.

Dean, Anita, *Nutrition for You,* Michigan State University Miscellaneous Series Circular E-640, July 1971, 15 pages.

Denman, Charles E., Huffine, Wayne W. and Arnold, James D., *Bermudagrass Forage Production Studies in Oklahoma, 1962–1970,* Oklahoma State University Bulletin B-692, April 1971, 15 pages.

Detwyler, Thomas R., *Man's Impact on Environment,* McGraw-Hill Book Co., Inc., Princeton Road, Hightstown, New Jersey 08520, 1971, 731 pages.

Dickson, Alex and Winch, Fred E., Jr., *Plantation Production of Christmas Trees in New York,* Cornell University Bulletin 1204, 1970, 28 pages.

Dinkel, D. H. and Epps, Alan C., *1971–72 Vegetable Varieties,* University of Alaska Publication No. 30, January 1971, 5 pages.

Directory—Agricultural Organizations in Connecticut, 1971, Bulletin 71-42, 1971, University of Connecticut, Storrs, Connecticut 06286, 19 pages.

Directory of Farmers' Organizations and Marketing Boards in Canada, Canada Department of Agriculture Publication 1365, 1971, 24 pages.

Diseases of Corn in the Midwest, State Universities of Illinois, Indiana, Iowa, Kansas, Michigan, Minnesota, Missouri, Ohio, Nebraska, North Dakota, South Dakota, and Wisconsin, Circular 967, November 1970, 24 pages.

Dodds, Duaine L., *Grazing Systems for Full Season Pastures,* North Dakota State University Circular R-559, May 1971, 4 pages.

Dodds, Duaine L. and Meyer, Dwain W., *Establishment of Dryland and Irrigated Forages,* North Dakota State University Circular R-563, September 1971, 6 pages.

Donahue, Roy L., *Our Soils and Their Management,* 3rd edition, 1970, Interstate Printers and Publishers, Danville, Illinois, 683 pages.

Donahue, Roy L., Shickluna, John C. and Robertson, Lynn S., *Soils: An Introduction to Soils and Plant Growth,* 3rd edition, June 1971, Prentice-Hall, Inc., Englewood Cliffs, New Jersey, 587 pages.

Dubetz, S. and Wilson, D. B., *Growing Irrigated Crops on the Canadian Prairies,* Canada Department of Agriculture, Publication No. 1400, 1969, 25 pages.

Ducks, Canada Department of Agriculture, Publication No. 1349, 1970, 9 pages.

Dunn, L. E., *Characteristics and Reclamation of Nevada Sodic Soils,* University of Nevada Bulletin R-35, January 1968, 36 pages.

Ecology, Price List 88, February 1972, Superintendent of Documents, Government Printing Office, Washington, D. C. 20402, 37 pages.

Effinger, Jerry L., *Land Use Controls,* Background Series 1, 1969, Office of Research and Development, 17 Grant Ave., West Virginia University, Morgantown, West Va. 25606

Epps, Alan C., *Gardening in Southeastern Alaska,* University of Alaska Publication No. 237, February 1971, 10 pages.

Erhardt, W. H., *Home Vegetable Gardening,* University of Maine Bulletin No. 544, February 1968, 19 pages.

Estimated Annual Costs, Production, and Income for Selected Livestock and Crop Enterprises, Eastern West Virginia, West Virginia University Bulletin 5947, June 1970, Morgantown, West Va. 26506, 92 pages.

Family Farm Records, Farmers Bulletin No. 2167, 1968, USDA, Washington, D. C. 20250, 22 pages.

FAO Filmstrips—In 1971 the Food and Agriculture Organization of the United Nations announced for sale a series of filmstrips on food and agriculture. A partial list of these films follows (available from Unipub, Inc., Box 443, New York, New York 10016):

Fish Food for Billions, 41 frames in color.

The New Rice Gives You a Choice, 66 frames in color.

To Survive, 41 frames in color.

The Protein Gap, 98 frames in color.

FAO in the Field, 90 frames in color.

That All the World May Eat, 47 frames in color.

Peasants into Farmers, 43 frames in color.

Extension Work and Teaching Aids, 46 frames in black and white.

Fertilizers in the War Against Hunger, 50 frames in color, double frames.

Fertilizers and Their Use, 62 frames in color, double frames.

Farm Crops Production Technology: A Suggested Post High School Curriculum, United States Department of Health, Education, and Welfare, No. OE-81016, February 1970, 179 pages, United States Government Printing Office, Washington, D. C. 20402.

Father-Son Agreements for Operating Farms, Farmers Bulletin No. 2179, 1970,

USDA, Washington, D. C. 20250, 18 pages.

Figurski, Leo and McReynolds, Kenneth, *Dairying on a Business Basis,* Kansas State University Circular C-339, 1969, 19 pages.

Finishing Beef Cattle, USDA Farmers Bulletin No. 2196, 1971, 30 pages.

Finkner, Ralph E., *Performance of Small Grain Cultivars on the High Plains,* New Mexico State University Bulletin 581, July 1971, 16 pages.

Fisheries in the Food Economy, FAO, 1968, Unipub, Inc., Box 443, New York, New York 10016, 79 pages.

Food. Report of the Second World Food Congress, FAO, The Hague, Netherlands, June 16-30, 1970, Volume 1, Unipub, Inc., Box 443, New York, New York 10016, 141 pages.

Forest Ecology and You, Society of American Foresters, Washington, D. C., 1971, 7 pages.

Foster, Albert B. and Fox, Adrian C., *Teaching Soil and Water Conservation—A Classroom and Field Guide,* USDA, PA-341, August 1970, 30 pages.

Fowl Tick and How to Control It, USDA Leaflet No. 382, February 1970, 5 pages.

Fox, R. T., *How to Make a Terrarium,* Cornell University Bulletin 1029, December 1970, 4 pages.

Freeman, Mervin L., *Answers to Questions About Land and Living in Alaska,* University of Alaska Publication No. 54, April 1967, 16 pages.

Fruit and Vegetable Marketing in Alabama, Auburn University Circular R-25, November 1970, 23 pages.

Gardening on the Contour, USDA Home and Garden Bulletin, No. 179, June 1970, 7 pages.

Garland, S. W., *Establishing a Farm in Canada,* Canada Department of Agriculture, Publication 1403, 1969, 32 pages.

Garrett, Wiley N., and Crawford, Johnny L., *Cotton Seedling Diseases,* University of Georgia Leaflet No. 49, February 1967, 4 pages.

Gay, Charles, *Range Management: Why and How,* New Mexico State University Circular 376, October 1965, 32 pages.

Mexico Range Plants, New Mexico State

Gay, Charles W. and Dwyer, Don D., *New* University Circular 374, August 1970, 86 pages.

Geiszler, G. N., Hoag, Ben K., Bauer, Armand, and Kucera, Henry L., *Influence of Seedbed Preparation on Some Soil Properties and Wheat Yields on Stubble,* Bulletin 488, North Dakota State University, Fargo, North Dakota 58102.

Geologic Quandrangle Maps of Alaska, U. S. Geological Survey, Washington, D. C.

George, Ernest J., *Shelterbelts for the Northern Great Plains,* USDA Farmers Bulletin No. 2109, 1966, 14 pages.

Georgia Agriculture, Georgia Department of Agriculture, Athens, Georgia, 1970, 27 pages.

Georgia's Beef Cattle Improvement Program, University of Georgia Circular 615, March 1970, 6 pages.

Goodspeed, Allen W. and Bell, John F., *The Practice of Forestry on the Island Creek Experimental Forest,* West Virginia University Bulletin 587, February 1970, 49 pages.

Graffis, D. W., Ross, G. L., and Dillon, J. E., *1970 Performance of Corn Hybrids in Illinois,* University of Illinois Circular 1032, January 1971, 52 pages.

Graham, Donald R., *Cattle, Hay and Soil—Relationships Between Cattle Nutrition, Hay Quality, and Soil Fertility,* Montana State University Bulletin 632, December 1970, 23 pages.

Grasses and Legumes for Soil Conservation in the Pacific Northwest and Great Basin States, USDA Handbook No. 339, April 1968, 69 pages.

Graves, Charles R., *1970 Performance of Field Crop Varieties,* University of Tennessee Bulletin 474, January 1971, 90 pages.

Gray, James A. and Groff, Jack L., *Keys to Finewool Sheep Production,* Texas A&M University Circular L-944, December 1970, 3 pages.

Gray, James A. and Groff, Jack L., *Keys to Profitable Angora Goat Production,* Texas A&M University Leaflet L-909, October 1970, 3 pages.

Gray, J. R. and Stucky, H. R., editors, *New*

Mexico Agriculture—1970, New Mexico State University Research Report 195, June 1971.

Greene, E. L., Rose, G. N., and Brooke, D. L., *Location of Agricultural Production in Florida,* University of Florida Bulletin 733, November 1969, 62 pages.

Griffing, M. E. and Thacker, W., *Enterprise Cost Report for Combination Irrigated Dryland Farm,* Montana State University Circular 1126, January 1971, 13 pages.

Grizzell, Roy A., Jr., Dillon, Alan W., Jr., and Sullivan, Edward G., *Catfish Farming —A New Farm Crop,* USDA Farmers Bulletin No. 2244, November 1969, 22 pages.

Hamor, Wade H., Uhlig, Hans G., and Compton, Lawrence V., *Ponds and Marshes for Wild Ducks on Farms and Ranches in the Northern Plains,* USDA Farmers Bulletin No. 2234, August 1968, 16 pages.

Hanway, J. J., Herrick, J. B., Willrich, T. L., Bennett, P. C. and McCall, J. T., *The Nitrate Problem,* Iowa State University Special Report No. 34, August 1963, 20 pages.

Hardin, James W., *Guide to the Literature on Plants of North Carolina,* North Carolina State University Miscellaneous Extension Publication No. 66, March 1971, 16 pages.

Harper, Harold B., *Instructions on Land Judging in Kansas,* Kansas State University Bulletin MF-224, December 1970, 15 pages.

Harter, Walter G. and Golden, W. I., *Catfish for Extra Farm Profit,* University of Georgia Circular 616, April 1970, 2 pages.

Harvey, L. H., *Performance of Cotton Varities in South Carolina,* Clemson University Circular 518, February 1971, 21 pages.

Hayes, Jack, editor, "A Good Life for More People," *The Yearbook of Agriculture 1971,* House Document No. 29, 1971, 391 pages. For sale by Superintendent of Documents, Washington, D. C., 20402.

Hayes, William A., *Mulch Tillage in Modern Farming,* USDA Leaflet No. 554, February 1971, 7 pages.

Hazardous Occupations in Agriculture for 4-H Members 14 to 16 years of Age, South Dakota State University, E.M.C. 593, April 1970, 63 pages.

Heid, Walter G., Jr., *Cost and Returns of Montana Dryland Wheat Production,* Montana State University Bulletin 653, November 1970, 25 pages.

Heinrichs, D. H., *Alfalfa in Canada,* Canada Department of Agriculture Publication No. 1377, 1969, 28 pages.

Helfman, Elizabeth S., *This Hungry World,* Lothrop, Lee, & Shephard Co., New York, 1970, 160 pages.

Herpich, Russell L., *Irrigating Soybeans,* Kansas State University Bulletin 1-5, December 1969, 3 pages.

Hill, Russell G. and Dersch, Eckhart, *Watersheds for Water Management,* Extension Bulletin 364, March 1971, Michigan State University, East Lansing, Michigan 48823, 14 pages.

Hohn, Charles M. and Wallace, E. H., *ABC's of Making Adobe Brick,* New Mexico State University Circular 429, February 1971, 4 pages.

Holyoke, V. H., Arno, J. R., and Hutchinson, F. E., *Soil Resources of Maine,* Bulletin 546, May 1968, 28 pages.

Hopen, H. J., Courter, J. W., Vandemark, J. S. and Nelson, W. R., Jr., *Vegetables for Minigardens,* University of Illinois Circular 1036, March 1971, 7 pages.

Hopp, Richard J., Way, Winston A. and Wiggans, Samuel C., *The Home Vegetable Garden,* University of Vermont Circular 138, September 1970, 27 pages.

Horn, Anton S., *Strawberry Growing in Idaho,* University of Idaho Bulletin No. 440, June 1970, 19 pages.

Horse Science, 4-H Horse Program, unnumbered, undated, 46 pages. Available from each State Extension Director.

Housing and Equipment for Sheep, USDA Farmers Bulletin No. 2242, 1970, 12 pages.

Hull, Jerome, Jr., *Strawberries in Home Gardens,* Michigan State University Extension Bulletin 521, June 1971, 6 pages.

Hutchinson, F. E., *The Nature of Water Pollution and its Relevance to Maine,*

Bulletin 561, March 1971, University of Maine, Orono, Maine, 18 pages.

Insecticides and Their Uses in Minnesota—1972, University of Minnesota Extension Bulletin 263, 1972, 43 pages.

Jackobs, J. A., Peck, T. R., and Walker, W. M., *Efficiency of Fertilizer Topdressings on Alfalfa,* University of Illinois Bulletin 738, December 1970, 22 pages.

Jackson, James E., *Soybeans in Georgia,* University of Georgia Bulletin 639, February 1970, 37 pages.

Jacobs, Hyde S. and Whitney, David A., *Determining Water Quality for Irrigation,* Kansas State University C-396, June 1971, 8 pages.

Jarvis, G. Jennings, *Ponds for Trout,* New Mexico State University Bulletin 400 B-106, March 1967, 2 pages.

Jaycox, Elbert R., *Pollination of Legume Seed in Illinois,* University of Illinois Circular 1039, May 1971, 5 pages.

Jefferies, Ned W., *4-H Leader's Range Management Guide,* Montana State University Bulletin 1026, April 1969, 20 pages.

Jefferies, Ned W., *Livestock Range—It's Nature and Use. Part III—Plant Relationships,* Montana State University Bulletin 1015, July 1968, 17 pages.

Jefferies, Ned W., *Livestock Range—It's Nature and Use. Part IV—A Management Plan,* Montana State University Bulletin 1018, September 1969, 10 pages.

Jefferies, Ned W., *Livestock Range—It's Nature and Use. Part V—Range Operation,* Montana State University Bulletin 1029, July 1971, 16 pages.

Jenkins, George H., Jr., *Septic Tanks for Homes,* University of Kentucky Circular 431-A, June 1971, 17 pages.

Johnson, P. R. and Hartman, C. W., *Environmental Atlas of Alaska,* University of Alaska, Environmental Engineering and Institute of Water Resources, College, Alaska, 1969.

Johnson, W. T., Dewey, J. E. and Kastl, H. J., editors, *A Guide to Safe Pest Control Around the Home,* Cornell University Miscellaneous Bulletin 74, 40 pages.

Jones, D. D., Day, D. L. and Dale, A. C., *Aerobic Treatment of Livestock Wastes,* University of Illinois Bulletin 737, April 1971, 55 pages.

Jones, J. R., *Swine Feeding for Profits,* North Carolina State University Extension Folder 281, October 1971, Raleigh, North Carolina, 27607, 6 pages.

Kahl, William C., editor, *Pollution: Problems, Projects, and Mathematical Exercises, Grades 6–9,* Department of Public Instruction, 126 Langdon St., Madison, Wisconsin 53702, 1970, 84 pages.

Kearney, Harold M., *Think Ahead: A High School Student's Guide to a Bright Future,* University of Maine Bulletin 552, September 1970, Orono, Maine 04473, 14 pages.

Keys to Profitable Range Management in Texas, Texas A&M University Bulletin MP-965, March 1971, 11 pages.

Keys to Profitable Rice Production, Texas A&M University Leaflet L-894, January 1970, 4 pages.

King, Claude L. and Willis, William G., *Soybean Diseases in the Midwest,* Kansas State University Bulletin MF-166, September 1966.

Kingsbury, John M., *Common Poisonous Plants,* Cornell University Bulletin 583, August 1971, 32 pages.

Kingsbury, John M., *Poison Ivy, Poison Sumac, and other Rash-Producing Plants,* Cornell University Bulletin 1154, July 1971, 16 pages.

Kraenzel, Carl F. and Macdonald, Frances H., *Social Forces in Rural Communities of Sparsely Populated Areas,* Montana State University Bulletin 647, February 1971, 17 pages.

Landforce, Andrew S., *Care and Culture of Earthworms,* Oregon State University Bulletin FS 140, February 1970, 2 pages.

Land Use and Wildlife Resources, National Resource Council, Washington, D. C. 20418, 1970, 262 pages.

Larsen, John E., *Planning Your Vegetable Garden,* Texas A&M University Fact Sheet L-911, March 1971, 4 pages.

Leedy, Clark D., *Control Iron Chlorosis,* New Mexico State University Circular 421, 4 pages.

Lentz, Austin N. and Hetzell, Albert A., A

Guide to Nature Study in New Jersey, Rutgers University Leaflet No. 433-A, January 1971, 16 pages.

List of Available Publications of the United States Department of Agriculture, July 1971, USDA, Washington, D. C. 20250.

Livestock on Small Farms, Canada Department of Agriculture Publication No. 1381, 1969, 24 pages.

Lodge, R. W., Campbell, J. B., Smoliak, S., and Johnston, A., *Management of the Western Range,* Canada Department of Agriculture Publication No. 1425, 1971, 34 pages.

Long, Roger B. and McEldowney, Bernard E., *Storage Facilities and Marketing Costs for Wheat in Southern Idaho,* University of Idaho Bulletin 515, June 1970, Moscow, Idaho 83843, 25 pages.

Magness, J. R., Markle, G. M., and Compton, C. C., *Food and Feed Crops for the United States,* New Jersey Agriculture Experiment Station, New Brunswick, New Jersey, 08903, Interregional Research Project IR-4, IR Bulletin No. 1 and No. 828, 1971, 255 pages.

Mark, G. G. and Dimmick, R. S., *Managing the Family Forest,* USDA Farmers Bulletin 2187, 1971, 61 pages, Superintendent of Documents, Washington, D. C., 20402.

Marketing: a Dynamic Force in Agricultural Development, 1970, FAO, Unipub, Inc., Box 443, New York, New York 10016, 40 pages.

Marr, Charles, *Kansas Garden Guide,* Kansas State University Circular C-436, February 1971, 23 pages.

Martin, E. C., *Basic Beekeeping,* Michigan State University Bulletin E625, September 1968, 12 pages.

Martin, James P. and Waksman, Selman A., *Synthetic Manure,* Rutgers University Circular 470, July 1971, 8 pages.

Massey, P. H., Jr., McNeil, Marshall, and Wilkins, L. B., *Plastic Greenhouses,* Circular 760, Revised, February 1960, Virginia Polytechnic Institute, Blacksburg, Virginia.

McAlister, J. T., *Mulch Tillage in the Southeast,* USDA Leaflet No. 512, June 1971, 8 pages.

McCall, Wade W., Shigeura, Gordon T., and Tamimi, Yusuf N., *Windbreaks for Hawaii,* University of Hawaii Circular 438, January 1970, 10 pages.

McCall, Wade W. and Nakawa Yukio, *Growing Plants Without Soil,* University of Hawaii Circular 400, March 1970, 20 pages.

McClain, E. F. and Jutras, M. W., *Performance of Forage Sorghum Hybrids for Silage Potential in the Piedmont of South Carolina, 1965–1970,* Clemson University Circular 521, March 1971, 16 pages.

McGill, J. Frank and Samples, L. E., *Peanuts in Georgia, Growing Peanuts in Georgia—A Package Approach,* University of Georgia Bulletin 640, April 1969, 39 pages.

McNeil, Richard J., *Deer in New York State,* Cornell University Extension Bulletin 1189, September 1969, 23 pages.

McNeill, Harold, *Perspectives on Air Pollution with Emphasis on Solid Waste Management,* Appalachian Center Information Series 2, 1970, Office of Environmental Development, 207 Coliseum, West Virginia University, Morgantown, West Virginia 26506.

Meggitt, William F., *Weed Control in Field Crops,* Michigan State University Bulletin E-434, May 1970, 8 pages.

Melvin, Stewart W., *How to Comply With Iowa's Feedlot Runoff Control Regulations,* Iowa State University Bulletin PM-511, June 1971, 4 pages.

Miller, James F. and Swann, Charles W., *Weed Control in Field Crops,* University of Georgia Bulletin 692, October 1971, 15 pages.

Miner, Ronald J., editor, *Farm Animal Waste Management,* North Central Regional Publication No. 206, Special Report No. 67, May 1971, 44 pages.

Minigardens, Montana State University Folder 115, May 1970, 4 pages.

Minigardens for Vegetables, USDA Home and Garden Bulletin 163, 1971, 12 pages.

Mitchell, W. H., Rahn, E. M., and McDaniel, John, *Chemical Weed Control in Field Crops,* University of Delaware, January 1971, 7 pages. (Unnumbered publication).

Moak, James E., *Forestry: It's Economic Importance,* Mississippi State University Bulletin 785, March 1971, 27 pages.

Montana's Water Conservancy Law—What it is—How it Works—What it Means, Montana State University Circular 295, February 1970, 8 pages.

Morrison, Denton E., editor, *Farmers' Organizations and Movements,* Michigan State University Research Bulletin 24, and North Central Regional Research Publication No. 200, 1970, 116 pages.

Morrison, Frank D., Willis, William G., and Brooks, Leroy, *Fruit Pest Control in Home Fruit Planting,* Kansas State University Leaflet L-254, July 1971.

Morrow, Robert R., Hamilton, Lawrence S., and Winch, Fred E., Jr., *Planting Forest Trees in Rural New York,* Cornell University Bulletin 1161, March 1971, 31 pages.

Muir, M. K., *Nutrient Status Survey of the Yellowstone Drainage of Montana,* Montana State University Bulletin 642, February 1971, 19 pages.

Mulches for Your Garden, USDA Home and Garden Bulletin No. 185, January 1971, 7 pages.

Musick, J. T. and Dusek, D. D., *Grain Sorghum Row Spacing and Planting Rates Under Limited Irrigation in the Texas High Plains,* Texas A&M University Bulletin MP-392, October 1969.

Myers, Kenneth H., *Facts for Prospective Farmers,* USDA Farmers Bulletin No. 2221, 1971, 22 pages.

Neely, William W. and Davison, Verne E., *Wild Ducks on Farmland in the South,* USDA Farmers Bulletin 2218, March 1971, 14 pages.

Nelson, A. B. and Neumann, A. L., *Alfalfa Hay, Cottonseed Hulls, Cottonseed, Cottonseed Meal, and Trace Minerals in the Rations for Finishing Steers,* New Mexico State University Bulletin 571, November 1970, 17 pages.

Newton, Michael, *Planting Forest Trees in the Pacific Northwest,* Universities of Oregon, Washington, and Idaho, PNW 120, April 1970, 8 pages.

Nitrogen Fertilization of Flue Cured Tobacco, North Carolina State University Extension Folder No. 279, June 1969, 11 pages.

Nolan, C. N., Smith, F. H., and Thomas, C. A., *1971 Peanut Production Guide for South Carolina,* Clemson University Information Card No. 115, February 1971, 6 pages.

Nutrient Disorders in Tree Fruits, Universities of Washington, Oregon, and Idaho, PNW 121, August 1970, 8 pages (color).

Nutritive Value of Foods, USDA Home and Garden Bulletin No. 72, January 1971, 41 pages.

Olson, Gerald W., *Using Soils as Ecological Resources,* Cornell University Information Bulletin 6, Biological Sciences, Agronomy No. 1. (Note: A set of 50 slides that are used to illustrate this bulletin are for sale by Visual Communication Section, Department of Commercial Arts, Roberts Hall, Cornell University, Ithaca, New York 14850).

Organization and Activities of the Canada Department of Agriculture, Canada Department of Agriculture Publication No. 1123, 1970, 70 pages.

Parker, K. G., *Livestock Range—It's Nature and Use. Part I—Range Plants,* Montana State University Bulletin 1014, July 1971, 32 pages.

Parker, K. G., *Livestock Range—It's Nature and Use. Part II—Forage Values,* Montana State University Bulletin 1028, March 1965, 16 pages.

Pastures and Grazing Systems for the Mountains of North Carolina, North Carolina State University Bulletin 437, May 1970, 49 pages.

Peck, Theodore P., *Pollution Control,* University of Minnesota Special Report 35, July 1970, 37 pages.

Pest Control Chart for Tobacco Fields, 1971, University of Connecticut and University of Massachusetts, Cooperating, No. 71-18, March 1971, 2 pages.

Pieper, Rex D., Montoya, James R., and Groce, V. Lynn, *Site Characteristics on Pinyon-Juniper and Blue Grama Ranges in South-Central New Mexico,* New Mexico State University Bulletin 573, January 1971, 21 pages.

Pinkston, Carthel T. and Gray, D. Leroy, *Guidelines for Program of Control Against Rodents,* University of Arkansas Circular 538, June 1971, 15 pages.

Planning for Profit, Circular 518, January 1971, University of North Carolina, Raleigh, North Carolina, 27607, 14 pages.

Plants: How They Improve Our Environment, Soil Conservation Society of America, unnumbered, undated.

Pollack, Bernard L., *New Jersey, 1971, Vegetable Production Recommendations,* Rutgers University Leaflet 437-B, February 1971, 42 pages.

Pretechnical Post High School Programs—A Suggested Guide, Technical Education Program Series No. 12, 1967, U.S. Dept. of Health, Education, and Welfare, Government Printing Office, Washington, D. C., 67 pages.

Principal Stored Grain Insects, Iowa State University PM 463-1, 1970.

Proctor, E. A., *Some Do's and Dont's in Marketing Fruits and Vegetables,* North Carolina State University Extension Folder No. 285, June 1970, 4 pages.

Producing Hay and Pasture for Horses in Michigan, Michigan State University Bulletin E-643, September 1971.

Production and Marketing of Sheep in New England, The, A New England Cooperative Extension Service Publication, October 1970, 49 pages, available from any Extension Service in New England.

Protecting the World Environment in the Light of Population Increase, U. S. Government Printing Office, Washington, D. C., 20402, 1970, 36 pages.

Putnam, A. R., *Chemical Weed Control for Horticultural Crops,* Michigan State University Bulletin 433, February 1971, 12 pages.

Rahn, E. M., Fisher, V. J., and Stevens, R. F., *Chemical Weed Control in Fruit and Vegetable Crops,* University of Delaware Mimeo Circular No. 84, February 1970, 16 pages.

Range Plant Series, Oregon State University. Single sheet descriptions of 77 principal range plants and 37 weeds.

Recommended Chemicals for Weed and Brush Control, Arkansas, 1972, University of Arkansas MP44, January 1972, 41 pages.

Reed, L. B. and Webb, Raymond E., *Insects and Diseases of Vegetables in the Home Garden,* USDA Home and Garden Bulletin No. 46, 50 pages.

Reid, R. L., Post, Amy J., and Jung, G. A., *Mineral Composition of Forages,* West Virginia University Bulletin 589T, February 1970, 35 pages.

Richardson, Curtis W., *Nutrient Requirements and Feed Composition Data,* Oklahoma State University, No. 4002, April 1971, 4 pages.

Roland, Charles, Seigler, W. E., and Goolsby, H. B., *Cotton Defoliation Guide,* University of Georgia Leaflet No. 86, July 1969, 4 pages.

Rossman, E. C. and Darling, Bary M., *Michigan Corn Production Hybrids Compared,* 1971, Michigan State University Bulletin 431, January 1971, 20 pages.

Savos, Milton G. and Schroeder, David B., *1971 Connecticut Vegetable Insect and Disease Control Guide,* University of Connecticut Bulletin 71-24, 1971.

Schermerhorn, R. W. and Cavett, K. D., *Economics of Fee Fishing Ponds,* Oklahoma State University, No. 9003, October 1970, 4 pages.

Schipper, I. A. and Alstad, A. D., *Anthrax,* North Dakota State University Circular A-561, August 1971, Fargo, North Dakota 58102, 4 pages.

Schoenfeld, Clay, editor, *Outlines of Environmental Education,* Dembar Educational Research Services, Inc., Madison, Wisconsin 43701, 1971, 256 pages.

Scholz, Earl W. and Askew, Robert G., *Improved Methods for Growing Tomatoes and Muskmelons in North Dakota,* North Dakota State University Circular H-555, March 1971, 4 pages.

Scott, H. E., Brett, Charles H., and Sorensen, K. A., *Insect Control for Vegetable Gardeners,* North Carolina State University Leaflet No. 169, February 1971, 16 pages.

Selecting Fertilizers for Lawns and Gardens, USDA Home and Garden Bulletin No. 89, June 1971, 7 pages.

Sheldrake, Raymond, Jr., *Planning, Con-*

structing, and Operating Plastic-Covered Greenhouses, Cornell University Miscellaneous Publication 72, February 1969, 15 pages.

Sidehill Gardening, University of Alaska FS-5, undated, 2 pages.

Six-year Experience with Point-Sample Cotton Insect Scouting, University of Arkansas Bulletin 754, April 1970, 39 pages.

Skog, Roy E., Koelling, Melvin R., and Bell, Lester E., Forests and the Environment, Michigan State University Extension Bulletin E-716, October 1971, 15 pages.

Slaughtering, Cutting and Processing Pork on the Farm, USDA Farmers Bulletin No. 2138, 1971, Washington, D. C. 20250, 48 pages.

Smaller Farmlands Can Yield More: Raising Agricultural Productivity by Technological Change, FAO, 1969, Unipub, Inc., Box 443, New York, New York 10016, 73 pages.

Smiley, William Lea, Forest Farming, University of Arkansas Circular 459, August 1971, 67 pages.

Smith, S. E. and Loosli, J. L., The Mineral and Vitamin Requirements of Livestock, Cornell University Extension Bulletin 1149, May 1970, 15 pages.

Smith, W. A. and Fate, D. W., Keys to Profitable Timber Production, Texas A&M University Leaflet L-923, December 1970, 4 pages.

Smith, Wil J., editor, The Poor and the Hard-Core Unemployed—Recommendations for New Approaches, 1970, Office of Research and Development, 17 Grant Ave., West Virginia University, Morgantown, West Virginia 26506.

Sod Industry Research, 1969–1970, Michigan State University Research Report 120, September 1970, 12 pages.

Soils Handbook, Miscellaneous Publication 383, April 1970, University of Kentucky, Lexington, Kentucky 40506, 41 pages.

Spivey, C. D., Blackberries, University of Georgia Leaflet 107, November 1970, 5 pages.

Splett, Philip J. and Miles, William R., Trees and Our Environment, University of Minnesota Extension Folder 253, 1971, 10 pages.

Spurgeon, W. I. and Cooke, Fred T., Jr., Cost Reduction Research for Cotton Production Systems in the Yazoo-Mississippi Delta, Mississippi State University Bulletin 783, January 1971, 12 pages.

Stanford, G., England, C. B., and Taylor, A. W., Fertilizer Use and Water Quality, ARS-41-168, October 1970, Agricultural Research Service, USDA, Washington, D. C. 20705, 19 pages.

Staten, Glen, Breeding Acala 1517 Cottons, 1926 to 1970, New Mexico State University, Memoir Series No. 4, 1971.

Steger, Robert E., Control Grass Tetany in Livestock, New Mexico State University Circular 400-B-809, June 1971, Las Cruces, New Mexico 88001.

Steger, Robert E., General Considerations for Noxious Plant Control on New Mexico Ranges, New Mexico State University Bulletin 400-B-806, February 1970, 3 pages.

Steger, Robert E., Grazing Systems for Range Care, New Mexico State University Circular 427, December 1970, 9 pages.

Steger, Robert E., Nitrate Poisoning from Forage, New Mexico State University Bulletin 400-B-807, March 1971, 3 pages.

Steger, Robert E., Prussic Acid Poisoning, New Mexico State University Bulletin 400-B-808, March 1971, 2 pages.

Stoner, Allan K., Growing Tomatoes in the Home Garden, USDA Home and Garden Bulletin No. 180, 1971, 12 pages.

Sudangrass and Sorghum—Sorghum-Sudangrass Hybrids for Forage, USDA Farmers Bulletin 2241, 1969, 12 pages.

Summary of Illinois Farm Business Records, 1970, Circular 1040, 1970, University of Illinois, Champaign, Illinois 61803, 23 pages.

Sutherland, Douglas, Sources of Information for Young Entomologists, University of Idaho Bulletin 519, September 1970, 11 pages.

Sutter, R. J. and Corey, G. L., Consumptive Irrigation Requirements for Crops in Idaho, University of Idaho Bulletin 516, July 1970, 97 pages.

Teague, Richard D., A Manual of Wildlife Conservation, The Wildlife Society,

Washington, D. C. 20016, 1971, 206 pages.

Texas Guide for Controlling External Parasites of Livestock and Poultry, Texas A&M University Bulletin MP-691, February 1971, 43 pages.

Thomas, W. B. and Tanksley, T. D., Jr., *Keys to Profitable Swine Production,* Texas A&M University Bulletin MP-953, January 1971, 8 pages.

Thorne, Wynne, *Agricultural Production in Irrigated Areas* and *Arid Lands in Transition,* American Association for the Advancement of Science, 1970, pages 31-56.

Tobacco, A Sensitive Monitor for Photochemical Air Pollution, U. S. Department of Health, Education, and Welfare Publication No. AP-55, June 1969, 23 pages.

Topoleski, Leonard D., *The Home Vegetable Garden,* Cornell University Extension Bulletin 1191, March 1971, 39 pages.

25 Technical Careers You Can Learn in Two Years or Less, U. S. Office of Education, The Conference Board and The Manpower Institute, Washington, D. C. (Department of Health, Education, and Welfare, Washington, D. C. 20201.)

Wade, Larkin and Buford, James A., Jr., *Forest Products Marketing in Alabama,* Auburn University Circular R-22, November 1970, 26 pages.

Wadleigh, Cecil H., *Wastes in Relation to Agriculture and Forestry,* USDA (Miscellaneous Publication No. 1065, March 1968, 112 pages.

Waelti, John J., *Understanding the Water Quality Controversy in Minnesota,* University of Minnesota Extension Bulletin 359, 1970, 27 pages.

Wagner, Richard H., *Environment and Man,* W. W. Norton and Co., 55 Fifth Avenue, New York, New York 10003, 1971, 491 pages.

Waldo, George F. and Bringhurst, Royce S., *Commercial Strawberry Growing in the Pacific Coast States,* USDA Farmers Bulletin 2236, 1971, 22 pages.

Walker, Robert D., *How Food Production Affects the Environment,* University of Illinois Circular 1037, June 1971, 5 pages.

Warnick, A. C., Koger, M., Martinez, A., and Cunha, T. J., *Productivity of Beef Cows As Influenced by Pasture and Winter Supplement During Growth,* University of Florida Bulletin 695, May 1969, 15 pages.

Way, Winston A., *Forage Crop Guide,* University of New Hampshire and University of Vermont Bulletin NEC-62, 1970, 4 pages.

What You Should Know About Coarse Grains, Canada Department of Agriculture Publication No. 1410, 1970, 11 pages.

What You Should Know About Seeds, Canada Department of Agriculture Publication No. 1412, 1970, 9 pages.

What You Should Know About Wheat, Canada Department of Agriculture Publication No. 1386, 1969, 12 pages.

Where and How to Get a Farm. Some Questions and Some Answers," Leaflet No. 432, August 1971, USDA, Washington, D. C. 20250, 8 pages.

Whittenburg, Bob, *Livestock Judging for Alabama 4-H Club Members,* Leaflet YA-13, September 1967, Auburn University, Auburn, Alabama 36830, 23 pages.

Williams, Harry E., *You Can Control Garden Insects,* University of Tennessee Publication 595, April 1971, 11 pages.

Wilson, Esther, *Your Food—and What it Means to You,* University of Idaho Bulletin No. 508, April 1971, Moscow, Idaho 83843, 8 pages.

Wilson, T. V. and Snell, A. W., *Irrigation of Corn and Cotton in South Carolina,* Clemson University Bulletin 540, August 1968, 15 pages.

Winkelblech, Carl S., *Farm Ponds in New York,* Cornell University Bulletin 949, April 1971, 31 pages.

Wilkins, Bruce T., *Using Your Land for Pleasant Living,* Cornell University Bulletin 1211, June 1970, 11 pages.

Your Future as an Extension Agent, Kansas State University, Manhattan, Kansas 66502, MF-94, December 1968.

Zeller, Frederick A. and Smith, Wil J., *Economic Development in West Virginia,* Information Series 6, 1968, Office of Research and Development, 17 Grant Ave., West Virginia University, Morgantown, West Virginia 26506.

YEARBOOKS OF AGRICULTURE

The following *Yearbooks of Agriculture,* when available, are for sale by the Superintendent of Documents, Government Printing Office, Washington, D. C. 20402.

Science in Farming, 1943–1947
Grass, 1948
Trees, 1949
Crops in Peace and War, 1950–1951
Insects, 1952
Plant Diseases, 1953
Marketing, 1954
Water, 1955
Animal Diseases, 1956
Soil, 1957
Land, 1958
Food, 1959
Power to Produce, 1960
Seeds, 1961
After a Hundred Years, 1962
A Place to Live, 1963
Farmer's World, 1964
Consumers All, 1965
Protecting Our Food, 1966
Outdoors, U.S.A., 1967
Science for Better Living, 1968
Food for Us All, 1969
Contours of Change, 1970
A Good Life for More People, 1971
Landscape for Living, 1972

STATE EXTENSION SERVICE DIRECTORS

ALABAMA — Auburn University, Auburn 36830

ALASKA — University of Alaska, College 99735

ARIZONA — University of Arizona, Tucson 85721

ARKANSAS — P. O. Box 391, Little Rock 72203

CALIFORNIA — 2200 University Avenue, Berkeley 94720

COLORADO — Colorado State University, Fort Collins 80521

CONNECTICUT — University of Connecticut, Storrs 06268

DELAWARE — University of Delaware, Newark 19711

DISTRICT OF COLUMBIA — Federal City College, 1424 K St., N. W., Washington, D. C. 20005.

FLORIDA — University of Florida, Gainesville 32603

GEORGIA — University of Georgia, Athens 30601

HAWAII — University of Hawaii, Honolulu 96822

IDAHO — University of Idaho, Moscow 83843

ILLINOIS — University of Illinois, Urbana 61803

INDIANA — Purdue University, Lafayette 47907

IOWA — Iowa State University of Science and Technology, Ames 50010

KANSAS — Kansas State University, Manhattan 66504

KENTUCKY — University of Kentucky, Lexington 40506

LOUISIANA — Louisiana State University, Baton Rouge 70803

MAINE — University of Maine, Orono 04473

MARYLAND — University of Maryland, College Park 20742

MASSACHUSETTS — University of Massachusetts, Amherst 01003

MICHIGAN — Michigan State University, East Lansing 48823

MINNESOTA — University of Minnesota, St. Paul 55101

MISSISSIPPI — Mississippi State University, State College 39762

MISSOURI — University of Missouri, Columbia 65202

MONTANA — Montana State University, Bozeman 59715

NEBRASKA — University of Nebraska, Lincoln 68503

NEVADA — University of Nevada, Reno 89507

NEW HAMPSHIRE — University of New Hampshire, Durham 03824

NEW JERSEY — Rutgers—The State University, Box 231, New Brunswick 08903

NEW MEXICO — New Mexico State University, Las Cruces 88001

NEW YORK — New York State College of Agriculture, Ithaca 14850

NORTH CAROLINA — North Carolina State University, Raleigh 27607

NORTH DAKOTA — North Dakota State University, Fargo 58103

OHIO — Ohio State University, 2120 Fyffe Road, Columbus 43210

OKLAHOMA — Oklahoma State University, Stillwater 74075

OREGON — Oregon State University, Corvallis 97331

PENNSYLVANIA — The Pennsylvania State University, University Park 16802

PUERTO RICO — University of Puerto Rico, Rio Piedras 00927

RHODE ISLAND — University of Rhode Island, Kingston 02881

SOUTH CAROLINA — Clemson University, Clemson 29631

SOUTH DAKOTA — South Dakota State University, Brookings 57007

TENNESSEE — University of Tennessee, P.O. Box 1071, Knoxville 37901

TEXAS — Texas A&M University, College Station 77841

UTAH — Utah State University, Logan 84321

VERMONT — University of Vermont, Burlington 05401

VIRGINIA — Virginia Polytechnic Institute, Blacksburg 24061

VIRGIN ISLANDS — Box 166, Kingshill, St. Croix 00850

WASHINGTON — Washington State University, Pullman 99163

WEST VIRGINIA — 294 Coliseum, West Virginia University, Morgantown 26506

WISCONSIN — University of Wisconsin, 432 N. Lake St., Madison 53706

WYOMING — University of Wyoming, Box 3354, University Station, Laramie 82071

SUGGESTIONS FOR ORGANIZING A COUNTY LIBRARY

1. Write to your state library extension agency (addresses below).
2. Talk to other people you think would be interested.
3. Organize an active committee from the whole county or region.
4. Invite a worker from the state library extension agency to a county meeting.

5. When all the steps have been taken to inform people, draw up a definite plan for an appropriation with the officials of the county or region.
6. Obtain formal action on the plan by the necessary governing body.

STATE LIBRARY EXTENSION AGENCIES

ALABAMA — Public Library Service, State of Alabama, Montgomery

ALASKA — State Library, Alaska Office Bldg., Juneau

ARIZONA — Department of Library and Archives, Phoenix

ARKANSAS — Arkansas Library Commission, 506½ Center St., Little Rock

CALIFORNIA — State Library, Library-Courts Bldg., Sacramento

COLORADO — State Library, Department of Education, 32 State Services Bldg., Denver

CONNECTICUT — Bureau of Library Services, Department of Education, P.O. Box 2219, Hartford

DELAWARE — Library Commission for the State of Delaware, Dover

FLORIDA — State Library, Tallahassee

GEORGIA — Division of Instructional Materials and Library Service, Department of Education, Atlanta

GUAM — Nieves M. Flores Memorial Library, Agana, Guam

HAWAII — Library of Hawaii, Honolulu.

IDAHO — Idaho State Library, 615 Fulton St., Boise

ILLINOIS — State Library, Centennial Memorial Bldg., Springfield

INDIANA — State Library, Indianapolis

IOWA — State Traveling Library, Historical Bldg., Des Moines

KANSAS — Traveling Libraries Commission, 801 Harrison, Topeka

KENTUCKY — Department of Libraries, Box 87, Berry Hill, Frankfort

LOUISIANA — State Library, Baton Rouge.

MAINE — State Library, Augusta

MARYLAND — Division of Library Extension, Department of Education, 301 W. Preston St., Baltimore

MASSACHUSETTS — Division of Library Extension, Department of Education, 200 Newbury St., Boston

MICHIGAN — State Library, 125 E. Shiawassee St., Lansing

MINNESOTA — Library Division, Department of Education, State Office Annex, St. Paul

MISSISSIPPI — Library Commission, 405 State Office Bldg., Jackson

MISSOURI — State Library, State Office Bldg., Jefferson City

MONTANA — State Library Commission, South Ave. and Middlesex, Missoula

NEBRASKA — Nebraska Public Library Commission, Lincoln

NEVADA — Nevada State Library, Carson City

NEW HAMPSHIRE — State Library, 20 Park St., Concord

NEW JERSEY — Division of the State Library, Archives and History, Department of Education, State House Annex, Trenton

NEW MEXICO — State Library, P.O. Box 1666, Santa Fe

NEW YORK — New York State Library, State Department of Education, Albany

NORTH CAROLINA — State Library, Raleigh

NORTH DAKOTA — State Library Commission, Liberty Memorial Bldg., Bismarck

OHIO — State Library, State Office Bldg., Columbus

OKLAHOMA — Oklahoma State Library, 109 State Capitol, Oklahoma City

OREGON — State Library, Salem

PENNSYLVANIA — State Library, Education Bldg., Harrisburg

PUERTO RICO — Library Division, Department of Education, Carnegie Library, San Juan

RHODE ISLAND — State Library, Providence

SOUTH CAROLINA — State Library Board, 1001-07 Main St., Columbia

SOUTH DAKOTA — State Library Commission, 322 South Fort St., Pierre

TENNESSEE — Public Libraries Division, State Library and Archives, Nashville

TEXAS — State Library, Austin

UTAH — Utah State Library, 609 East South Temple, Salt Lake City

VERMONT — Free Public Library Commission, State Library Bldg., Montpelier

VIRGINIA — Virginia State Library, Richmond

VIRGIN ISLANDS — Libraries and Museums, Department of Education, Box 390, Charlotte Amalie, St. Thomas

WASHINGTON — State Library, Olympia

WEST VIRGINIA — Library Commission, 2004 Quarrier St., Charleston

Wisconsin — Wisconsin Free Library Commission, Madison

WYOMING — State Library, Supreme Court Bldg., Cheyenne

REGIONAL OFFICES OF THE UNITED STATES FOREST SERVICE

Northern Region
Federal Building
Missoula, Montana 59801

Rocky Mountain Region
Federal Center, Building 85
Denver, Colorado 80225

Southwestern Region
517 Gold Avenue S. W.
Albuquerque, New Mexico 87101

Intermountain Region
324 25th Street
Ogden, Utah 84401

California Region
630 Sansome Street
San Francisco, California 94111

Pacific Northwest Region
Post Office Box 3623
Portland, Oregon 97208

Eastern Region
633 West Wisconsin Avenue
Milwaukee, Wisconsin 53203

Southern Region
1720 Peachtree Road N. W.
Atlanta, Georgia 30309

Alaska Region
Post Office Box 1628
Juneau, Alaska 99801

LAND-GRANT UNIVERSITIES AND COLLEGES

All requests for publications and information from the land-grant universities and colleges should be made by the teacher.

ALABAMA — Auburn University, Auburn 36830

ALASKA — University of Alaska, College 99735

ARIZONA — University of Arizona, Tucson 85721

ARKANSAS — University of Arkansas, Fayetteville 72701

CALIFORNIA — University of California, Berkeley 94720

COLORADO — Colorado State University, Fort Collins 80521

CONNECTICUT — University of Connecticut, Storrs 06268

DELAWARE — University of Delaware, Newark 19711

FLORIDA — University of Florida, Gainesville 32603

GEORGIA — University of Georgia, Athens 30601

HAWAII — University of Hawaii, Honolulu 96822

IDAHO — University of Idaho, Moscow 83843

ILLINOIS — University of Illinois, Urbana 61803

INDIANA — Purdue University, Lafayette 47907

IOWA — Iowa State University of Science and Technology, Ames 50010

KANSAS — Kansas State University, Manhattan 66504

KENTUCKY — University of Kentucky, Lexington 40506

LOUISIANA — Louisiana State University, Baton Rouge 70803

MAINE — University of Maine, Orono 04473

MARYLAND — University of Maryland, College Park 20742

MASSACHUSETTS — University of Massachusetts, Amherst 01003

MICHIGAN — Michigan State University, East Lansing 48823

MINNESOTA — University of Minnesota, University Farm, St. Paul 55101

MISSISSIPPI — Mississippi State University, State College 39762

MISSOURI — University of Missouri, Columbia 65202

MONTANA — Montana State University, Bozeman 59715

NEBRASKA — University of Nebraska, Lincoln 68503

NEVADA — University of Nevada, Reno 89507

NEW HAMPSHIRE — University of New Hampshire, Durham 03824

NEW JERSEY — Rutgers University, New Brunswick 08903

NEW MEXICO — New Mexico State University, Las Cruces 88001

NEW YORK — Cornell University, Ithaca 14850

NORTH CAROLINA — University of North Carolina, Raleigh 27607

NORTH DAKOTA — North Dakota State University, Fargo 58103

OHIO — Ohio State University, Columbus 43210

OKLAHOMA — Oklahoma State University, Stillwater 74075

OREGON — Oregon State University, Corvallis 97331

PENNSYLVANIA — Pennsylvania State University, University Park 16802

PUERTO RICO — University of Puerto Rico, Rio Piedras 00927

RHODE ISLAND — University of Rhode Island, Kingston 02881

SOUTH CAROLINA — Clemson University, Clemson 29631

SOUTH DAKOTA — South Dakota State University, Brookings 57007

TENNESSEE — University of Tennessee, Knoxville 37901

TEXAS — Texas A&M University, College Station 77841

UTAH — Utah State University, Logan 84321

VERMONT — University of Vermont, Burlington 05401

VIRGINIA — Virginia Polytechnic Institute, Blacksburg 24061

WASHINGTON — Washington State University, Pullman 99163

WEST VIRGINIA — West Virginia University, Morgantown 26506

WISCONSIN — University of Wisconsin, Madison 53706
WYOMING — University of Wyoming, Laramie 82071

SOIL CONSERVATION SERVICE STATE OFFICES

The following State Offices will supply publications, motion pictures, charts, and posters on soil and water conservation and insofar as possible furnish speakers, tour guides, and other assistance to schools in their conservation work.

State Conservationist
138 South Gay St.
P.O. Box 311
Auburn, Alabama 36830

State Conservationist
P.O. Box F
Palmer, Alaska 99645

State Conservationist
230 N. First Ave.
Phoenix, Arizona 85025

State Conservationist
5401 Federal Office Bldg.
Little Rock, Arkansas 72201

State Conservationist
Tioga Bldg., Room 203
2020 Milvia St.
Berkeley, California 94704

State Conservationist
Box 17107
Denver, Colorado 80217

State Conservationist
Mansfield Professional Park
Storrs, Connecticut 06268

State Conservationist
9 East Loockerman St.
Dover, Delaware 19901

State Conservationist
P.O. Box 1208
Gainesville, Florida 32601

State Conservationist
Old Post Office Bldg.
P.O. Box 832
Athens, Georgia 30601

State Conservationist
440 Alexander Young Bldg.
Honolulu, Hawaii 96813

State Conservationist
304 North 8th St.
Boise, Idaho 83702

State Conservationist
200 W. Church St.
P.O. Box 678
Champaign, Illinois 61820

State Conservationist
5610 Crawfordsville Rd.
Indianapolis, Indiana 46224

State Conservationist
210 Walnut St.
Des Moines, Iowa 50309

State Conservationist
760 S. Broadway, Box 600
Salina, Kansas 67401

State Conservationist
1409 Forbes Road
Lexington, Kentucky 40505

State Conservationist
3737 Government St.
P.O. Box 1630
Alexandria, Louisiana 71301

State Conservationist
University of Maine
U.S.D.A. Bldg.
Orono, Maine 04473

State Conservationist
4321 Hartwick Rd.
College Park, Maryland 20740

State Conservationist
27-29 Cottage St.
Amherst, Massachusetts 01002

State Conservationist
1405 S. Harrison Road
East Lansing, Michigan 48823

State Conservationist
316 N. Robert St.
St. Paul, Minnesota 55101

State Conservationist
Milner Bldg.
Lamar & Pearl Sts., Box 610
Jackson, Mississippi 39205

State Conservationist
P.O. Box 459
Columbia, Missouri 65201

State Conservationist
Federal Bldg.
P.O. Box 970
Bozeman, Montana 59715

State Conservationist
134 South 12th Street
Lincoln, Nebraska 68508

State Conservationist
P.O. Box 4850
Reno, Nevada 89505

State Conservationist
Federal Bldg.
Durham, New Hampshire 03824

State Conservationist
1370 Hamilton St.
P.O. Box 219
Somerset, New Jersey 08873

State Conservationist
517 Gold Ave., SW
P.O. Box 2007
Albuquerque, New Mexico 87103

State Conservationist
700 E. Water St.
Syracuse, New York 13210

State Conservationist
Federal Office Bldg.
P.O. Box 27307
Raleigh, North Carolina 27611

State Conservationist
P.O. Box 1458
Bismarck, North Dakota 58501

State Conservationist
311 Old Federal Bldg.
Third and State Sts.
Columbus, Ohio 43215

State Conservationist
Farm Road & Brumley St.
Stillwater, Oklahoma 74074

State Conservationist
1218 Washington St. S.W.
Portland, Oregon 97205

State Conservationist
P.O. Box 985, Federal Square Station
Harrisburg, Pennsylvania 17108

Director, Caribbean Area
G.P.O. Box 4868
Santurce Station
San Juan, Puerto Rico 00936

State Conservationist
U.S. Post Office Bldg.
East Greenwich, Connecticut 02818

State Conservationist
Federal Bldg.
901 Sumter St.
Columbia, South Carolina 29201

State Conservationist
239 Wisconsin Ave. S.W.
P.O. Box 1357
Huron, South Dakota 57350

State Conservationist
561 U.S. Court House
Nashville, Tenn. 37203

State Conservationist
First National Bank Bldg.
16-20 S. Main St., Box 648
Temple, Texas 76501

State Conservationist
4012 Federal Bldg.
125 South State St.
Salt Lake City, Utah 84111

State Conservationist
96 College Rd.
Burlington, Vermont 05401

State Conservationist
P.O. Box 10026, Federal Bldg.
Richmond, Virginia 23240

State Conservationist
360 U.S. Courthouse
W. 920 Riverside Ave.
Spokane, Washington 99201

State Conservationist
209 Prairie Ave.
P.O. Box 865
Morgantown, West Virginia 26505

State Conservationist
P.O. Box 4248
Madison, Wisconsin 53711

State Conservationist
Federal Office Bldg.
Box 2440
Casper, Wyoming 82601

WHERE TO GET COMMUNICATION AIDS

Motion Pictures. Write to the Motion Picture Service, Office of Information, U.S.D.A., for *Agriculture Handbook No. 14,* "Motion Pictures of the U.S. Department of Agriculture."

The films of the U.S.D.A. are on loan from the cooperating film libraries in the 50 states, the District of Columbia, and the territory of Puerto Rico.

Persons wishing to borrow a film for educational purposes should consult their respective state film library listed here. The films are loaned free of charge except for transportation costs.

STATE FILM LIBRARIES THAT LEND U.S.D.A. MOTION PICTURES

ALABAMA — Agricultural Extension Service, Auburn University, Auburn 36830

ALASKA — Agricultural Extension Service, University of Alaska, College 99735

ARIZONA — Visual Aids Bureau, Extension Division, University of Arizona, Tucson 85721

ARKANSAS — Audio-Visual Aids, Department of Public Relations, Arkansas State College, Conway

CALIFORNIA — Extension Division, University of California, Berkeley 94720
Extension Division, University of California, 10851 LeConte Ave., Los Angeles 90024

COLORADO — Bureau of Audio-Visual Instruction, University of Colorado, Boulder
Agricultural Extension Service, Colorado State University, Fort Collins 80521

CONNECTICUT — Audio-Visual Aids Center, University of Connecticut, Storrs 06268

DELAWARE — Department of Rural Communications, University of Delaware, Newark 19711

DISTRICT OF COLUMBIA — District of Columbia Public Library, Eight & K Streets, N.W., Washington, D. C.

FLORIDA — Dept. of Audio-Visual Instruction, General Extension Division, University of Florida, Gainesville 32603

GEORGIA — Agricultural Extension Service, Athens 30601
Audio-Visual Aids Department, University of Georgia, Athens 30601

HAWAII — Agricultural Extension Service, University of Hawaii, Honolulu 96822

IDAHO — Agricultural Extension Service, State House, Boise

ILLINOIS — Visual Aids Service, University of Illinois, Urbana 61803

INDIANA — Audio-Visual Center, Indiana University, Bloomington
Purdue Film Library, Purdue University, Lafayette 47907

IOWA — Visual Instruction Service, Iowa State University of Science and Technology, Ames 50010

KANSAS — Bureau of Visual Instruction, University Extension Division, University of Kansas, Lawrence
Agricultural Extension Service, Kansas State University, Manhattan

KENTUCKY — Bureau of Audio-Visual Aids, University of Kentucky, Lexington 40506

LOUISIANA — Agricultural Extension Service, Louisiana State University, Baton Rouge 70803

MAINE — Agricultural Extension Service, University of Maine, Orono 04473

MARYLAND — Agricultural Extension Service, University of Maryland, College Park 20742

MASSACHUSETTS — Audio-Visual Center, University of Massachusetts, Amherst 01003

MICHIGAN — Audio-Visual Center, Michigan State University, East Lansing 48823
Audio-Visual Education Center, University of Michigan, Ann Arbor

MINNESOTA — Institute of Agriculture, University of Minnesota Farm, St. Paul 1

MISSISSIPPI — Agricultural Extension Service, Mississippi State University, State College 39762
Audio-Visual Education, Department of Education, Jackson

MISSOURI — Visual Education Service, University Extension, University of Missouri, Columbia 65202

MONTANA — Publications Department, Agricultural Extension Service, Montana State University, Bozeman 59715
Montana State Film Library, Sam Mitchell Building, Helena

NEBRASKA — Bureau of Audio-Visual Instruction, University of Nebraska, Lincoln 68503

NEVADA — Agricultural Extension Service, University of Nevada, Reno 89507

NEW HAMPSHIRE — Audio-Visual Center, University of New Hampshire, Durham 03824

NEW JERSEY — New Jersey State Museum, State Department of Education, State House Annex, Trenton 7

NEW MEXICO — Museum Film Service, Museum of New Mexico, Sante Fe
Agricultural Extension Service, New Mexico State University, Las Cruces 88001

NEW YORK — Film Library, State Department of Commerce, 40 Howard St., Albany

NORTH CAROLINA — Bureau of Visual Instruction, University of North Carolina, Chapel Hill
Agricultural Extension Service, State College Station, Raleigh

NORTH DAKOTA — Department of Information, Agricultural Extension Service, North Dakota State University, Fargo 58103

OHIO — Department of Audio-Visual Education, State Department of Education, State Office Bldg., Columbus 43215
Agricultural Extension Service, College of Agriculture, Ohio State University, Columbus 43210
Columbus Public Library, 96 S. Grant Ave., Columbus 43215

OKLAHOMA — Audio-Visual Education Department, University of Oklahoma, Norman
Agricultural Extension Service, Oklahoma State University, Stillwater 74075

OREGON — Department of Visual Instruction, Oregon State University, Corvallis 97331

PENNSYLVANIA — Audio-Visual Aids Library, Pennsylvania State University, University Park 16802

PENNSYLVANIA — PCW Audio-Visual Materials Center, 1500 Woodland Road, Pittsburgh

PUERTO RICO — Agricultural Extension Service, College of Agriculture, Rio Piedras 00927

RHODE ISLAND — University of Rhode Island Library, Kingston 02881

SOUTH DAKOTA — Extension Division, University of South Dakota, Vermillion Agricultural Extension Service, South Dakota State University, Brookings 57007

TENNESSEE — Division of University Extension, University of Tennessee, Box 1071, University Station, Knoxville 37901

TEXAS — Visual Instruction Bureau, University of Texas, Austin 78712
Agricultural Extension Service, Texas A & M University, College Station

UTAH — Audio-Visual Aids Dept., The Libraries, Utah State University, Logan

VERMONT — Vermont State Film Library, University of Vermont, Burlington 05401

VIRGINIA — Agricultural Extension Service, Virginia Polytechnic Institute, Blacksburg 24061

WASHINGTON — Audio-Visual Center, Washington State University 99163

WEST VIRGINIA — Audio-Visual Aids Department, The Library, West Virginia University, Morgantown 26506

WISCONSIN — Bureau of Visual Instruction, University of Wisconsin, Madison 53706

WYOMING — Wyoming Film Library, University of Wyoming, Box 3354, Laramie 82071

FILM STRIPS

Film strips on almost 150 subjects are available for purchase from Photo Laboratory, Inc., 3825 Georgia Avenue, N. W., Washington, D. C. The prices vary from approximately one to four dollars, depending upon the length of the film strip. A complete list of subjects and a brief description of each, with the selling price, may be obtained by writing to Office of Information, USDA, Washington, D. C. 20250 for a copy of "Film Strips of the USDA," *Agriculture Handbook No. 87.*

CAREERS IN AGRICULTURE

For complete information, write to one of the 70 colleges or universities offering courses leading to these professions:

Aerial Plant Nutrient Applicator
Agriculture Advertising Writer
Agricultural Advertiser
  Account Executive
Agricultural Attache
Agricultural Chemical Salesman
Agricultural Chemist
Agricultural Commodities Broker
Agricultural Commodity Grader
Agricultural Commodity Inspector
Agricultural Commodity Warehousing
  Examiner
Agricultural Consultant
Agricultural Broker

Agricultural Business Administrator
Agricultural Economist
Agricultural Writer, Editor
Agricultural Educator
Agricultural Engineer
Agricultural Geographer
Agricultural Instructor
Agricultural Journalist
Agricultural Management Specialist
Agricultural Market Reporter
Agricultural Marketing Specialist
Agricultural Missionary
Agricultural Program Specialist
Agricultural Researcher
Agricultural Statistician
Agricultural Trade Magazine Editor
Agriculturist
Agronomist
Animal Behaviorist
Animal Breeder
Animal Specialist
Aquatic Weed Specialist
Arachnologist (zoology of spiders)
Area Economist
Associate Buyer
Avian Specialist
Bacterial Pesticide Specialist
Bacteriologist
Baker Scientist
  Technologist, Manager
Bank Agriculture Representative
Banking Official
Beekeeper
Biochemist
Biological Pesticide Specialist
Biologist
Biophysicist
Biostatistician
Botanist
Breeding Technician
Cattle Manager
Clay Mineralogist
Climatologist
College Agricultural Researcher
College Faculty Member
Commercial Cattle Feeder
Conservationist—forest, range, soil, water, wildlife
Consumer Marketing Specialist
Cooperative Manager
County Extension Agent

County Extension 4-H Agent
Crop Insurance Specialist
Crop Physiologist
Crop Researcher
Crop Specialist
Cytogeneticist (heredity through cells and genetics)
Dairyman
Dairy Plant Manager
Dairy Technologist
Dendrologist (tree rings to date past events)
Director of Agricultural Research
Earth Scientist
Ecologist
Electric Co-Op Manager
Elevator Manager
Entomologist
Environmental Scientist
Extension Specialist
Farm Appraiser for a bank or agricultural lending institution
Farm Building Designer
Farm Credit Manager
Farmer
Farm Equipment Specialist
Farm Machinery Dealer
Farm Machinery Designer
Farm Manager
Farm Planner
Farm Realtor
Farm Store Manager
Farm Superintendent
Farm Supply Cooperative Salesman
Feed Dealer
Feedlot Manager
Feed Mill Manager
Feed Salesman
Feed Technologist
Field Crop Grower
Field Representative of:
  Beef Breed Association
  Beef/Pork Packing or Processing Co.
  Dairy Breed Association
  Feed Company
  Fruit Packing or Processing Company
  Fuel Company
  Insurance Company
  Lumber Processing Company
  Machinery Company
  Plant Nutrient Company
  Poultry Packing or Processing Company
  Veterinary Supply Company

Fish Culturist
Fish and Wildlife Specialist
Fishery Biologist
Fishery Manager
Floriculturist
Florist
Food Inspector
Food Processor
Food Retailer
Food Technologist
Foreign Agricultural Affairs Officer
Forester
Forest Products Technologist
Forest Ranger
Fruit Grower
Game Warden
Geneticist—plant, animal
Geochemist
Geomorphologist (genesis of earth forms)
Golf Course Superintendent
Grain Buyer
Grain Processor
Greenhouse Grower
Greenhouse Owner
Ground-water Geologist
Horticulturist
Husbandman
Hydrologist
Ichthyologist (zoology of fish)
Industrial Agriculturist
Information Specialist
Inspector—food, feed
International Agriculture Specialist
Insurance Broker
Irrigation Engineer
Irrigation Manager
IVS Volunteer
Laboratory Technician
Land Appraiser
Land Economist
Land Surveyor
Landscape Architect
Livestock Breeder
Livestock Buyer
Livestock Feeder
Loan Specialist
Market Analyst
Market Manager
Meat Department Manager
Meat Grader
Meat Inspector
Microbiologist

Microscopist
Miller
Molecular Biologist
Mycologist (botany of fungi)
Naturalist
Nematologist (zoology of nematodes)
Nurseryman
Nursery Owner or Operator
Nutritionist—plant, animal
Organizational Fieldman
Ornithologist (zoology of birds)
Outdoor Recreation Specialist
Packinghouse Manager
Parasitologist
Park Manager
Park Naturalist
Park Ranger
Park Superintendent
Pathologist—plant, animal
Peace Corps Administrator
Peace Corps Volunteer, especially in Africa,
    Asia, or Latin America
Pest Control Inspector
Pesticide-residue Analyst
Pet Food Processor
Physiologist
Plant Breeder
Plant Pathologist
Plant Physiologist
Plant Propagator
Plant Quarantine Inspector
Poultryman
Poultry Inspector
Poultry Scientist
Produce Department Manager
Product Development Specialist
Production Credit Fieldman
Production Manager
Public Relations Manager
Purchasing Agent
Quality Control Specialist
Radio Farm Director
Rancher
Ranch Manager
Range Conservationist
Range Scientist
Range Specialist
Ranger—forest, park
Recreation Development Planner

Resource Economist
Rural Sociologist
Salesman
Sales Representative
Sanitarian
Securities Salesman
Science Editor
Seed Broker
Seed Grower
Seed Technologist
Soil Analyst
Soil Chemist
Soil Conservationist
Soil Fertility Scientist
Soil Physicist
Soil Scientist
Soil Surveyor
Statistician
Taxonomist (classifies plants, animals)
Technical Editor
Technical Writer
Textile Researcher
Textile Technologist
Timber Manager, Specialist
Transportation Manager
Turf Producer
Turf Specialist
TV Farm Director
Vegetable Grower
Veterinarian (small animal)
Virologist (viruses)
VISTA Volunteer
Vocational Agriculture Teacher
Waterfowl Specialist
Water-life Management Specialist
Water Economist
Water Resources
    Administrator
    Engineer
    Researcher, Specialist
Water Resources Development Official
Weed Science Specialist
Wildlife Administrator
Wildlife Biologist
Wildlife Specialist
Wildlife Writer, Editor
Youth Corps Conservation Director
Zoologist

SCHOOLS OFFERING A 4-YEAR DEGREE IN AGRICULTURE

Seventy Colleges and Universities offer degrees in agriculture. Address correspondence to: Director of Instruction, College of Agriculture, at the address below:

AUBURN UNIVERSITY
Auburn, Alabama 36830

ALABAMA AGRICULTURAL AND MECHANICAL COLLEGE
Normal, Alabama 35762

UNIVERSITY OF ALASKA
College, Alaska 99701

ARIZONA STATE UNIVERSITY
Tempe, Arizona 85281

UNIVERSITY OF ARIZONA
Tucson, Arizona 85721

UNIVERSITY OF ARKANSAS
Fayetteville, Arkansas 72701

AGRICULTURAL, MECHANICAL & NORMAL COLLEGE
Pine Bluff, Arkansas 71601

UNIVERSITY OF CALIFORNIA
Berkeley, California 94720
Davis, California 95616
Los Angeles, California 90024
Riverside, California 92502

COLORADO STATE UNIVERSITY
Fort Collins, Colorado 80521

UNIVERSITY OF CONNECTICUT
Storrs, Connecticut 06268

UNIVERSITY OF DELAWARE
Newark, Delaware 19711

DELAWARE STATE COLLEGE
Dover, Delaware 19901

UNIVERSITY OF FLORIDA
Gainesville, Florida 32601

FLORIDA A&M UNIVERSITY
Tallahassee, Florida 32307

UNIVERSITY OF GEORGIA
Athens, Georgia 30601

FORT VALLEY STATE COLLEGE
Fort Valley, Georgia 31030

UNIVERSITY OF HAWAII
Honolulu, Hawaii 96822

UNIVERSITY OF IDAHO
Moscow, Idaho 83843

SOUTHERN ILLINOIS UNIVERSITY
Carbondale, Illinois 62903

UNIVERSITY OF ILLINOIS
Urbana, Illinois 61801

PURDUE UNIVERSITY
Lafayette, Indiana 47907

IOWA STATE UNIVERSITY
Ames, Iowa 50010

KANSAS STATE UNIVERSITY
Manhattan, Kansas 66502

UNIVERSITY OF KENTUCKY
Lexington, Kentucky 40506

KENTUCKY STATE COLLEGE
Frankfort, Kentucky 40601

LOUISIANA STATE UNIVERSITY
Baton Rouge, Louisiana 70803

SOUTHERN UNIVERSITY & AGRICULTURAL & MECHANICAL COLLEGE
Baton Rouge, Louisiana 70813

UNIVERSITY OF MAINE
Orono, Maine 04473

UNIVERSITY OF MARYLAND
College Park, Maryland 20742

MARYLAND STATE COLLEGE
Princess Anne, Maryland 21853

UNIVERSITY OF MASSACHUSETTS
Amherst, Massachusetts 01002

MICHIGAN STATE UNIVERSITY
East Lansing, Michigan 48823

UNIVERSITY OF MINNESOTA
St. Paul, Minnesota 55101

MISSISSIPPI STATE UNIVERSITY
State College, Mississippi 39762

ALCORN AGRICULTURAL & MECHANICAL
COLLEGE
Lorman, Mississippi 39096

UNIVERSITY OF MISSOURI
Columbia, Missouri 65201

LINCOLN UNIVERSITY
Jefferson City, Missouri 65102

MONTANA STATE UNIVERSITY
Bozeman, Montana 59715

UNIVERSITY OF NEBRASKA
Lincoln, Nebraska 68503

UNIVERSITY OF NEVADA
Reno, Nevada 89507

UNIVERSITY OF NEW HAMPSHIRE
Durham, New Hampshire 03824

RUTGERS UNIVERSITY
New Brunswick, New Jersey 08903

NEW MEXICO STATE UNIVERSITY
Las Cruces, New Mexico 88001

CORNELL UNIVERSITY
Ithaca, New York 14850

UNIVERSITY OF NORTH CAROLINA
Raleigh, North Carolina 27607

AGRICULTURAL & TECHNICAL COLLEGE
OF NORTH CAROLINA
Greensboro, North Carolina 27411

NORTH DAKOTA STATE UNIVERSITY
Fargo, North Dakota 58102

THE OHIO STATE UNIVERSITY
Columbus, Ohio 43210

OKLAHOMA STATE UNIVERSITY
Stillwater, Oklahoma 74074

LANGSTON UNIVERSITY
Langston, Oklahoma 73050

OREGON STATE UNIVERSITY
Corvallis, Oregon 97331

PENNSYLVANIA STATE UNIVERSITY
University Park, Pennsylvania 16802

UNIVERSITY OF PUERTO RICO
COLLEGE OF AGRICULTURE
Rio Piedras, Puerto Rico 00928

UNIVERSITY OF RHODE ISLAND
Kingston, Rhode Island 02881

CLEMSON UNIVERSITY
Clemson, South Carolina 29631

SOUTH CAROLINA STATE COLLEGE
Orangeburg, South Carolina 29115

SOUTH DAKOTA STATE UNIVERSITY
Brookings, South Dakota 57006

UNIVERSITY OF TENNESSEE
Knoxville, Tennessee 37901

TENNESSEE AGRICULTURAL &
INDUSTRIAL STATE UNIVERSITY
Nashville, Tennessee 37203

TEXAS A & M UNIVERSITY
College Station, Texas 77843

TEXAS TECHNOLOGICAL COLLEGE
Lubbock, Texas 79409

PRAIRIE VIEW AGRICULTURAL &
MECHANICAL COLLEGE
Prairie View, Texas 77445

UTAH STATE UNIVERSITY
Logan, Utah 84321

UNIVERSITY OF VERMONT
Burlington, Vermont 05401

VIRGINIA POLYTECHNIC INSTITUTE
Blacksburg, Virginia 24061

VIRGINIA STATE COLLEGE
Petersburg, Virginia 23800

WASHINGTON STATE UNIVERSITY
Pullman, Washington 99163

WEST VIRGINIA UNIVERSITY
Morgantown, West Virginia 26505

UNIVERSITY OF WISCONSIN
Madison, Wisconsin 53706

UNIVERSITY OF WYOMING
Laramie, Wyoming 82070

POST-SECONDARY EDUCATION IN AGRIBUSINESS AND NATURAL RESOURCES OCCUPATIONS

For education beyond secondary school level, write for information at the following state addresses:

ALABAMA: State Supervisor, Agribusiness Education, Department of Education
State Office Building
Montgomery, AL 36104

ALASKA: Superintendent, Trades, Industries, and Fisheries, State Department of Education
Alaska Office Building
Juneau, AK 99801

ARIZONA: State Supervisor
Agricultural Education
1333 W. Camelback Road, Suite No. 111
Phoenix, AZ 85013

ARKANSAS: Director, Agricultural Education, Division of Vocational, Technical, and Adult Education
Arch Ford Education Building
Little Rock, AR 72201

CALIFORNIA: Division of Occupational Education, California Community Colleges
825 Fifteenth Street
Sacramento, CA 95814

COLORADO: Supervisor, Agricultural Education, State Board for Community Colleges and Occupational Education
207 State Services Building
Denver CO 80203

CONNECTICUT: State Supervisor, Agricultural Education, State Department of Education
P.O. Box 2219
Hartford, CT 06115

DELAWARE: State Supervisor
Agricultural Education
University of Delaware
College of Agricultural Sciences
Agricultural Experiment Station
Newark, DE 19711

FLORIDA: Administrator
Agricultural Education
State Department of Education
Knott Building
Tallahassee, FL 32304

GEORGIA: State Supervisor
Agricultural Education
State Department of Education
325 State Office Building
Atlanta, GA 30334

HAWAII: Program Specialist
Agricultural Education Section
Office of Instructional Services
State Department of Education
Post Office Box 2360
Honolulu, HI 96804

IDAHO: State Supervisor
Agricultural Education
State Board for Vocational Education
518 Front Street
Boise, ID 83702

ILLINOIS: Head Consultant, Applied Biological and Agricultural Occupations
Board of Vocational Ed. and Rehabilitation
1035 Outer Park Drive
Springfield, IL 62706

INDIANA: Chief Consultant
Agribusiness Education
Division of Vocational Education
700 North High School Road
Indianapolis, IN 46224

IOWA: Consultant, Post-secondary Service
State Department of Public Instruction
Area Schools & Career Education Branch
Des Moines, IA 50319

KANSAS: State Supervisor
Agricultural Education
Division of Vocational Education
Kansas State Department of Education
Kansas State Education Building
120 East 10th Street
Topeka, KS 66612

KENTUCKY: Vice President, University of
   Kentucky Community College System
Breckinridge Hall
Lexington, KY 40506

Director, Agricultural Education
Bureau of Vocational Education
Commonwealth of Kentucky Education De-
   partment
Frankfort, KY 40601

LOUISIANA: State Director
Vocational Agriculture
State Department of Education
State Capitol Building
Baton Rouge, LA 70804

MAINE: State Consultant
Agricultural Education
State Department of Education
Augusta, ME 04330

MARYLAND: State Supervisor
Agricultural Education
State Department of Education
600 Wyndhurst Avenue
Baltimore, MD 21210

MASSACHUSETTS: Supervisor of Agricultural
   Education
Division of Occupational Education
The Commonwealth of Massachusetts
Department of Education
182 Tremont Street
Boston, MA 02111

MICHIGAN: Supervisor, Post-Secondary Unit
State Department of Education
Division of Vocational Education, Box 928
Lansing, MI 48904

MINNESOTA: State Supervisor
Agricultural Education
State Department of Education
Capitol Square, 550 Cedar Street
St. Paul, MN 55101

MISSISSIPPI: State Supervisor
Vocational Agriculture Department
State Department of Education
Vocational Education Division
Post Office Box 771
Jackson, MS 39205

MISSOURI: Director, Agricultural Education
State Department of Education
Division of Public Schools
Post Office Box 480
Jefferson City, MO 65101

MONTANA: State Supervisor
Agricultural Education
State Department of Public Instruction
State Capitol
Helena, MT 59601

NEBRASKA: State Director
Agricultural Education
State Department of Education
223 South 10th Street
Lincoln, NB 68508

NEVADA: State Supervisor
Agricultural Education
State Department of Education
Heroes Memorial Building
Carson City, NV 89701

New Hampshire: Consultant
Agricultural Education
State Department of Education
Stickney Avenue
Concord, NH 03301

New Jersey: State Director of Agricultural
Education
New Jersey State Department of Education
Vocational Division
225 West State Street
Trenton, NJ 08625

New Mexico: State Supervisor
Agricultural Education
State Department of Vocational Education
200 DeVargas Street
Santa Fe, NM 87501

New York: Associate in Higher Occupa-
tional Education
University of the State of New York
The State Education Department
99 Washington Avenue
Albany, NY 12210

Vice Chancellor for Community Colleges
and Provost for Vocational and Technical Education
State University of New York
Albany, NY 12201

North Carolina: Educational Consultant
Agricultural and Biological Education
Division of Occupational Education
Department of Community Colleges
State Board of Education
Raleigh, NC 27602

North Dakota: State Supervisor
Agricultural Education
State Department of Education
State Office Building
900 East Boulevard
Bismark, ND 58501

Ohio: State Supervisor
Agricultural Education
Agricultural Administration Building
Ohio State University
2120 Fyffe Road
Columbus, OH 43210

Oklahoma: State Supervisor
Oklahoma State Department of Vocational
and Technical Education
1515 West Sixth Avenue
Stillwater, OK 74074

Oregon: Agribusiness Specialist
Oregon Board of Education
942 Lancaster Drive N. E.
Salem, OR 97310

Pennsylvania: State Supervisor
Agricultural Education
Pennsylvania Department of Education
Post Office Box 911
Harrisburg, PA 17126

Puerto Rico: Director, Vocational Agricul-
ture Program
Commonwealth of Puerto Rico
Area of Vocational & Technical Education
Post Office Box 759
Hato Rey, PR 00919

Rhode Island: University of Rhode Island
Kingston, RI 02881

South Carolina: Coordinator
Agricultural Technology Programs
State Committee for Technical Education
1429 Senate Street
Columbia, SC 29201

South Dakota: State Supervisor
Agricultural Education, Division of Voca-
tional and Technical Education
Agricultural Education Service
State Department of Public Instruction
Pierre, SD 57501

Tennessee: Supervisor
Agricultural Occupations
State Department of Education
210 Cordell Hull Building
Nashville, TN 37219

Texas: Director
Vocational Agricultural Education
Texas Education Agency
201 East Eleventh Street
Austin, TX 78701

UTAH: State Specialist
Agricultural Education
State Board for Vocational Education
1300 University Club Building
136 East South Temple
Salt Lake City, UT 84111

VERMONT: State Consultant
Agricultural Education
State Department of Education
State Office Building
Montpelier, VT 05602

WASHINGTON: Program Director
Agricultural Education, Coordinating Council for Occupational Education
Post Office Box 248
Olympia, WA 98501

WEST VIRGINIA: State Supervisor
Vocational Agriculture
Department of Education
Building 6, Unit B, State Capitol
Charleston, WV 25305

WISCONSIN: Supervisor
Vocational Agriculture
State Board of Vocational, Technical and Adult Education
137 East Wilson Street
Madison, WI 53703

WYOMING: State Director
Agricultural Education
State Department of Education
Capitol Building
Cheyenne, WY 82001

For further information on United States post-secondary agricultural programs, write:

Educational Program Specialist
Agribusiness and Natural Resources Occupations
Division of Vocational and Technical Education
United States Office of Education
Department of Health, Education and Welfare
Washington, D. C. 20202

FEDERAL JOB INFORMATION CENTERS

Address: Executive Secretary, Interagency Board of United States Civil Service Examiners, at the location nearest you. Telephone numbers may be found under U. S. Government listing in the directory in cities where Federal Job Information Centers are located. (Source: "Federal Career Directory —A Guide for College Students," U. S. Civil Service Commission, October 1968, Washington, D. C.)

Sutherland Building
806 Governors Dr. SW.
Huntsville, Ala. 35801

107 Francis St.
Mobile, Ala. 36602

Hill Building
632 Sixth Ave.
Anchorage, Alaska 99501

Balke Building
44 W. Adams St.
Phoenix, Ariz. 85003

923 W. Fourth St.
Little Rock, Ark. 72201

851 S. Broadway
Los Angeles, Calif. 90014

455 Capitol Mall, Suite 125
Sacramento, Calif. 95814

380 W. Court St.
San Bernardino, Calif. 92401

1400 Fifth Ave., Suite 100
San Diego, Calif. 92101

450 Golden Gate Ave.
Box 36122
San Francisco, Calif. 94102

Post Office Building, Room 203
18th and Stout Sts.
Denver, Colo. 80202

Federal Building, Room 716
450 Main St.
Hartford, Conn. 06103

U.S. Post Office and Courthouse
11th and King Sts.
Wilmington, Del. 19801

123 S. Court Ave.
Orlando, Fla. 32801

275 Peachtree St. NE.
Atlanta, Ga. 30303

Federal Building
451 College St.
Macon, Ga. 31201

Federal Building
Honolulu, Hawaii 96813

Federal Building-U.S. Court House, Room
663
550 W. Fort St.
Boise, Idaho 83702

219 S. Dearborn St., Room 1322
Chicago, Ill. 60604

36 S. Pennsylvania St., Room 102
Indianapolis, Ind. 46204

Federal Building, Room 191
210 Walnut St.
Des Moines, Iowa 50309

Beacon Building, Room 2
114 S. Main St.
Wichita, Kans. 67202

Heyburn Building
721 S. Fourth St.
Louisville, Ky. 40202

Federal Building South
600 South St.
New Orleans, La. 70130

Federal Building
Augusta, Maine 04330

Federal Office Building
Lombard St. and Hopkins Pl.
Baltimore, Md. 21201

Post Office and Courthouse Building
Boston, Mass. 02109

144 W. Lafayette St.
Detroit, Mich. 48226

Building 57, Fort Snelling
Minneapolis-St. Paul, Minn. 55111

802 N. State St.
Jackson, Miss. 39201

Federal Building, Room 134
601 E. 12th St.
Kansas City, Mo. 64106

Federal Building, Room 1712
1520 Market St.
St. Louis, Mo. 63103

I.B.M. Building
130 Neill Ave.
Helena, Mont. 59601

U.S. Courthouse and Post Office Building,
Room 1014
215 N. 17th St.
Omaha, Nebr. 68102

300 Booth St.
Reno, Nev. 89502

Federal Building-U.S. Post Office
Daniel and Penhallow Sts.
Portsmouth, N.H. 03801

Federal Building
970 Broad St.
Newark, N.J. 07102

Federal Building
421 Gold Ave. S.W.
Albuquerque, N. Mex. 87101

Federal Building
26 Federal Plaza
New York, N.Y. 10007

O'Donnell Building
301 Erie Blvd. W.
Syracuse, N.Y. 13202

415 W. Hillsborough St.
Raleigh, N.C. 27603

Manchester Building, Room 206
112 N. University Dr.
Fargo, N. Dak. 58102

New Federal Building
1240 E. Ninth St.
Cleveland, Ohio 44199

Knott Building
21 E. Fourth St.
Dayton, Ohio 45402

210 NW. Sixth St.
Oklahoma City, Okla. 73102

319 SW. Pine St.
Portland, Oreg. 97204

Federal Building
128 N. Broad St.
Philadelphia, Pa. 19102

Federal Building
1000 Liberty Ave.
Pittsburgh, Pa. 15222

Federal Building and Post Office
Kennedy Plaza
Providence, R.I. 02903

Federal Office Building
334 Meeting St.
Charleston, S.C. 29403

Dusek Building, Room 118
919 Main St.
Rapid City, S. Dak. 57701

Federal Office Building
167 N. Main St.
Memphis, Tenn. 38103

1114 Commerce St., Room 103
Dallas, Tex. 75202

El Paso National Bank Building
411 N. Stanton St.
El Paso, Tex. 79901

702 Caroline St.
Houston, Tex. 77002

U.S. Post Office and Courthouse
615 E. Houston St.
San Antonio, Tex. 78205

Federal Building Annex
135 S. State St.
Salt Lake City, Utah 84111

Federal Building
Elmwood Ave. and Pearl St.
Burlington, Vt. 05401

Rotunda Building
415 St. Paul Blvd.
Norfolk, Va. 23510

Federal Office Building
First and Madison Sts.
Seattle, Wash. 98104

Federal Building
500 Quarrier St.
Charleston, W. Va. 25301

161 W. Wisconsin Ave., Room 215
Milwaukee, Wis. 53203

2005 Warren Ave.
Cheyenne, Wyo. 82001

Hato Rey Building
255 Ponce de Leon Ave.
San Juan, P.R. 00917

Civil Service Commission Building
1900 E St. NW.
Washington, D.C. 20415

INFORMATION FROM THE UNITED STATES DEPARTMENT OF AGRICULTURE

The USDA has vast resources of information that are made available on request. The USDA Bulletin "How to Get Information from the USDA" lists the various subdivisions of the Department with the names and telephone exchanges of the heads of each subdivision and office. This bulletin may be obtained by writing to the Office of Information, USDA, Washington, D. C. 20250. The following list defines various subdivisions of the Department.

OFFICE OF INFORMATION

This office directs and coordinates information work with the various agencies and has final review of all informational materials involving departmental policy. It provides assistance and facilities in the production of motion pictures, still photography, exhibits, and art and graphics.

ECONOMIC RESEARCH SERVICE

Research is done in agricultural economics and marketing; factors affecting agricultural production, supplies, prices, and income are analyzed; studies United States trade in agricultural products; analyzes farm productivity costs, financing, use of resources, potentials of low-income areas.

FOREIGN ECONOMIC DEVELOPMENT SERVICE

This office is concerned with economic development through technical agricultural assistance and training of foreign agricultural experts.

STATISTICAL REPORTING SERVICE

Prepares monthly, quarterly, or annual estimates of production, supply, prices of agricultural commodities, farm labor, etc., nationally and by states.

AGRICULTURAL STABILIZATION AND CONSERVATION SERVICE

This office is responsible for "set-aside" and related commodity production adjustment and support programs; administration of the Sugar Act and the National Wool Act; drought and other natural disaster assistance; certain national emergency preparedness programs, and the Rural Environmental Assistance Program.

EXPORT MARKETING SERVICE

Recommends policies and programs to maximize exports of agricultural commodities, with particular emphasis on export dollars.

FEDERAL CROP INSURANCE CORPORATION

Insures farmers against loss of crop investment due to risk beyond their control.

FOREIGN AGRICULTURAL SERVICE

Promotes the export of United States farm products; protects domestic agricultural markets from unfair foreign competition; is a basic source of information to American agriculture on world crops, policies, and markets.

COMMODITY EXCHANGE AUTHORITY

Supervises futures trading on commodity exchanges to maintain fair trading practices and competitive pricing; prevents price manipulation and other violations of the Commodity Exchange Act.

CONSUMER AND MARKETING SERVICE

Responsible for meat and poultry inspection and label approval; standardization and grading, market news, commodity purchase, plentiful foods promotion, and other marketing services; marketing agreements and orders; specified marketing regulatory programs.

FOOD AND NUTRITION SERVICE

Administers the Federal food-help programs within the United States and its territories (including food stamp, school lunch, school breakfast, special milk, and other child nutrition or family-assistance food programs).

PACKERS AND STOCKYARDS ADMINISTRATION

Responsible for maintaining competitive interstate livestock and poultry marketing and meat packing under the provisions of the Packers and Stockyards Act.

FARMER COOPERATIVE SERVICE

Helps farmers and other rural Americans use cooperatives most effectively by conducting research, advising cooperative leaders and others; provides publications and educational materials.

FARMERS HOME ADMINISTRATION

Provides a supplementary source of credit for strengthening family farms, constructing rural housing and developing rural community facilities; makes loans for farm operating expenses, farm enlargement, improvement and purchase (including loans for individual homes, rental units, self-help homes, and farm housing); loans and grants for developing rural water systems, and sewage and disposal systems; emergency credit needs, soil and water conservation, shifts in land use to grazing areas and watershed development.

FOREST SERVICE

Applies sound conservation and utilization practices to natural resources of the National Forests and National Grasslands in cooperation with states and private land owners; carries out extensive forest and range research.

RURAL ELECTRIFICATION ADMINISTRATION

Makes loans to cooperatives and other qualified organizations to bring electric power and modern telephone service on a continuing basis to rural areas; works with borrowers to stimulate economic development in their areas.

SOIL CONSERVATION SERVICE

Cooperates with land owners and operators in local and national soil and water conservation programs, and with other government agencies; administers the Great Plains Conservation Program, Watershed Protection, and the Flood Prevention Program; carries on the National Cooperative Soil Survey; assists land owners and local groups in Resource Conservation and Development projects; assists in the establishment of income-producing recreation enterprises.

AGRICULTURAL RESEARCH SERVICE

Conducts basic and applied research on: the production of animals and plants, and the prevention of environmental pollution by agricultural practices; conservation of soil and water; nutrition of humans, animals, and plants; protection of animals and plants from diseases, parasites, insects, and other pests; machinery, buildings, and facilities needed for more economical and efficient production and marketing of agricultural products. Conducts plant and animal regulatory, quarantine, and control activities to protect the nation's agriculture.

COOPERATIVE STATE RESEARCH SERVICE

Administers federal-grant funds for research at State Agricultural Experiment Stations; coordinates agricultural research among the states and between the states and the USDA.

EXTENSION SERVICE

Has primary responsibility for and leadership in USDA educational programs, and coordination of all educational activities of the Department.

NATIONAL AGRICULTURAL LIBRARY

Collects and maintains worldwide publications in agricultural, biological, and chemical sciences; serves the USDA and state agricultural agencies, other libraries, institutions, and the general public; sells microfilm and photocopies of material in its collection; makes available, on magnetic tape, bibliographic information previously included in the *Bibliography of Agriculture, National Agricultural Library Monthly Catalog,* and *Pesticides Documentation Bulletin.*

HOW TO PURCHASE PHOTOGRAPHS FROM THE U.S.D.A.

Reproductions of photographs, maps, and charts used in authorized work of the USDA may be purchased from the Department. Requests for news and general illustration photographs and price schedules should be addressed to the Photography Division, Office of Information, USDA, Washington, D. C. 20250.

1. Only reproductions of photographs, maps, and charts used in authorized work of the Department are available.
2. Since regular Department activities must be given precedence, there may be some delay in filling orders. Normally orders are filled within 15 days.
3. USDA photographic reproductions may not be used in advertising or otherwise which would indicate that the Department either directly or by implication endorses any commercial product. The Department will not furnish reproductions that may be used for such purpose.
4. Price quotations will be furnished for all items. An extra charge may be made for excessive laboratory time caused by unusual instructions from the purchaser.
5. Advance payment is required before any photographic reproductions will be made. Payment should be made by check, draft, or money order, payable to the Office of Information, USDA.
6. If for any reason the Department does not furnish reproductions ordered, all money received for items not furnished will be refunded.

HOW TO GET INFORMATION FROM THE DEPARTMENT OF THE INTERIOR

The Department of the Interior is responsible for a wide variety of natural resource activities. Your requests for information can be handled more rapidly if they are directed to the office concerned with the specific subjects in which you are interested. The following list will help you locate the best Department source for information materials.

BUREAU OF SPORT FISHERIES AND WILDLIFE: (Hunting and fishing license sales statistics; fish and wildlife research; Federal aid to fish and wildlife restoration; duck stamp data; migratory bird hunting regulations; recreational use of national fish hatcheries and national wildlife refuges; river basin studies; rare and endangered wildlife; photographs.)
Office of Conservation Education
Bureau of Sport Fisheries and Wildlife
Room 3242, Interior Building
Washington, D. C. 20240

GEOLOGICAL SURVEY: (General and detailed geologic and topographic maps; reports on mining districts, mineral occurrences; laboratory investigations; water resources studies; stream-flow, quality, sediment, and ground-water level measuring stations; aerial photographs.)
Information Officer, Geological Survey
Department of the Interior
5232 GSA Building
Washington, D. C. 20240

NATIONAL PARK SERVICE: (Information on 35 national parks, 13 national recreation areas, 7 national seashores, and other areas; park usage—camping, swimming, boating, hiking, etc.; pack trips; archeological studies; winter sports; wildlife management in parks.)
Information Officer
National Park Service, Room 2023
Interior Building
Washington, D. C. 20240

BUREAU OF LAND MANAGEMENT: (Wildlife use and management on public land; camping, hunting, fishing, etc., on public land; obtaining public lands for state and community parks; forest and watershed practices aiding wildlife and recreation.)
Information Officer
Bureau of Land Management, Room 5643
Interior Building
Washington, D. C. 20240

BUREAU OF INDIAN AFFAIRS:
(Information on Indians and their relationship to the Federal Government; various tribes; ceremonies and celebrations of interest to visitors; locations of specific reservations.)
Information Officer
Bureau of Indian Affairs, Room 222
1951 Constitution Avenue
Washington, D. C. 20242

BUREAU OF OUTDOOR RECREATION:
(National recreation needs and plans; Federal, state and private programs for recreational development; coordination of various Federal recreation programs; forecasts on recreational needs.)
Information Officer
Bureau of Outdoor Recreation, Room 4125
Interior Building
Washington, D. C. 20240

BUREAU OF MINES:
(Booklets and technical literature on mineral production and consumption; abatement of environmental pollution associated with minerals; free-loan 16 mm. films on minerals and the resources of the States.)
Chief, Office of Mineral Information
Bureau of Mines, Room 2611
Interior Building
Washington, D. C. 20240

OFFICE OF SALINE WATER:
(Information on the program for desalting sea or brackish water through the operation of demonstration plants; plant visits welcome: Wrightsville Beach, N. C.; Freeport, Texas; Roswell, N. M.; Webster, S. D.; and San Diego, Cal.)
Information Officer
Office of Saline Water, Room 5024
Interior Building
Washington, D. C. 20240

BUREAU OF RECLAMATION:
(Information on use of more than 200 reservoirs for recreation—camping, fishing, boating, swimming; scenic tours and camping areas; sightseeing at dams and related works.)
Information Officer
Bureau of Reclamation, Room 7642
Interior Building
Washington, D. C. 20240

OFFICE OF TERRITORIES:
(Travel, sightseeing, recreation, sports, etc., in the Virgin Islands, Trust Territory of Pacific Islands, Guam, American Samoa.)
Director, Office of Territories
Room 6514, Interior Building
Washington, D. C. 20240

For any additional information, contact the Director of Information, Department of the Interior, Washington, D. C. 20240. Information about environmental problems not directly related to the Department of the Interior may be obtained from the Public Affairs Office, Environmental Protection Agency, Washington, D. C. 20460.

UNITED STATES DEPARTMENT OF COMMERCE FIELD OFFICES

ALABAMA, Birmingham—Title Building, 2028 Third Avenue, N. 35203

ALASKA, Anchorage—Room 60, United States Post Office and Court House 99501

ARIZONA, Phoenix—New Federal Building, 230 North First Avenue 85025

CALIFORNIA, Los Angeles—Room 450, Western Pacific Building 1031 S. Broadway 90015

CALIFORNIA, San Francisco—Room 419, Customhouse, 555 Battery Street 94011

COLORADO, Denver—142 New Custom House, 19th and Stout Street 80202

CONNECTICUT, Hartford—18 Asylum Street 06103

FLORIDA, Jacksonville—512 Greenleaf Building, 204 Laura Street 32202

FLORIDA, Miami—408 Ainsley Building, 14 N. E. First Avenue 33132

GEORGIA, Atlanta—4th Floor, Home Savings Building, 75 Forsyth Street, N.W. 30303

GEORGIA, Savannah—235 United States Courthouse, 125-129 Bull Street 31402

HAWAII, Honolulu—202 International Savings Building, 1022 Bethel Street 96813

ILLINOIS, Chicago—Room 1302, 226 West Jackson Boulevard 60606

LOUISIANA, New Orleans—1508 Masonic Temple Building, 333 St. Charles Avenue 70130

MASSACHUSETTS, Boston—Room 230, 80 Federal Street 02110

MICHIGAN, Detroit—438 Federal Building 48226

MINNESOTA, Minneapolis—Room 304, Federal Building, 110 S. Fourth Street 55401

MISSOURI, Kansas City—Room 2011, 911 Walnut Street 64106

MISSOURI, St. Louis—2511 New Federal Building, 1520 Market Street 63103

NEVADA, Reno—1479 Wells Avenue 89502

NEW MEXICO, Albuquerque—United States Court House 87101

NEW YORK, Buffalo—504 Federal Building, 117 Ellicott Street 14203

NEW YORK, New York—61st Floor, Empire State Building, 350 Fifth Avenue 10001

NORTH CAROLINA, Greensboro—Room 407, United States Post Office Building 27402

OHIO, Cincinnati—809 Fifth Third Bank Building, 36 East Fourth Street 45202

OHIO, Cleveland—4th Floor, Federal Reserve Bank Building, East 6th Street and Superior Avenue 44101

OREGON, Portland—217 Old United States Court House, 520 S. W. Morrison Street 97204

PENNSYLVANIA, Philadelphia — Jefferson Building, 1015 Chestnut Street 19107

PENNSYLVANIA, Pittsburgh — 2201 Federal Building, 1000 Liberty Avenue 15222

PUERTO RICO, Santurce—605 Condado Avenue, Room 628 00907

SOUTH CAROLINA, Charleston—No. 4 North Atlantic Wharf 29401

TENNESSEE, Memphis—212 Falls Building, 22 North Front Street 38103

TEXAS, Dallas—Room 3-104 Merchandise Mart, 500 S. Ervay Street 75201

TEXAS, Houston—5102 Federal Building, 515 Rusk Avenue 77002

UTAH, Salt Lake City—222 S. W. Temple Street 84101

VIRGINIA, Richmond—2105 Federal Building, 400 North 8th Street 23240

WASHINGTON, Seattle—809 Federal Office Building, 909 First Avenue 98104

WISCONSIN, Milwaukee—1201 Straus Building, 238 W. Wisconsin Avenue 53203

WYOMING, Cheyenne—207 Majestic Building, 16th and Capitol Avenue 82001

FORESTRY TECHNICIAN SCHOOLS IN THE UNITED STATES AND CANADA

The following institutions have been recognized by the Society of American Foresters, having met the Society's *Minimum Guidelines for Forest Technician Training.* Forest technician graduates of these institutions are eligible for SAF's membership category of Technician Member. Full-time students in forest technology at these institutions are eligible for the Student Member category. (Reprinted by special permission of the *Journal of Forestry,* magazine of the Society of American Foresters.)

CALIFORNIA — Lassen Community College, 1100 Main Street, Susanville 96130

CALIFORNIA — College of the Redwoods, 1040 Del Norte Street, Eureka 95501

CALIFORNIA — Sierra College, 5000 Rocklin Road, Rocklin 95677

FLORIDA — Lake City Community College, Lake City 32055

GEORGIA — Savannah Area Vocational-Technical School, Savannah 31401

ILLINOIS — Southeastern Illinois College, 333 West College Street, Harrisburg 62946

KENTUCKY — Forestry and Wood Technology, University of Kentucky, Quicksand 41363

MAINE — Forest Technician Program, University of Maine, Orono 04473

MARYLAND — Allegany Community College, Cumberland 21502

MICHIGAN — Alpena Community College, Alpena 49707

MICHIGAN — Ford Forestry Center, Michigan Technological University, L'Anse 49946

MINNESOTA — North Central School, University of Minnesota, Grand Rapids 55744

MONTANA — Flathead Valley Community College, Box 1174, Kalispel 59901

MONTANA — Missoula Technical Center, 909 South Avenue West, Missoula 59801

NEW HAMPSHIRE — Forest Technician Curriculum, University of New Hampshire, Durham 03824

NEW YORK — Ranger School Forest Technician Program, State University College of Forestry, Wanakena 13695

NEW YORK — Paul Smith's College of Arts and Sciences, Paul Smiths 12970

NORTH CAROLINA — Haywood Technical Institute, P. O. Box 427, Clyde 28721

NORTH CAROLINA — Martin Technical Institute, P. O. Drawer 866, Williamson 27530

NORTH CAROLINA — Wayne Community College, Drawer 1878, Goldsboro 27530

OHIO — Tri-County Technical Institute, Nelsonville 45764

OKLAHOMA — Eastern Oklahoma State College, Wilburton 74578

OREGON — Clatstop Community College, Astoria 97103

OREGON — Central Oregon Community College, College Way, Bend 97701

OREGON — Lane Community College, 4000 East 30th Avenue, Eugene 97405

OREGON — Southwestern Oregon Community College, P. O. Box 518, Coos Bay 97420

OREGON — Umpqua Community College, Roseburg 97470

PENNSYLVANIA — The Pennsylvania State University, Mont Alto 17237

PENNSYLVANIA — Williamsport Area Community College, 1005 West Third Street, Williamsport 17701

SOUTH CAROLINA — Horry-Georgetown Technical Education Center, P. O. Box 317, Conway 29526

VIRGINIA — Dabney S. Lancaster Community College, Clifton Forge 24422

WASHINGTON — Centralia College, Centralia 98531

WASHINGTON — Everett Community College, 801 Wetmore Avenue, Everett 98201

WASHINGTON — Green River Community College, 12401 S. E. 320th Street, Auburn 98002

WASHINGTON — Peninsula College, Port Angeles 98362

WASHINGTON — Shoreline Community College, 161st Street and Greenwood Avenue, Seattle 98133

WASHINGTON — Spokane Falls Community College, W3410 Fort George Wright Drive, Spokane 99204

WASHINGTON — Wenatchee Valley College, Wenatchee 98801

WEST VIRGINIA — Glenville State College, Glenville 26351

CANADA

ALBERTA — Forest Technology School, Hinton

ONTARIO — Forest Technician Program, Lakehead University, Thunder Bay

ONTARIO — Sir Sandford Fleming College, Lindsay

NEW BRUNSWICK — Maritime Forest Ranger School, Fredericton

NATIONAL LIVESTOCK REGISTRY ASSOCIATIONS

BEEF CATTLE

American Angus Association, Flynn W. Stewart, Pres., Lloyd D. Miller, Secy., 3201 Frederick Blvd., St. Joseph, Missouri 65401.

American Brahman Breeders Association, Leon Locke, Pres., Box 166, Hingerford, Texas 77448; Harry P. Gayden, Exec.-Secy., 4815 Gulf Freeway, Houston, Texas 77023.

American Devon Cattle Club, Senator Wayne Morse, Pres., 209 Senate Office Bldg., Washington, D.C. 20510; Kenneth Hinshaw, Secy., Agawam, Mass. 01001.

American Galloway Breeders Association, Box 1424, Billings, Montana 59103.

American Hereford Association, Wayne Naugle, Pres., Naugle Hereford Ranch, Nampa, Idaho 83651; Dr. W. T. Berry, Jr., Secy., Hereford Drive, Kansas City, Missouri 64105.

American-International Charolais Association, Edward P. Shurick, Pres., Nutmeg Farms, Northrup Road, Bridgwater, Conn. 06752; J. Scott Henderson, Exec.-Secy., 923 Lincoln Liberty Life Building, Houston, Texas 77002.

American Polled Hereford Association, Walter M. Lewis, Pres., Larned, Kansas 67550; Orville K. Sweet, Secy., 4700 E. 63 St., Kansas City, Missouri 64130.

American Polled Shorthorn Society, Paul T. Loyd, Pres., 1225 Washington Rd., Newton, Kansas 67114; Charlotte Ekness, Secy., 8288 Hascall St., Omaha, Nebraska 68124.

American Scotch Highland Breeders' Association, Keith Crew, Pres., Interior, South Dakota 57750; Mrs. Margaret Manke, Secy., Edgemont, South Dakota 57735.

American Shorthorn Association, John C. Ragsdale, Pres., Prospect, Kentucky 40059; C. D. Swaffar, Secy., 8288 Hascall St., Omaha, Nebraska 68124.

American Simmental Association, Jack Winninger, Pres., Meeteetse, Wyoming 82433; Dale J. Lynch, Secy., 270 Country Commons Rd., Cary, Illinois 60013.

Beefmaster Breeders Universal, W. T. Hix, Pres., Third Floor, Gunter Hotel, San Antonio, Texas 78206; Mrs. Maxine E. Brown, Exec.-Secy., Third Floor-Gunter Hotel, San Antonio, Texas 78206.

International Brangus Breeders Association, Inc., Charles L. Cobb, Pres., Box 888, Bay City, Texas 77414; Roy W. Lilley, Exec.-Secy., 908 Livestock Exchange Building, Kansas City, Missouri 64102.

Performance Registry International, Roy Beeby, Pres., Prairie City Farms, Marshall, Oklahoma 73056; Glenn Butts, Secy., 201-204 Frisco Building, Joplin, Missouri 64801.

Red Angus Association of America, Bert Crane, Pres., Merced, California 95340; Mrs. Sybil Parker, Denton, Texas 76201.

Santa Gertrudis Breeders International, Winthrop Rockefeller, Pres., Winrock Farms, Rt. 3, Morrilton, Arkansas 72110; R. P. Marshall, Secy., Box 1257, Kingsville, Texas 78363.

DAIRY CATTLE

American Guernsey Cattle Club, Frank D. Brown, Jr., Pres., Port Deposit, Maryland 21904; F. X. Chapman, Secy., Peterborough, New Hampshire 03458.

American Jersey Cattle Club, C. Scott Mayfield, Pres., Athens, Tennessee 37303;

J. F. Cavanaugh, Exec.-Secy., 1521 East Broad Street, Columbus, Ohio 43205.

Ayrshire Breeders' Association, Phillip Schuyler, Pres., 15 Grand St., Cobleskill, New York 12043; David Gibson, Jr., Secy., Brandon, Vermont 05733.

Brown Swiss Cattle Breeder's Association of America, George W. Opperman, Pres., Manning, Iowa 51455; Marvin L. Kruse, Secy., Box 1019, Beloit, Wisconsin 53511.

Holstein-Friesian Association of America, R. DeWitt Mallary, Pres., Lower Plain Mallary Farm, Bradford, Vermont 05033; Robert H. Rumler, Exec.-Secy., Box 808, Brattleboro, Vermont 05301.

DUAL-PURPOSE CATTLE

American Milking Shorthorn Society, J. M. White, Pres., Marathon, New York 13803; Harry Clampitt, Secy., 313 S. Glenstone Ave., Springfield, Missouri 65802.

Red Poll Cattle Club of America, Elmer J. Miller, Pres., Rt. 3, Columbus, Wisconsin 53925; Wendell H. Severin, Secy., 3275 Holdrege St., Lincoln, Nebraska 68503.

GOATS

American Angora Goat Breeders Association, Mrs. Thomas L. Taylor, Secy., Rocksprings, Texas 78880.

American Dairy Goat Association, Wesley Nordfelt, Pres., Box 577, Ripon, California 95366; Dan Wilson, Secy., Box 186, Spindale, North Carolina 28160.

American Goat Society, Inc., Carl W. Romer, Secy., 7900 E. 66 St., Kansas City, Missouri, 64133.

HORSES

American Albino Horse Association, Inc., Ruby Shumaker, Pres., Rt. 3, Box 113,

Scio, Oregon 97374; Ruth White, Secy., Box 79, Crabtree, Oregon 97335.

American Paint Horse Association, Edgar Robinson, Pres., Box 153, Abilene, Texas 79604; Ralph Dye, Secy., Box 12487, Fort Worth, Texas 76116.

American Quarter Horse Association, E. H. Honnen, Pres., Box 5188 T.A., Denver, Colorado 80217; Don Jones, Secy., Box 200, Amarillo, Texas 79105.

American Saddle Horse Breeders Association, Inc., Thomas J. Morton, Jr., Pres., Old Stone House Farms, Newburgh, Indiana 47630; Charles J. Cronan, Jr., Secy., 929 S. 4th St., Louisville, Kentucky 40203.

American Shetland Pony Club, T. R. Huston, Pres., Box 236, Hanna City, Illinois 61536; Burton J. Zuege, Secy., Box 1250, Lafayette, Indiana 47906.

Appaloosa Horse Club, Inc., Bill Moore, Pres., Box 44, Hansen, Idaho 83334; George B. Hatley, Secy., Box 403, Moscow, Idaho 83843.

Arabian Horse Club Registry of America, Daniel C. Gainey, Pres., Owatonna, Minnesota 55060; Ward B. Howland, Secy., 332 S. Michigan Ave., Chicago, Illinois 60604.

Belgian Draft Horse Corporation of America, Herbert Schneckloth, Pres., Rt. 3, Davenport, Iowa 52804; Blanche A. Schmalzried, Secy., 189 West Hill St., Wabash, Indiana 46992.

Clydesdale Breeders Association of the U.S., Charles W. Willhoit, Secy., Batavia, Iowa 52533.

Jockey Club, Ogden Phipps, Chairman, 300 Park Avenue, New York, New York 10022; John F. Kennedy, Exec.-Secy., 300 Park Avenue, New York, New York 10022.

Morgan Horse Club, Inc., Deane C. Davis, Pres., Dyer Avenue, Montpelier, Vermont 05602; Seth P. Holcombe, Secy., Box 2157, West Hartford, Connecticut 06117.

Palomino Horse Association, Willard R. Beanland, Secy., 22249 Saticoy Street, Canoga Park, California 91304.

Palomino Horse Breeders of America, Jay Forrest Kratz, Pres., 113-115 Main Street, North Wales, Pennsylvania 19454; Howard Grekel, Secy., Rt. 1, Box 540, Claremore, Oklahoma 74017.

Percheron Horse Association of America, Ray H. Bast, Pres., Rt. 1, Box 114, Richfield, Wisconsin 53076; Dale Gossett, Secy., Rt. 1, Belmont, Ohio 43718.

Pinto Horse Association of America, Inc., Helen Smith, Secy., Box 3984, San Diego, California 92103.

Pony of the Americas Club, Inc., James S. Bicknell, III, Pres., Box 67, Clare, Michigan 48617; Leslie L. Boomhower, Secy., Exec.-Secy., Rt. 4, Mason City, Iowa 50401.

Tennessee Walking Horse Breeders' Association of America, Sen. J. T. Kelley, Pres., Rt. 7, Columbia, Tennessee 38401; Mrs. Sharon Brandon, Secy., Box 286, Lewisburg, Tennessee 37091.

U. S. Pony Trotting Association, Walter Butcher, Pres., William Burns, Secy., Box 1250, Lafayette, Indiana 47902.

U. S. Trotting Association, Walter J. Michael, Pres., Box 511, Bucyrus, Ohio 44820; Perley J. Gale, Secy., 750 Michigan Avenue, Columbus, Ohio 43215.

Welsh Pony Society of America, Mrs. Robert Pirie, Pres., Aquila Farm, Hamilton, Mass. 01936; Louise F. Gehret, Secy., 1770 Lancaster, Paoli, Pennsylvania 19301.

SHEEP

American Cheviot Sheep Society, Vern Williams, Pres., 11924 Braddock Rd., Fairfax, Virginia 22630; Stan Gates, Secy., 5051 Flourtown Rd., Lafayette Hill, Pennsylvania 19444.

American Corriedale Association, Inc., W. J. Marshall, Pres., 6810 Mondova Rd., Maumee, Ohio 43537; Mrs. Gail Sublett, Clerk, Box 809, Columbia, Missouri 65201.

American and Delaine-Merino Record Association, Harold E. Simms, Secy., Aleppo, Pennsylvania 15310.

American Hampshire Sheep Association, George R. Scott, Pres., Rt. 2, Box 82 E, Ft. Collins, Colorado 80521; Ron A. Gilman, Secy., Box 190, Stuart, Iowa 50250.

American Oxford Down Record Association, C. E. Puffenberger, Secy., Rt. 1, Eaton Rapids, Michigan 48827.

American Rambouillet Sheep Breeders' Association, 2709 Sherwood Way, San Angelo, Texas 76901; Murl Patrick, Pres., White, South Dakota 57276; Mrs. A. D. Harvey, Secy., Wall St., San Angelo, Texas 76901.

American Romney Breeders Association, Erwin Riddell, Pres., Rt. 1, Box 324, Independence, Oregon 97351; Dr. John H. Landers, Jr., Secy., 212 Withycombe Hall, Corvallis, Oregon 97331.

American Shropshire Registry Association, Inc., William A. McKerrow, Pres., Rt. 1, Box 175, Pewaukee, Wisconsin 53072; Mrs. Jessie F. Ritenour, Secy., Box 678, Lafayette, Indiana 47902.

American Southdown Breeders Association, Aime F. Real, Pres., Kerrville, Texas 78028; W. L. Henning, Secy.-Treas., University Park, Pennsylvania 16801.

American Suffolk Sheep Society, J. Alden Olsen, Pres., Spanish Fork, Utah 84660; Allan Jenkins, Secy., Newton, Utah 84327.

Black-Top National Delaine-Merino Sheep Association, I. Y. Hamilton, Pres., 23 W. Pitt St., Canonsburg, Pa. 15417; Jane R. McCarty, Secy., Box 23, Hickory, Pa. 15340.

Columbia Sheep Breeders Association of America, Otho Whitefield, Pres., Friona, Texas 79035; Richard L. Gerber, Secy., Box 272, Upper Sandusky, Ohio 43351.

Continental Dorset Club, Kenneth L. Young, Pres., RFD#2, Box 1601, Medina, Ohio 44256; J. R. Henderson, Secy., Box 97, Hickory, Pennsylvania 15340.

Debouillet Sheep Breeders Association, Joe Rawlings, Pres., Bronte, Texas 76933; Mrs. A. D. Jones, Secy., 300 South Kentucky, Roswell, New Mexico 88201.

Karakul Fur Sheep Registry, Robert Harris, Pres., Rt. 1, Fabius, New York 13063; Theodore R. Stultz, Secy., Rt. 1, Fabius, New York 13063.

Montadale Sheep Breeders Association, E. H. Mattingly, Secy., 4103 N. Broadway, St. Louis, Missouri 63147.

National Lincoln Sheep Breeders Association, E. N. Stemen, Pres., Ft. Jennings, Ohio 45844; Ralph O. Shaffer, Secy., 5284 S. Albaugh Rd., West Milton, Ohio 45383.

National Suffolk Sheep Association, John Shonkwiler, Pres., Rt. 2, Box 200, Neosho, Missouri 64850; Betty J. Biellier, Secy., Box 324, Columbia, Missouri 65201.

Texas Delaine Association, F. G. Jones, Jr., Pres., Rt. 1, Lackney, Texas 79241; Mrs. G. A. Glimp, Secy., Rt. 1, Burnet, Texas 78611.

U. S. Targhee Sheep Association, Lawrence Capra, Pres., Boyes, Montana 59316; Gene Coombs, Secy.-Treas., Security Bank, Box 2513, Billings, Montana 59103.

SWINE

American Berkshire Association, Milo Wolrab, Pres., Mt. Vernon, Iowa 52314; Gene Mason, Secy., 601 W. Monroe, Springfield, Illinois 62704.

American Landrace Association, Inc., Dr. Marvin Clark, Pres., Rt. 2, Monmouth, Illinois 61462; Eugene G. Benedict, Secy., Box 111, Culver, Indiana 46511.

American Yorkshire Club, Inc., Hal B. Clark, Pres., Rt. 1, Clarks Hill, Indiana 47930; Wilbur L. Plager, Secy., 1001 South Street, Box 878, Lafayette, Indiana 47902.

Hampshire Swine Registry, A. Ruben Edwards, Pres., Middletown, Missouri 63359; Harold Boucher, Secy., 1111 Main Street, Peoria, Illinois 61606.

National Chester White Swine Record Association, Donald W. Brown, Pres., Battleground, Indiana 47920; J. Marvin Garner, Secy., Box 228, Rochester, Indiana 46975.

National Hereford Hog Record Association, Leslie Schrecengost, Pres., Trent, South Dakota 57065; Sylvia Schulte, Secy., Norway, Iowa 52318.

National Spotted Swine Record, Norman Pothast, Pres., Spencer, Iowa 51301; Duane Fort, Secy., Bainbridge, Indiana 46105.

OIC Swine Breeders Association, Merlyn G. Oates, Pres., Rt. 2, Bellville, Ohio 44813; Thomas R. Hendricks, Secy., Box 111, Greencastle, Indiana 46135.

Poland China Record Association, Milt Friedow, Pres., Rt. 1, Kanawha, Iowa 50447; C. W. Mitchell, Secy., 501 E. Losey, Box 71, Galesburg, Illinois 61401.

Tamworth Swine Association, Wm. E. Strain, Pres., 4809 Mayfair, Bellaire, Texas 77401; Wm. R. Horne, Secy., Rt. 2, Cedarville, Ohio 45314.

United Duroc Swine Registry, Henry Hamer, Pres., Ft. Atkinson, Wisconsin 53538; Bruce Henderson, Secy., 239 N. Monroe, Peoria, Illinois 61614.

COMMERCIAL LIVESTOCK—CATTLE

American National Cattlemen's Association, Bill House, Pres., 801 E. 17th Ave., Denver, Colorado 80218; R. E. Sneddon, Secy., 801 E. 17th Ave., Denver, Colorado 80218.

National Livestock Feeders Association, Charles Phelps, Pres., Hastings, Iowa 51540; Don F. Magdanz, Secy., 309 Livestock Exchange Bldg., Omaha, Nebraska 68107.

COMMERCIAL LIVESTOCK—SHEEP

National Lamb Feeders Association, Reed C. Culp, Sr., Pres., 133 Virginia Street, Salt Lake City, Utah 84103; Roy A. Hanson, Secy., Box 867, Miles City, Montana 59301.

National Wool Growers Association, W. E. Overton, Pres., Box 61, Yeso, New Mexico 88136; Edwin E. Marsh, Secy., 600 Crandall Bldg., Salt Lake City, Utah 84101.

COMMERCIAL LIVESTOCK—SWINE

National Pork Producers Council, Albert Gehlbach, Pres., Rt. 3, Lincoln, Illinois 62656; Rolland Paul, Exec. V.P., 3101 Ingersoll Ave., Des Moines, Iowa 50312.

APPENDIX B: WEIGHTS, MEASURES, AND CONVERSION FACTORS

WEIGHTS, MEASURES, AND CONVERSION FACTORS *
Avoirdupois and Troy Weights

U.S. equivalents			Metric equivalent
Avoirdupois weight:[1]			
1 grain - - - - - - - - - -	- - - - - - - - - - - - - -	- - - - - - - - - - - - - -	64.799 milligrams
437.5 grains - - - - - - -	1 ounce, avoirdupois -	0.911 ounce, troy - - - -	28.35 grams
16 ounces, avoirdupois - -	1 pound avoirdupois -	1.215 pounds, troy - - - - -	453.59237 grams
100 pounds, avoirdupois -	1 short hundredweight -	- - - - - - - - - - - - - - -	45.3592 kilograms
2,000 pounds			
20 short hundredweights -	1 short ton - - - - - -	- - - - - - - - - - - - - -	0.907 metric ton
112 pounds, avoirdupois -	1 long hundredweight -	- - - - - - - - - - - - - -	50.802 kilograms
2,240 pounds - - - - - -			
20 long hundredweights	1 long ton - - - - - -	- - - - - - - - - - - - - -	1.016 tons
Troy weight:[3]			
1 grain - - - - - - - - - -	- - - - - - - - - - - - - -	- - - - - - - - - - - - - -	64.799 milligrams
24 grains - - - - - - - - - -	1 pennyweight - - - - -	- - - - - - - - - - - - - - -	1.555 grams
20 pennyweights - - - -	1 ounce, troy - - - - -	1.097 ounces, avoirdupois -	31.103 grams
12 ounces, troy - - - - - -	1 pound, troy - - - - -	0.823 pound, avoirdupois -	373.242 grams

[1]The system used in weighing all commodities except precious stones, precious metals, and drugs.
[2]The system used in weighing precious stones and precious metals such as silver and gold.

Weights of the United States Bushel

Item	Pounds	Kilograms
Wheat, beans, peas, potatoes (Irish or white) - - - - - - - - - - - - - - - -	60	27.22
Rye, corn (maize), linseed (flatseed), malin (mixed grain) - - - - - - - - - - - - -	56	25.40
Barley, buckwheat -	48	21.77
Onions -	57	25.86
Rough rice -	45	20.41
Malt -	34	15.42
Oats -	32	14.51
Peanuts, green, in shell -	22	9.98
Castor beans -	50	22.68
Sorghums for grain -	56	25.40

Weights of the Barrel of Nonliquid Products

	Pounds	Kilograms
Wheat flour, barley flour, rye flour, and corn meal (net)[1] - - - - - - - - - - - - - - - - -	196	88.90
Rosin, tar and pitch (gross) -	500	226.80
Fish, pickled (net) -	200	90.72
Lime (net) -	200	90.72
Cement (4 bags counted as 1 barrel) (net)[1] -	376	170.55

[1]Except as noted in the tables.

Capacities or Volumes

U.S. equivalents			
1 barrel (liquid)[1] - - - -	31 to 42 gallons - - - -	- - - - - - - - - -	- -
1 barrel (dry)[2] - - - - - -	7,056 cubic inches - -	105 dry quarts - -	3,281 bushels (struck measure).[3]

* Source: <u>Statistical Abstract of the United States, 1965</u>, 86th Edition, United States Department of Commerce, Washington, D.C.

[1]There are various "barrels" established by law or usage. Federal law recognizes a 31-and a 40-gallon barrel depending on the type of liquor. In addition, the number of gallons recognized as making up a barrel differs among various States.

[2]Standard for fruits and vegetables and other dry commodities, except cranberries, recognized by the Dept. of Agriculture.

[3]Standard bushel as measured by U.S. Government.

Approximate Weight of Petroleum and Products

(In the United States petroleum and its products are measured by bulk and weight. Quantities are customarily reduced to the equivalent of barrels of 42 U.S. gallons (158.984 liters). In many foreign countries these commodities are measured by weight. The specific gravity of the different grades of crude petroleum and of the finished products varies materially. On the basis of approximate averages, the Department of Commerce used the following factors for converting foreign weight statistics to gallons or to barrels of 42 gallons.)

ITEM	U.S. Gallons		Barrel of 42 Gallons	
	U.S. equivalent	Metric equivalent	U.S. equivalent	Metric equivalent
Crude petroleum- - - - - - - - -	7.3 pounds -	3.311 kilograms -	330.6 pounds -	139.07 kilograms
Lubricating oils - - - - - - - -	7.0 pounds -	3.175 kilograms -	294.0 pounds -	133.36 kilograms
Illuminating oils (kerosene) -	6.6 pounds -	2.994 kilograms -	277.2 pounds -	125.74 kilograms
Gasoline and related products motor spirit, benzine, etc. -	6.1 pounds -	2.767 kilograms -	256.2 pounds -	116.21 kilograms
Fuel and gas oils - - - - - - - -	7.7 pounds -	3.493 kilograms -	323.4 pounds -	146.69 kilograms

Liquid Measures

U.S. equivalents		Metric equivalent
1 fluid dram - - - - - - - - -	- -	3.697 milliliters
8 fluid drams - - - - - - - -	1 fluid ounce -	29.573 milliliters
4 fluid ounces - - - - - - - - -	1 gill -	0.118 liter
4 gills - - - - - - - - - - - - -	1 pint -	0.473 liter
2 pints - - - - - - - - - - - -	1 quart -	0.9463 liter
4 quarts - - - - - - - - - - -	1 gallon -	3.785 liters

Dry Measures

U.S. equivalents			Metric equivalent
1 pint - - - - - - - - - - -	- - - - - - - - - - - - - - - - - - -	33.600 cubic inches -	0.551 liter
2 pints - - - - - - - - - - -	1 quart - - - - - - - - - - -	67.2006 cubic inches	1.101 liters
8 quarts - - - - - - - - - - -	1 peck - - - - - - - - - - -	537.605 cubic inches	8.810 liters
4 pecks - - - - - - - - - - -	1 bushel (struck measure)[1] - - -	2,150.42 cubic inches	35.238 liters

[1]Standard bushel as measured by U.S. Government.

Length Measures

U.S. equivalents		Metric equivalent
1 inch - - - - - - - - - - - - -	- -	2.540 centimeters
12 inches - - - - - - - - - -	1 foot -	0.3048 meter
3 feet - - - - - - - - - - - - -	1 yard -	0.9144 meter
5½ yards - - - - - - - - - - - -	1 rod or pole -	5.029 meters
5,280 feet - - - - - - - - - - -	1 statute (land) mile - - - - - - - - - - - - - - - - -	1.609 kilometers

Volume of sphere = .5238 x diameter x diameter x diameter
Volume of cylinder = .7854 x height x diameter x diameter
Volume of pyramid = 1/3 x area of base x altitude
Volume of cone (like a stack of grain) = .2618 x height x diameter x diameter

Agricultural Commodity Conversion Factors *

Commodity	Unit	Approximate Equivalent
Apples	1 pound dried	8 pounds fresh (beginning 1943)
Barley	1 metric ton[1]	45.93 bushels
Barley flour	100 pounds	4.59 bushels barley
Beans, snap or wax	1 case canned[2]	0.010 tons fresh
Calves	1 pound live weight	0.555 pounds dressed weight (since 1952)
Cattle	1 pound live weight	0.549 pounds dressed weight (since 1952)
Chickens	1 pound live weight	0.72 pound ready-to-cook weight
Corn	1 metric ton[1]	39.368 bushels
Corn, shelled	1 bushel (56 pounds)	2 bushels (70 pounds) of husked corn
Cotton	1 pound ginned	3.26 pounds seed cotton including trash[3]
Cottonseed meal	1 pound	2.10 pounds cottonseed
Cottonseed oil	1 pound	5.88 pounds cottonseed
Dairy products		
Butter	1 pound	21.1 pounds milk
Cheese	1 pound	10 pounds milk
Nonfat dry milk	1 pound	11 pounds liquid skim milk
Eggs	1 case	47 pounds
Flaxseed	1 bushel	About 2 1/2 gallons oil
Hogs	1 pound live weight	0.569 pounds dressed weight excluding lard (since 1952)
Oats	1 metric ton[1]	68.8944 bushels
Peanuts	1 pound shelled	1 1/2 pounds unshelled
Rye	1 metric ton[1]	39.368 bushels
Rye flour	100 pounds	2.23 bushels rye (beginning 1947)
Sheep and lambs	1 pound live weight	0.477 pounds dressed weight (since 1952)
Soybeans	1 metric ton[1]	36.744 bushels
Soybean meal	1 pound	1.28 pounds soybeans
Soybean oil	1 pound	5.45 pounds soybeans
Tomatoes	1 case canned[2]	0.027 ton fresh
Turkeys	1 pound live weight	0.80 pound ready-to-cook weight
Wheat	1 metric ton[1]	36.7437 bushels
Wheat flour	100 pounds	2.30 bushels wheat[4]

*Source: Agriculture Handbook No. 230, U.S. Department of Agriculture, 1964

[1]A metric ton is equivalent to 2,204.6 pounds.

[2]Case of 24 number 2 cans.   [3]Varies widely by method of harvest.

[4]This is equivalent to 4.51 bushels of wheat per barrel (196 pounds) of flour and has been used in conversions beginning July 1, 1957.

Area Measures

U.S. equivalents		Metric equivalent
1 square inch - - - - - - - -	- -	6.452 square centimeters
144 square inches - - - - - -	1 square foot -	929.030 square centimeters
9 square feet - - - - - - - -	1 square yard -	0.8361 square meter
30¼ square yards - - - - - -	1 square rod -	25.293 square meters.
160 square rods - - - - - -	1 acre -	0.4047 hectare.
4,840 square yards - - - - - - -		
650 acres - - - - - - - - - - -	1 square mile -	259.000 hectares.

Cubic Measures

U.S. equivalents		Metric equivalent
1 cubic inch - - - - - - - - - - -	- -	16.387 cubic centimeters
1,728 inches - - - - - - - - - - -	1 cubic foot -	28.317 cubic decimeters
27 cubic feet - - - - - - - -	1 cubic yard -	0.7646 cubic meter

AGRICULTURAL CONVERSION EQUIVALENTS AND MEASUREMENT FORMULAS *

Unit	Equivalent Units
Acre - - - - - - - - - -	43,560 square feet; 4,840 square yards; 160 square rods
Barrel - - - - - - - - - -	31.5 gallons; 196 pounds (flour)
Bushel - - - - - - - -	1.244 cubic feet; 2,150.42 cubic inches; 32 quarts (dry); 4 pecks
Cord - - - - - - - - - -	128 cubic feet
Cubic foot - - - - - -	0.8 bushel; 1,728 cubic inches; 7.481 gallons; 62.4 pounds (water)
Cubic yard - - - - - -	27 cubic feet; 202 gallons
Feet per second - - - - - -	(22/15) x miles per hour
Foot - - - - - - - - - -	12 inches; 1/3 yard; 0.305 meter
Gallon - - - - - - - -	231 cubic inches (liquid); 268.8 cubic inches (dry); 0.1337 cubic foot; 4 quarts; 8.345 pounds (water)
Inch - - - - - - - - - -	2.54 centimeter
Meter - - - - - - - - - -	39.37 inches; 1.094 yards; 3.281 feet
Mile - - - - - - - - - -	5,280 feet; 1,760 yards; 320 rods; 1.609 kilometers
Peck - - - - - - - - - -	8 quarts; 1/4 bushel
Pint - - - - - - - - - -	1/2 quart; 16 ounces
Quart - - - - - - - - - -	2 pints; 32 ounces; 67.2 cubic inches (dry); 57.75 cubic inches (liquid)
Rod - - - - - - - - - -	16.5 feet; 5.5 yards
Square foot - - - - - -	1/9 square yard; 144 square inches
Square mile - - - - - -	640 acres
Square yard - - - - - -	9 square feet
Ton - - - - - - - - - -	2,240 pounds (long); 2,205 pounds (metric); 2,000 pounds (short)
Yard - - - - - - - - - -	3 feet; 0.9144 meter

Measurement Formulas

Circumference of circle = 3.1416 x diameter = $\dfrac{22 \text{ x diameter}}{7}$

Area of circle = .7854 x diameter x diameter

Area of rectangle = length x width

Area of triangle = 0.5 x base x altitude

Area of curved surface of cylinder (like a silo) = 3.1416 x diameter x height

* Source: Agriculture Handbook No. 230, U.S. Department of Agriculture, 1964.

INDEX